PEM Fuel Cell Testing and Diagnosis

PEM Fuel Cell Testing and Diagnosis

Jianlu Zhang
Huamin Zhang
Jinfeng Wu
Jiujun Zhang

AMSTERDAM • BOSTON • HEIDELBERG • LONDON
NEW YORK • OXFORD • PARIS • SAN DIEGO
SAN FRANCISCO • SINGAPORE • SYDNEY • TOKYO

ELSEVIER

Elsevier
Radarweg 29, PO Box 211, 1000 AE Amsterdam, The Netherlands
The Boulevard, Langford Lane, Kidlington, Oxford OX5 1GB, UK

Notice
No responsibility is assumed by the publisher for any injury and/or damage to persons or property
as a matter of products liability, negligence or otherwise, or from any use or operation of any
methods, products, instructions or ideas contained in the material herein. Because of rapid advances
in the medical sciences, in particular, independent verification of diagnoses and drug dosages
should be made

British Library Cataloguing in Publication Data
A catalogue record for this book is available from the British Library

Library of Congress Cataloging-in-Publication Data
A catalog record for this book is available from the Library of Congress

ISBN: 978-0-444-53688-4

Contents

Proton exchange membrane (PEM) fuel cells are considered to be a viable solution to the challenges of reducing environmental pollution and global warming, as they are unique among energy-converting devices in producing low or zero emissions. They possess several other advantages, including high efficiency and power density, which make them attractive in a wide range of application areas, such as portable, stationary, and transportation power sources. However, large-scale commercialization of PEM fuel cells requires intensive R&D to overcome several challenges, including high cost and inadequate reliability/durability. In the effort to address these issues, fuel cell testing and diagnostics have been recognized as playing a critical role in material characterization, performance optimization, design validation, and in fundamental understanding for further progress.

Several factors have motivated the writing of this book. As PEM fuel cells are being developed by fuel cell researchers and practicing engineers at an increasingly fast pace, particularly in the area of electric vehicles, there is a pressing need for an overview of fuel cell testing and diagnosis that describes the basic principles, measurements, and applications of the technology. In addition, as fuel cells grow in popularity among undergraduate students and M.Sc. and Ph.D. candidates in mechanical, chemical, and electrical engineering; environmental studies and engineering; and material science and engineering, a textbook or reference book on fuel cell testing and diagnosis is definitely desirable.

This book is the direct result of decades of experience in PEM fuel cell testing and diagnosis. All the contributing authors have been working in PEM fuel cell technology for many years in the areas of design, materials, components, operation, diagnostics, and systems. Our aim is to provide a general understanding of the relevant techniques, and detailed guidance in their applications for PEM fuel cells. We also hope that the readers, especially those who are not electrochemists, will find the book's descriptions of testing and diagnosis techniques to be accessible and sufficient preparation for experimental result analysis.

The following twelve chapters contain comprehensive information on the fundamentals of PEM fuel cells, as well as the basic principles and practical implementation of testing and diagnosis. Chapter 1 introduces the readers to the general field of PEM fuel cells, including fuel cell fundamentals and electrochemical approaches in fuel cell testing and diagnosis; Chapter 2 describes the design and assembly of the membrane electrode assembly, single fuel cell, and fuel cell stack; Chapter 3 gives a brief overview of techniques used in PEM fuel cell testing and diagnosis; Chapter 4 describes how the temperature can affect PEM fuel cell kinetics and performance; Chapters 5 covers membrane/ionomer proton conductivity measurements; Chapter 6 discusses hydrogen crossover in PEM fuel cells and associated measurements; Chapter 7 provides a detailed description of fuel cell open circuit voltage; Chapters 8 and 9 delve into the effects of humidity and pressure on fuel cell performance; Chapter 10 describes high-temperature PEM fuel cells; Chapter 11 discusses fuel cell degradation and failure analysis, and their associated testing and diagnosis; and Chapter 12 discusses the use of electrochemical half-cells for evaluating PEM fuel cell catalysts and catalyst layers. In the writing of these chapters, the authors have cited the open literature available in scientific journals up to the time of the final draft.

We would like to thank the family members of all the authors for their continued patience, understanding, encouragement, and support throughout the writing of this book.

If this book contains technical errors, the authors would deeply appreciate the readers' constructive comments for correction and further improvement.

**Jianlu Zhang, Huamin Zhang,
Jinfeng Wu and Jiujun Zhang**
January 2013

Dr. Jianlu Zhang is currently a Senior Engineer at Palcan Energy Corporation, Canada. He has eleven years of R&D experience in renewable energy conversion and storage technologies. In addition, Dr. Zhang has nine years of hands-on experience in PEMFC research, including membrane electrode assembly fabrication, electrocatalyst synthesis and characterization, PEMFC design, testing and diagnosis, and fuel cell contamination studies.

Dr. Zhang received his B.S. degree in Chemistry in 1998 from Liaocheng University, his M.Sc. in Applied Chemistry in 2001 from the Dalian University of Technology, and his Ph.D. in Physical Chemistry in 2005 from the Dalian Institute of Chemical Physics, Chinese Academy of Sciences (CAS-DICP), where he worked on catalyst synthesis and catalyst layer design for PEMFCs. Subsequently, he conducted his postdoctoral research at the National Research Council of Canada's Institute for Fuel Cell Innovation (NRC-IFCI) for four years and at the Pacific Northwest National Laboratory, US Department of Energy, for one year. In 2011, Dr. Zhang worked as an associate professor at CAS-DICP, where his research was focused on PEMFCs and technologies for energy storage. To date, Dr. Zhang has authored/coauthored more than sixty publications, including forty-nine peer-reviewed journal articles, nine conference proceeding articles, and six book chapters. He has also been a major inventor in seven patents.

Professor Huamin Zhang is the Head of the Energy Storage Division at the Dalian Institute of Chemical Physics, Chinese Academy of Sciences (CAS-DICP). Professor Zhang received his B.S. from the Shandong University in 1982, and his M.Sc. and Ph.D. from the Kyushu University in Japan in 1985 and 1988, respectively, under the supervision of Professor Noboru Yamazoe. After completing his Ph.D., Professor Zhang worked in Japan in the fields of functional materials, catalytic materials, solid oxide materials, and fuel cells. As a professor and the Director of the Fuel Cell Center at CAS-DICP since 2000, he has been in charge of the Fuel Cell Engine project of China's national "863" high-technology project, the Fuel Cell Vehicle program supported by the National Ministry of Science and Technology, the Knowledge Innovation Program's Large-Scale Proton Exchange Membrane Fuel Cell Engine and Hydrogen Source, supported by the Chinese Academy of Sciences, the international cooperation project Key Technology of Fuel Cells, supported by China's National Ministry of Science and Technology, the Orientated Knowledge Innovation Program Fluoride Ionomer for PEMFC, supported by the Chinese Academy of Sciences, and the national "863" high-technology project Research on Testing the Key Materials, Components, and Environmental Adaptability of Fuel Cells. Currently, he is the chief scientist responsible for the national "973" project Basic Research on Large-Scale and High-Efficiency Flow Batteries for Energy Storage.

Under his leadership, Professor Zhang's team successfully developed a 30-kW fuel cell engine in 2001, as well as a 150-kW fuel cell engine for China's first fuel cell minibus and first fuel cell bus in 2004. Recently, Professor Zhang's team also succeeded in developing and demonstrating 10-kW and 5-MW vanadium flow battery energy storage systems.

Professor Zhang's current research interests are in key materials and components of flow batteries, including vanadium electrolyte, bipolar plates, and advanced ion exchange membranes, as well as in system integration of flow batteries, high specific energy lithium-based batteries, low Pt and non-noble metal electrocatalysts, self-humidified membranes, advanced MEA fabrication, and process technologies for PEMFCs. Professor Zhang is responsible for fuel cell and flow battery standardization in China. He is also actively involved in European and international flow battery standardization.

Professor Zhang holds four ministry-level awards for outstanding scientific and technological achievement (from the Ministry of Science and Technology, the Ministry of Finance, the State Planning Commission, and the State Economic and Trade Commission), two first prizes and two second prizes for technological inventions in Liaoning province, and one second prize in the GM China Science and Technology Achievement Awards.

To date, Professor Zhang has coauthored >260 research articles published in refereed journals, >150 patents, and two book chapters.

Dr. Jinfeng Wu is a Senior Engineer at UniEnergy Technologies, LLC. Dr. Wu has been involved in fuel cell and renewable energy research since 1998, and in 2004, he was awarded a Ph.D. in Electrochemical Engineering from the Dalian Institute of Chemical Physics, Chinese Academy of Science. He then became a full-time researcher at the Casaccia Research Centre in the Italian National Agency for New Technologies, Energy and the Environment, in Rome. In November 2005, he was awarded a Natural Sciences and Engineering Research Council of Canada postdoctoral fellowship at the Institute for Fuel Cell Innovation, National Research Council of Canada (NRC-IFCI). From November 2007 to October 2010, Dr. Wu acted as an assistant research officer with the Low Temperature Fuel Cell Group at NRC-IFCI. After that, he worked as a Senior Engineer and technical leader at Palcan Energy Corporation, Canada from October 2010 to August 2012. Dr. Wu has coauthored >70 journal and conference articles, five book chapters, plus many technical reports, as well as one patent in the area of fuel cell water management. Dr. Wu is an active member of the Electrochemical Society.

Dr. Jiujun Zhang is a Senior Research Officer and Technical Leader at the National Research Council of Canada's Institute for Fuel Cell Innovation (NRC-IFCI, now has been changed to Energy, Mining and Environment Portfolio (NRC-EME) since April 2012). Dr. Zhang received his B.S. and M.Sc. in Electrochemistry from Peking University in 1982 and 1985, respectively, and his Ph.D. in Electrochemistry from Wuhan University in 1988. After completing his Ph.D., he took the position as an associate professor at the Huazhong Normal University for two years. Starting in 1990, he carried out three terms of postdoctoral research at the California Institute of Technology, York University, and the University of British Columbia. Dr. Zhang has over thirty years of R&D experience in theoretical and applied electrochemistry, including over fourteen years of fuel cell R&D (among these 6 years at Ballard Power Systems and 9 years at NRC-IFCI), and 3 years of electrochemical sensor experience. Dr. Zhang holds several adjunct professorships, including one at the University of Waterloo, one at the University of British Columbia, and one at Peking University. Up to now,

Dr. Zhang has coauthored 290 publications, including 200 refereed journal articles with approximately 5700 citations, 11 edited/coauthored books, 11 conference proceedings articles, 14 book chapters, as well as 80 conference and invited oral presentations. He also holds >10 US/EU/WO/JP/CA patents, 9 US patent publications, and has produced in excess of eighty industrial technical reports. Dr. Zhang serves as the editor/editorial board member for several international journals, as well as Editor for book series (Electrochemical Energy Storage and Conversion, CRC Press). Dr. Zhang is an active member of The Electrochemical Society, the International Society of Electrochemistry, and the American Chemical Society.

PEM Fuel Cell Fundamentals

PEM Fuel Cell Testing and Diagnosis. http://dx.doi.org/10.1016/B978-0-444-53688-4.00001-2

1.1. INTRODUCTION

Proton exchange membrane (PEM) fuel cells, which directly convert chemical energy to electrical energy, have attracted great attention due to their numerous advantages, such as high power density, high energy conversion efficiency, fast startup, low sensitivity to orientation, and environmental friendliness. Figure 1.1 shows the schematic of a typical single PEM fuel cell [1], in which the anode and cathode compartments are separated by a piece of PEM such as Nafion®. This Nafion® membrane serves as the electrolyte and helps conduct protons from the anode to the cathode and also separates the anode and the cathode. During fuel cell operation, the fuel (e.g. H_2) is oxidized electrochemically within the anode catalyst layer (CL), and this produces both protons and electrons. The protons then get transported across the membrane to the cathode side, while the electrons move through the outer circuit and thereby also reach the cathode side. These protons and electrons electrochemically react with the oxidant (i.e. oxygen in the feed air) within the cathode CL and produce both water and heat. The whole process of a H_2/air PEM fuel cell produces electricity, water, and heat, without any polluting byproducts.

To better understand how a PEM fuel cell works, it is necessary to grasp the fundamentals of PEM fuel cells, including their cell structure and the thermodynamics and kinetics of fuel cell electrochemical reactions. In the following sections of this chapter, the fundamentals of H_2/air PEM fuel cells will be discussed in detail.

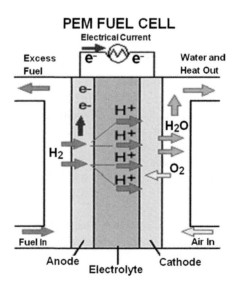

FIGURE 1.1 Schematic of a typical H_2/air PEM fuel cell [1]. (For color version of this figure, the reader is referred to the online version of this book.)

Several other types of fuel cells also belong to the PEM fuel cell family; these include the direct methanol fuel cell, direct ethanol fuel cell, and direct formic acid fuel cell. However, the scope of this book is such that we will only focus on the H_2/air PEM fuel cell.

1.2. ELECTROCHEMICAL REACTION THERMODYNAMICS IN A H_2/AIR FUEL CELL

1.2.1. Thermodynamic Electrode Potential and Cell Voltage of a H_2/Air Fuel Cell

A H_2/air PEM fuel cell converts chemical energy stored in the fuel (hydrogen) into electrical energy through electrochemical reactions between H_2 and O_2. These electrochemical reactions can be written as follows:

$$\text{Anode:} \quad H_2 \leftrightarrow 2H^+ + 2e^- \tag{1.I}$$

$$\text{Cathode:} \quad O_2 + 4H^+ + 4e^- \leftrightarrow 2H_2O \tag{1.II}$$

$$\text{Overall:} \quad H_2 + \frac{1}{2}O_2 \leftrightarrow H_2O \tag{1.III}$$

Note that the two-directional arrows in these reaction expressions indicate that all these reactions are chemically or electrochemically reversible, although they are not thermodynamically reversible due to their limited reaction rate in both reaction directions. Assuming that these reactions are in equilibrium states, the thermodynamic electrode potentials for the half-electrochemical Reactions (1.I) and (1.II) and the overall Reaction (1.III) can be expressed using the following Nernst equations:

$$E^r_{H_2/H^+} = E^o_{H_2/H^+} + 2.303\frac{RT}{n_H F}\log\left(\frac{a^2_{H^+}}{a_{H_2}}\right) \tag{1.1}$$

$$E^r_{O_2/H_2O} = E^o_{O_2/H_2O} + 2.303\frac{RT}{n_O F}\log\left(\frac{a_{O_2}a^4_{H^+}}{a^2_{H_2O}}\right) \tag{1.2}$$

In Eqn (1.1), $E^r_{H_2/H^+}$ is the reversible anode potential (V) at temperature T; a_{H_2} and a_{H^+} are the respective activities of H_2 and H^+; $E^o_{H_2/H^+}$ is the electrode potential of the H_2/H^+ redox couple under standard conditions (1.0 atm, 25 °C), which is defined as zero voltage; n_H is the electron transfer number (= 2 for the H_2/H^+ redox couple); R is the universal gas constant (8.314 J K^{-1} mol^{-1}); and F is Faraday's constant (96,487 C mol^{-1}). In Eqn (1.2), $E^r_{O_2/H_2O}$ is the reversible cathode potential (V) at temperature T; a_{O_2} and

a_{H^+} are the respective activities of O_2 and H^+; $E^o_{O_2/H_2O}$ is the electrode potential of the O_2/H_2O redox couple under standard conditions (1.0 atm, 25 °C), which is 1.229 V (vs. the standard hydrogen electrode (SHE)); n_O is the electron transfer number ($= 4$ for the O_2/H_2O redox couple); and a_{H_2O} is the activity of H_2O. Combining Eqn (1.1) and (1.2) yields a thermodynamic or theoretical fuel cell voltage (V^{OCV}_{cell}), which can be expressed as Eqn (1.3):

$$V^{OCV}_{cell} = E^r_{O_2/H_2O} - E^r_{H_2/H^+}$$

$$= E^o_{O_2/H_2O} - E^o_{H_2/H^+} + 2.303 \frac{RT}{2F} \log \left(\frac{a_{H_2} a_{O_2}^{\frac{1}{2}}}{a_{H_2O}} \right) \quad (1.3)$$

where $E^o_{O_2/H_2O} - E^o_{H_2/H^+}$ equals 1.229 V. Therefore, if T, a_{O_2}, a_{H_2}, and a_{H_2O} are known, the theoretical cell voltage under different conditions can be calculated according to Eqn (1.3). Note that for an approximate evaluation of theoretical electrode potentials and fuel cell voltage, the variables in Eqn (1.3), a_{O_2}, a_{H_2}, and a_{H_2O}, can be replaced by their partial pressures:P_{O_2}, P_{H_2}, and P_{H_2O}, respectively.

In fact, the theoretical cell voltage, V^{OCV}_{cell}, is the fuel cell open circuit voltage (OCV). However, in practice, the measured OCV value is always lower than the theoretical value calculated by Eqn (1.3). This is because several factors can affect the OCV, including Pt/PtO catalyst mixed potential and hydrogen crossover, which will be discussed in Chapter 7.

It is worth mentioning that Eqn (1.3) for calculating the OCV or theoretical cell voltage are derived from the thermodynamic concept called the change in Gibbs free energy (ΔG). For an electrochemical system, ΔG is the thermodynamic potential measuring the "useful" electrical work that can be obtained from an isothermal and isobaric electrochemical reaction. For example, the change in Gibbs free energy (ΔG_{cell}) for fuel cell Reaction (1.III) can be related to the fuel cell voltage by the following equation:

$$\Delta G_{cell} = -2FV^{OCV}_{cell} \quad (1.4)$$

where 2 is the electron number when each H_2 is oxidized, and F has the same meaning as in Eqn (1.3). Using this equation, if the change in Gibbs free energy is known, the corresponding fuel cell OCV can be calculated. Of course, if the fuel cell OCV is known, the change in the Gibbs free energy of a reaction can be calculated using Eqn (1.4). Note that the calculated fuel cell OCV is not the real OCV in a practical fuel cell and is normally lower than the theoretically expected value due to the catalyst mixed potential and hydrogen crossover; therefore, Eqn (1.4), a thermodynamic equation, may not be applicable to a real situation. However, as a very rough estimation, this equation may be usable, depending on the user's own opinion.

1.2.2. Fuel Cell Electrical Work and Heat

As a general observation, during fuel cell operation, both electrical work and heat are generated. This can be theoretically understood based on the thermodynamic driving force of the fuel cell reactions. For example, at a constant temperature and pressure, if O_2 and H_2 in an atomic ratio of 1:2 were directly reacted in a calorimeter, the reaction energy, expressed as the total reaction heat ΔH, could be obtained. However, if the same reaction was carried out in the same calorimeter by using the fuel cell device shown in Fig. 1.1, two parts of the reaction energy could be measured: (1) the electrical energy output through the fuel cell's load resistance and (2) the heat released from the fuel cell reaction. However, the sum of these two portions of energy will be exactly the same as the total reaction heat energy (ΔH) released by the direct reaction of O_2 and H_2 at the same temperature and pressure, if the energy used to heat the fuel cell hardware is not considered. If the load resistance value of the fuel cell is infinitely great (and therefore the fuel cell reaction is infinitely slow), these two portions of energy can be related to the overall reaction heat energy ΔH by the following equation:

$$\Delta H = \Delta G_{cell} + T\Delta S \tag{1.5}$$

where ΔH is the total heat energy of the fuel cell reaction (also called the change in reaction enthalpy), ΔG_{cell} is the maximum electrical work that the fuel cell can generate, and $T\Delta S$ is the maximum heat the fuel cell can release; here, T is the temperature and ΔS is the change in reaction entropy. For example, at 25 °C and 1.0 atm, the values of ΔH, ΔG_{cell}, and ΔS for fuel cell Reaction (1.III) are -285.8 kJ mol^{-1}, -237.2 kJ mol^{-1}, and -48.7 J mol^{-1}, respectively. These numbers exactly obey Eqn (1.5).

1.2.3. Fuel Cell Electrical Energy Efficiency

In Eqn (1.5), the electrical work (ΔG_{cell}) produced by the fuel cell is practically desirable, whereas the heat ($T\Delta S$) produced is not as useful as the electrical work. In practical fuel cell operations, this heat has to be removed through a cooling system, which places an extra burden on the fuel cell system.

The theoretical electrical energy efficiency (η_e^o) of a fuel cell can be defined as follows:

$$\eta_e^o = \frac{\Delta G_{cell}}{\Delta H} 100\% \tag{1.6}$$

For example, at 25 °C and 1.0 atm, the fuel cell electrical efficiency can be as high as 83%, based on $\Delta G_{cell} = 237.2$ kJ mol^{-1}, and $\Delta H = 285.8$ kJ mol^{-1}, indicating that of the overall energy generated by the fuel cell reaction, 83% is converted into electrical energy and the other

17% is released as heat. In addition, this fuel cell electrical efficiency is a function of temperature. For example, in the temperature range of 25–1000 °C, this electrical efficiency will be reduced almost linearly from 83 to 66% at a reduction rate of 0.0174% per degree centigrade [2]. Note that this fuel cell electrical efficiency is a thermodynamic concept or a predicted maximum efficiency. For practical fuel cell operation, where the fuel cell reaction drifts significantly from the ideal thermodynamic situation, one should be especially cautious about using Eqn (1.6) to evaluate the electrical energy efficiency of a fuel cell.

1.3. ELECTROCHEMICAL REACTION KINETICS IN A H₂/AIR FUEL CELL

In the previous section, we discussed fuel cell thermodynamics. However, in reality, fuel cell operation with an external load is much more practical than in a thermodynamic state. When a H₂/air PEM fuel cell outputs power, the half-electrochemical reactions will proceed simultaneously on both the anode and the cathode. The anode electrochemical reaction expressed by Reaction (1.I) will proceed from H₂ to protons and electrons, while the oxygen from the air will be reduced at the cathode to water, as expressed by electrochemical Reaction (1.II). For these two reactions, although the hydrogen oxidation reaction (HOR) is much faster than the oxygen reduction reaction (ORR), both have limited reaction rates. Therefore, the kinetics of both the HOR and the ORR must be discussed to achieve a better understanding of the processes occurring in a PEM fuel cell.

1.3.1. Kinetics of the Hydrogen Oxidation Reaction

For a H₂/air PEM fuel cell, the anode HOR described by Reaction (1.I) is an overall reaction expression, which contains several steps that form the HOR mechanism. When Pt particles are used as the catalyst in the anode CL, a generally recognized reaction mechanism can be expressed as follows [3–8]:

$$Pt + H_2 \underset{\overleftarrow{k_{1H}}}{\overset{\overrightarrow{k_{1H}}}{\rightleftharpoons}} Pt - H_2 \tag{1.IV}$$

$$Pt - H_2 + Pt \underset{\overleftarrow{k_{2H}}}{\overset{\overrightarrow{k_{2H}}}{\rightleftharpoons}} 2Pt - H \quad \text{(Rate-determining step)} \tag{1.V}$$

$$2Pt - H \underset{\overleftarrow{k_{3H}}}{\overset{\overrightarrow{k_{3H}}}{\rightleftharpoons}} 2Pt + 2H^+ + 2e^- \tag{1.VI}$$

Reaction (1.IV) is the adsorption of H_2 on the platinum surface and is fast. Reaction (1.V) is the slow dissociative chemical adsorption of adsorbed H_2 and is considered as the rate-determining step for H_2 oxidation, and Reaction (1.VI) is the fast electrochemical oxidation of the dissociated hydrogen atom. In all these reaction equations, $\overrightarrow{k}_{iH}$ and $\overleftarrow{k}_{iH}$ represent the reaction rate constants in the forward and backward directions, respectively. For half–electrochemical Reaction (1.VI), $\overrightarrow{k}_{3H}$ and $\overleftarrow{k}_{3H}$ are dependent on the electrode potential and can be expressed as follows:

$$\overrightarrow{k}_{3H} = \overrightarrow{k}_{3H}^{\,o}\exp\left(\frac{\alpha_H n_{\alpha H}FE_a}{RT}\right) \tag{1.7}$$

$$\overleftarrow{k}_{3H} = \overleftarrow{k}_{3H}^{\,o}\exp\left(-\frac{(1-\alpha_H)n_{\alpha H}FE_a}{RT}\right) \tag{1.8}$$

The concentrations of Pt-related surface species or surface reaction sites, such as Pt–H_2, Pt–H, and Pt in Reactions (1.IV)–(1.VI), may be expressed as the corresponding surface coverage, such as θ_{Pt-H_2}, θ_{Pt-H}, and θ_{pt}, respectively. They have the following relationship:

$$\theta_{Pt-H_2} + \theta_{Pt-H} + \theta_{Pt} = 1 \tag{1.9}$$

Assuming the fast Reaction (1.IV) is always in an equilibrium state, the following equation will apply:

$$\theta_{Pt-H_2} = \frac{\overrightarrow{k}_{1H}}{\overleftarrow{k}_{1H}}C_{H_2}\theta_{Pt} \tag{1.10}$$

where C_{H_2} is the concentration or partial pressure of hydrogen in the CL. Combining Eqns (1.9) and (1.10) yields

$$\theta_{Pt-H} + \left(1 + \frac{\overrightarrow{k}_{1H}}{\overleftarrow{k}_{1H}}C_{H_2}\right)\theta_{Pt} = 1 \tag{1.11}$$

Because Reaction (1.V) is the slowest reaction and Reaction (1.VI) is a fast reaction, it is reasonable to assume that the surface coverage of the Pt–H site during the whole reaction process remains constant (steady-state assumption), that is, the amount of Pt–H produced is equal to the amount consumed. The following equation will then apply:

$$\overleftarrow{k}_{2H}\theta^2_{Pt-H} = \overrightarrow{k}_{2H}\theta_{Pt-H_2}\theta_{Pt} - \overrightarrow{k}_{3H}\theta^2_{Pt-H} + \overleftarrow{k}_{3H}C^2_{H^+}\theta^2_{Pt} \tag{1.12}$$

where C_{H^+} is the concentration of protons in the aqueous phase. Combining Eqns (1.11) and (1.12) yields

$$\theta_{Pt-H} = \left(\frac{\overrightarrow{k_{2H}} \frac{\overrightarrow{k_{1H}}}{\overleftarrow{k_{1H}}} C_{H_2} + \overleftarrow{k_{3H}} C_{H^+}^2}{\overleftarrow{k_{2H}} + \overrightarrow{k_{3H}}} \right)^{1/2} \theta_{Pt} \qquad (1.13)$$

By substituting Eqns (1.7), (1.8), and (1.11) into (1.9), both θ_{Pt-H} and θ_{Pt} become functions of the hydrogen concentration:

$$\theta_{Pt} = \left\{ 1 + \frac{\overrightarrow{k_{1H}}}{\overleftarrow{k_{1H}}} C_{H_2} \right.$$

$$\left. + \left[\frac{\overrightarrow{k_{2H}} \frac{\overrightarrow{k_{1H}}}{\overleftarrow{k_{1H}}} C_{H_2} + \overleftarrow{k_{3H}}^o C_{H^+}^2 \exp\left(-\frac{(1-\alpha_H)n_{\alpha H} F E_a}{RT} \right)}{\overleftarrow{k_{2H}} + \overrightarrow{k_{3H}}^o \exp\left(\frac{\alpha_H n_{\alpha H} F E_a}{RT} \right)} \right]^{1/2} \right\}^{-1} \qquad (1.14)$$

$$\theta_{Pt-H} = \left[\frac{\overrightarrow{k_{2H}} \frac{\overrightarrow{k_{1H}}}{\overleftarrow{k_{1H}}} C_{H_2} + \overleftarrow{k_{3H}}^o C_{H^+}^2 \exp\left(-\frac{(1-\alpha_H)n_{\alpha H} F E_a}{RT} \right)}{\overleftarrow{k_{2H}} + \overrightarrow{k_{3H}}^o \exp\left(\frac{\alpha_H n_{\alpha H} F E_a}{RT} \right)} \right]^{1/2}$$

$$\times \left\{ 1 + \frac{\overrightarrow{k_{1H}}}{\overleftarrow{k_{1H}}} C_{H_2} + \left[\frac{\overrightarrow{k_{2H}} \frac{\overrightarrow{k_{1H}}}{\overleftarrow{k_{1H}}} C_{H_2} + \overleftarrow{k_{3H}}^o C_{H^+}^2 \exp\left(-\frac{(1-\alpha_H)n_{\alpha H} F E_a}{RT} \right)}{\overleftarrow{k_{2H}} + \overrightarrow{k_{3H}}^o \exp\left(\frac{\alpha_H n_{\alpha H} F E_a}{RT} \right)} \right]^{1/2} \right\}^{-1} \qquad (1.15)$$

Equations (1.14) and (1.15) are fairly complicated, and they contain all reaction constants, the concentrations of both hydrogen and protons, as well as the anode potential. However, these two equations indicate that both θ_{Pt-H} and θ_{Pt} can be expressed as functions of the hydrogen and proton concentrations, which are practically controllable and measurable.

As mentioned previously, both $\overrightarrow{k}_{3H}$ and $\overleftarrow{k}_{3H}$ are electrode potential dependent and can be expressed as follows:

$$\overrightarrow{k}_{3H} = \overrightarrow{k}_{3H}^{o} \exp\left(\frac{\alpha_H n_{\alpha H} F E_a}{RT}\right) \tag{1.16}$$

$$\overleftarrow{k}_{3H} = \overleftarrow{k}_{3H}^{o} \exp\left(-\frac{(1-\alpha_H) n_{\alpha H} F E_a}{RT}\right) \tag{1.17}$$

The reaction rates in the forward and backward directions for Reaction (1.VI) can be expressed as Eqns (1.18) and (1.19), respectively:

$$\overrightarrow{r}_{H_2/H^+} = \overrightarrow{k}_{3H}\theta_{Pt-H}^2 A = \left[\overrightarrow{k}_{3H}^{o}\exp\left(\frac{\alpha_H n_{\alpha H} F E_a}{RT}\right)\right]\theta_{Pt-H}^2 A \tag{1.18}$$

$$\overleftarrow{r}_{H_2/H^+} = \overleftarrow{k}_{3H}\theta_{Pt}^2 C_{H^+}^2 A = \left[\overleftarrow{k}_{3H}^{o}\exp\left(-\frac{(1-\alpha_H) n_{\alpha H} F E_a}{RT}\right)\right]\theta_{Pt}^2 C_{H^+}^2 A \tag{1.19}$$

where $\overrightarrow{r}_{H_2/H^+}$ and $\overleftarrow{r}_{H_2/H^+}$ are the forward and backward HOR rates; $\overrightarrow{k}_{3H}^{o}$ and $\overleftarrow{k}_{3H}^{o}$ are the forward and backward reaction rate constants, which are independent of the electrode potential; A is the electrode surface area; α_H and $n_{\alpha H}$ are, respectively, the electron transfer coefficient and the electron transfer number (here equals to 2) in Reaction (1.VI); E_a is the anode potential; and R and T have the same meaning as in Eqn (1.3). The net reaction rate (r_{H_2/H^+}) can be expressed as in Eqn (1.20):

$$r_{H_2/H^+} = \overrightarrow{r}_{H_2/H^+} - \overleftarrow{r}_{H_2/H^+} = \overrightarrow{k}_{3H}^{o}\theta_{Pt-H}^2 A \exp\left(\frac{\alpha_H n_{\alpha H} F E_a}{RT}\right)$$
$$\tag{1.20}$$
$$- \overleftarrow{k}_{3H}^{o}\theta_{Pt}^2 C_{H^+}^2 A \exp\left(-\frac{(1-\alpha_H) n_{\alpha H} F E_a}{RT}\right)$$

The net reaction rate can also be expressed as the net current density (i_{H_2/H^+}):

$$i_{H_2/H^+} = \frac{r_{H_2/H^+}}{A} = n_H F\left[\overrightarrow{k}_{3H}^{o}\theta_{Pt-H}^2\exp\left(\frac{\alpha_H n_{\alpha H} F E_a}{RT}\right)\right.$$

$$\left. - \overleftarrow{k}_{3H}^{o}\theta_{Pt}^2 C_{H^+}^2\exp\left(-\frac{(1-\alpha_H) n_{\alpha H} F E_a}{RT}\right)\right] \tag{1.21}$$

where n_H is the overall electron transfer number for H_2 electro-oxidation to protons, which is 2. In electrochemistry, Eqn (1.21) is a form of the Butler–Volmer equation.

We now introduce another important parameter for electrode kinetics, called the exchange current density $(i^0_{H_2/H^+})$. This parameter is defined as the current density when the forward and backward electrochemical reaction rates reach an equilibrium state—that is, the forward current density is equal to the backward current density, or the net current density i_{H_2/H^+} in Eqn (1.21) equals zero $(i_{H_2/H^+} = 0)$. In this case, the electrode potential should reach the thermodynamic or reversible or equilibrium electrode potential (E_a becomes $E^r_{H_2/H^+}$). Then Eqn (1.21) becomes Eqn (1.22):

$$nF\overset{\rightarrow o}{k}_{3H}(\theta^o_{Pt-H})^2\exp\left(\frac{\alpha_H n_{\alpha H}FE^r_{H_2/H^+}}{RT}\right)$$

$$= nF\overset{\leftarrow o}{k}_{3H}(\theta^o_{Pt})^2C^2_{H^+}\exp\left(-\frac{(1-\alpha_H)n_{\alpha H}FE^r_{H_2/H^+}}{RT}\right)\right] \tag{1.22}$$

where θ^o_{Pt-H} and θ^o_{Pt} are the surface coverage of the Pt–H and Pt sites at the equilibrium electrode potential. Therefore, the exchange current density can be expressed as in Eqn (1.23):

$$i^0_{H_2/H^+} = nF\overset{\rightarrow o}{k}_{3H}(\theta^o_{Pt-H})^2\exp\left(\frac{\alpha_H n_{\alpha H}FE^r_{H_2/H^+}}{RT}\right)$$

$$= nF\overset{\leftarrow o}{k}_{3H}(\theta^o_{Pt})^2C^2_{H^+}\exp\left(-\frac{(1-\alpha_H)n_{\alpha H}FE^r_{H_2/H^+}}{RT}\right)\right] \tag{1.23}$$

This exchange current density can also be expressed in another form. For example, Eqn (1.23) can be rewritten as Eqn (1.24):

$$E^r_{H_2/H^+} = \frac{2.303RT}{n_{\alpha H}F}\log\left(\frac{\overset{\leftarrow o}{k}_{3H}}{\overset{\rightarrow o}{k}_{3H}}\right) + \frac{2.303RT}{n_{\alpha H}F}\log\left[\frac{(\theta^o_{Pt})^2C^2_{H^+}}{(\theta^o_{Pt-H})^2}\right] \tag{1.24}$$

Combining Eqns (1.23) and (1.24), an alternative expression for the exchange current density can be obtained:

$$i^0_{H_2/H^+} = nF(\overset{\rightarrow o}{k}_{3H})^{1-\alpha_H}(\overset{\leftarrow o}{k}_{3H})^{\alpha_H}(\theta^o_{Pt})^{2\alpha_H}(C_{H^+})^{2\alpha_H}(\theta^o_{Pt-H})^{2(1-\alpha_H)} \tag{1.25}$$

It can be seen that Eqns (1.22)–(1.24), and (1.25) all contain the parameters θ^o_{Pt} and θ^o_{Pt-H}, which actually can be expressed as functions of measurable and controllable parameters such as the concentrations of hydrogen and protons. For example, if the electrode potential E_a in Eqns (1.14) and (1.15) is replaced by the equilibrium electrode potential $E^r_{H_2/H^+}$, θ^o_{Pt} and θ^o_{Pt-H} can be alternatively expressed as Eqns (1.26) and (1.27), respectively:

$$
\theta^o_{Pt} = \left\{ 1 + \frac{\overrightarrow{k}_{1H}}{\overleftarrow{k}_{1H}} C_{H_2} \right.
$$
$$
\left. + \left[\frac{\overrightarrow{k}_{2H} \dfrac{\overrightarrow{k}_{1H}}{\overleftarrow{k}_{1H}} C_{H_2} + \overleftarrow{k}^o_{3H} C^2_{H^+} \exp\left(-\dfrac{(1-\alpha_H)n_{\alpha H}FE^r_{H_2/H^+}}{RT} \right)}{\overleftarrow{k}_{2H} + \overrightarrow{k}^o_{3H} \exp\left(\dfrac{\alpha_H n_{\alpha H} E^r_{H_2/H^+}}{RT} \right)} \right]^{1/2} \right\}^{-1}
$$

(1.26)

$$
\theta^o_{Pt-H} = \left[\frac{\overrightarrow{k}_{2H} \dfrac{\overrightarrow{k}_{1H}}{\overleftarrow{k}_{1H}} C_{H_2} + \overleftarrow{k}^o_{3H} C^2_{H^+} \exp\left(-\dfrac{(1-\alpha_H)n_{\alpha H}FE^r_{H_2/H^+}}{RT} \right)}{\overleftarrow{k}_{2H} + \overrightarrow{k}^o_{3H} \exp\left(\dfrac{\alpha_H n_{\alpha H} E^r_{H_2/H^+}}{RT} \right)} \right]^{1/2}
$$
$$
\times \left\{ 1 + \frac{\overrightarrow{k}_{1H}}{\overleftarrow{k}_{1H}} C_{H_2} + \left[\frac{\overrightarrow{k}_{2H} \dfrac{\overrightarrow{k}_{1H}}{\overleftarrow{k}_{1H}} C_{H_2} + \overleftarrow{k}^o_{3H} C^2_{H^+} \exp\left(-\dfrac{(1-\alpha_H)n_{\alpha H}FE^r_{H_2/H^+}}{RT} \right)}{\overleftarrow{k}_{2H} + \overrightarrow{k}^o_{3H} \exp\left(\dfrac{\alpha_H n_{\alpha H} E^r_{H_2/H^+}}{RT} \right)} \right]^{1/2} \right\}^{-1}
$$

(1.27)

By inserting θ^o_{Pt} and θ^o_{Pt-H} into Eqn (1.25), an alternative expression for exchange current density $\left(i^o_{H_2/H^+}\right)$ can be derived, which contains all reaction constants as well as the measurable and controllable concentrations of both hydrogen and protons.

In fact, this exchange current density can be directly measured by experiments. To do this, another alternative expression for Eqn (1.21) is very useful. First, let us introduce a concept called overpotential. The anode overpotential, η_a, is the difference between the anode potential and its equilibrium potential, and it can be expressed as in Eqn (1.28):

$$
\eta_a = E_a - E^r_{H_2/H^+}
$$

(1.28)

By combining Eqns (1.21), (1.23), and (1.28), one can obtain an alternative Butler–Volmer equation:

$$
i_{H_2/H^+} = i^o_{H_2/H^+} \left[\exp\left(\frac{\alpha_H n_{\alpha H} F \eta_a}{RT} \right) - \exp\left(-\frac{(1-\alpha_H)n_{\alpha H}F \eta_a}{RT} \right) \right]
$$

(1.29)

If the overpotential is large enough, for example, $\eta_a > 60$ mV, the second term on the right-hand side of Eqn (1.29) can be omitted when compared to the first term, and this equation then becomes

$$i_{H_2/H^+} = i^o_{H_2/H^+} \exp\left(\frac{\alpha_H n_{\alpha H} F \eta_a}{RT}\right) \qquad (1.30)$$

By taking the log of both sides of Eqn (1.30), we get

$$E_a = E^r_{H_2/H^+} - \frac{2.303RT}{\alpha_H n_{\alpha H} F} \log(i^o_{H_2/H^+}) + \frac{2.303RT}{\alpha_H n_{\alpha H} F} \log(i_{H_2/H^+}) \qquad (1.31)$$

Equation (1.31) are also known as the Tafel equation. In experiments, the electrode potential (E_a) at different current densities (i_{H_2/H^+}) can be measured. The plot of $\log(i_{H_2/H^+})$ vs. E_a can then be obtained, as shown in Fig. 1.2. From the slope of this plot $\left(= \dfrac{2.303RT}{\alpha_H n_{\alpha H} F},\right.$ called the Tafel slope), the kinetic parameter $\alpha_H n_{\alpha H}$ can be calculated according to Eqn (1.31), and from the intercept $= \left(E^r_{H_2/H^+} - \dfrac{2.303RT}{\alpha_H n_{\alpha H} F} \log(i^o_{H_2/H^+})\right)$ of this plot at $i_{H_2/H^+} = 1.0$, the exchange current density ($i^o_{H_2/H^+}$) can be calculated if $E^r_{H_2/H^+}$ is known. Therefore, three important electrode kinetic parameters—exchange current density ($i^o_{H_2/H^+}$), electron transfer coefficient (α_H), and electron transfer number ($n_{\alpha H}$)—can be experimentally measured based on Tafel Eqn (1.31).

An alternative way to obtain the exchange current density from Eqn (1.29) is to use the data within a very small η_a range near the open circuit potential.

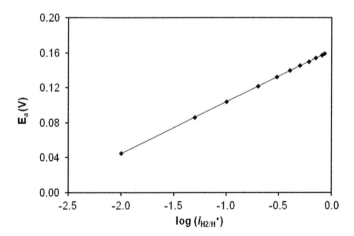

FIGURE 1.2   Tafel plot for the HOR on a Pt/C electrocatalyst at 23 °C.

In this small η_a range, Eqn (1.29) can be approximately simplified as Eqn (1.32):

$$i_{H_2/H^+} = i^o_{H_2/H^+}\left[\left(1 + \frac{\alpha_H n_{\alpha H} F \eta_a}{RT}\right) - \left(1 - \frac{(1-\alpha_H) n_{\alpha H} F \eta_a}{RT}\right)\right]$$

$$= i^o_{H_2/H^+} \frac{n_{\alpha H} F \eta_a}{RT} \tag{1.32}$$

By rearranging Eqn (1.32), we can obtain an alternative expression:

$$R^a_{ct} = \left|\frac{\partial \eta_a}{\partial i_{H_2/H^+}}\right| = \frac{\partial(E_a - E^r_{H_2/H^+})}{\partial i_{H_2/H^+}} = \frac{\partial E_a}{\partial i_{H_2/H^+}} = \frac{RT}{n_{\alpha H} i^o_{H_2/H^+}} \tag{1.33}$$

where R^a_{ct} is defined as the charge transfer resistance for the HOR, which can be measured using electrochemical impedance spectroscopy (EIS) at the open circuit potential. From Eqn (1.33), if the R^a_{ct} and $n_{\alpha H}$ are known, the exchange current density can be calculated. A detailed discussion of EIS measurements will be presented in Chapter 3.

Note that both the exchange current density $(i^o_{H_2/H^+})$ and the electron transfer coefficient (α_H) are temperature dependent, particularly $i^o_{H_2/H^+}$. This is because $i^o_{H_2/H^+}$ is a function of θ^o_{Pt} and θ^o_{Pt-H}, both of which contain all the reaction constants, which are all temperature dependent. The temperature dependency of $i^o_{H_2/H^+}$ can be expressed as the Arrhenius form:

$$\log(i^o_{H_2/H^+}) = \log(I^o_{H_2/H^+}) - \frac{E^o_{H_2}}{2.303R}\frac{1}{T} \tag{1.34}$$

where $I^o_{H_2/H^+}$ is the exchange current density at infinite temperature, and $E^o_{H_2}$ is the reaction activation energy for the HOR. If one can measure the exchange current densities at different temperatures, the plot of $\log(i^o_{H_2/H^+})$ vs. $1/T$ will yield both the reaction activation energy and the exchange current density at infinite temperature.

In the above section, all the equations we derived are based on pure electron transfer kinetics. Unfortunately, in reality, mass transfer (e.g. hydrogen diffusion inside a porous fuel cell CL) will have an effect on the overall reaction rate, and sometimes can become the rate-determining step. To address this mass transfer effect, we need to introduce another concept, called limiting diffusion current density, which can be expressed as in Eqns (1.35) and (1.36) [9]:

$$\overrightarrow{I}^l_{H_2/H^+} = n_H F \overrightarrow{m}_{H_2/H^+} C_{H_2} \tag{1.35}$$

$$\overleftarrow{I}^l_{H_2/H^+} = n_H F \overleftarrow{m}_{H_2/H^+} C_{H^+} \tag{1.36}$$

where $\overrightarrow{I}_{H_2/H^+}^{\,l}$ and $\overleftarrow{I}_{H_2/H^+}^{\,l}$ are the mass transfer limiting current densities for the HOR (forward) and the proton reduction reaction (backward), respectively, while $\overrightarrow{m}_{H_2/H^+}$ and $\overleftarrow{m}_{H_2/H^+}$ are the corresponding mass transfer coefficients. These limiting current densities represent the maximum current densities achieved at a given hydrogen concentration (C_{H_2}) and proton concentration (C_{H^+}). When the current density reaches the limiting current density, the hydrogen or proton concentration at the reaction surface (or catalyst surface) is fully exhausted. The hydrogen or proton supply is totally controlled by the mass transfer process (normally a diffusion process) of this reactant from the bulk phase to the reaction surface. For a fuel cell CL, the reaction zone is a porous matrix structure rather than a planar surface, so the mass transfer process should be more complicated than on a planar electrode surface. For example, the hydrogen concentration C_{H_2} should be considered the average concentration inside the gas-diffusion layer (GDL), and the corresponding mass transfer coefficient $\overrightarrow{m}_{H_2/H^+}$ may be the average mass transfer coefficient inside the CL. To reflect this mass transfer effect of the reactant, Eqn (1.29) can be modified to Eqn (1.37):

$$i_{H_2/H^+} = i^o_{H_2/H^+}\left[\left(1 - \frac{i_{H_2/H^+}}{\overrightarrow{I}^{\,l}_{H_2/H^+}}\right)\exp\left(\frac{\alpha_H n_{\alpha H} F \eta_a}{RT}\right)\right.$$

$$\left. - \left(1 - \frac{i_{H_2/H^+}}{\overleftarrow{I}^{\,l}_{H_2/H^+}}\right)\exp\left(-\frac{(1 - \alpha_H)n_{\alpha H} F \eta_a}{RT}\right)\right] \quad (1.37)$$

Equations (1.37) is the current density after the H_2 and H^+ mass transfer effects are considered. From this equation, it can be seen that if the current density in the forward direction reaches a limiting value $(i_{H_2/H^+} = \overrightarrow{I}^{\,l}_{H_2/H^+})$, the forward kinetic current density will become zero; this suggests that the forward reaction is totally controlled by the mass transfer process. In the backward direction, the same situation is valid.

If the overpotential is large enough, an equation similar to Eqn (1.31), with the mass transfer considered, can be derived as Eqn (1.38):

$$E_a = E^r_{H_2/H^+} - \frac{2.303RT}{\alpha_H n_{\alpha H} F}\log(i^o_{H_2/H^+}) + \frac{2.303RT}{\alpha_H n_{\alpha H} F}\log\left(\frac{\overrightarrow{I}^{\,l}_{H_2/H^+}\, i_{H_2/H^+}}{\overrightarrow{I}^{\,l}_{H_2/H^+} - i_{H_2/H^+}}\right)$$

$$(1.38)$$

1.3.2. Kinetics of the Oxygen Reduction Reaction

As is generally recognized [10], the ORR catalyzed by a Pt catalyst follows a complex reaction mechanism. The oxygen adsorbed on the Pt surface can be electrochemically reduced through several elementary steps, as shown in Reactions (1.VII)–(1.XIII):

$$Pt + O_2 \leftrightarrow Pt - O_{2(ads)} \qquad (1.VII)$$

$$Pt - O_{2(ads)} + H^+ + e^- \leftrightarrow Pt - O_2H_{(ads)} \quad \text{(Rate-determining step)} \qquad (1.VIII)$$

$$xPt - O_2H_{(ads)} + xe^- + xH^+ \leftrightarrow xPt - O_2H_{2(ads)} \qquad (1.IX)$$

$$xPt - O_2H_{2(ads)} \leftrightarrow xPt + xH_2O_2 \qquad (1.X)$$

$$(1 - x)Pt + (1 - x)Pt - O_2H_{(ads)} \leftrightarrow (1 - x)Pt - O_{(ads)}$$

$$+ (1 - x)Pt - OH_{(ads)} \qquad (1.XI)$$

$$(1 - x)Pt - O_{(ads)} + (1 - x)H^+ + (1 - x)e^- \leftrightarrow (1 - x)Pt - OH_{(ads)} \qquad (1.XII)$$

$$2(1 - x)Pt - OH_{(ads)} + 2(1 - x)H^+ + 2(1 - x)e^-$$

$$\leftrightarrow 2(1 - x)Pt + 2(1 - x)H_2O \qquad (1.XIII)$$

In this mechanism, Reaction (1.VII) represents the adsorption/desorption of O_2 at the Pt surface, which should be a fast reaction. Reaction (1.VIII) represents the first electron transfer of the ORR, which is considered the slowest reaction in the whole mechanism (i.e. the rate-determining step), and it produces the intermediate $Pt - H_2O_{(ads)}$. One part of this intermediate can be further reduced to produce H_2O_2 through Reactions (1.IX) and (1.X). The other part will be further reduced to produce H_2O through reactions (1.XI)–(1.XIII). The first part of the reaction forms a 2-electron transfer ORR pathway to produce H_2O_2, and the other part forms a 4-electron transfer ORR pathway to produce H_2O. In reality, the ORR normally has a mixed 2- and 4-electron transfer pathway and gives an overall electron transfer number of <4. The relative proportions of 2- and 4-electron transfer pathways are dependent on Reactions (1.IX) and (1.XI). The proportion of Reaction (1.IX) can be expressed as x, and the proportion of Reaction (1.XI) can be expressed as $(1-x)$. If $x = 1$, the mechanism will be a 2-electron transfer pathway, whereas if $x = 0$, the mechanism will become a totally 4-electron pathway. In practical situations on a Pt catalyst, the x value is much <1, and this gives a 4-electron dominated transfer pathway to produce H_2O, with $<1\%$ H_2O_2.

The ORR mechanism expressed by Reactions (1.VII)–(1.XIII) is very complicated. So, to obtain the relationship between the current and the electrode potential, we may want to make some reasonable assumptions to simplify this mechanism. We may assume that (1) $x = 0$ so that the whole mechanism only goes through a 4-electron pathway to produce water; and (2) the fast Reactions (1.XI)–(1.XIII) can be combined together into one reaction. Therefore, the ORR mechanism may be simplified as in the following three reactions:

$$Pt + O_2 \underset{\overleftarrow{k_{1O}}}{\overset{\overrightarrow{k_{1O}}}{\leftrightarrow}} Pt - O_{2(ads)} \qquad (1.XIV)$$

$$Pt - O_{2(ads)} + H^+ + e^- \underset{\overleftarrow{k_{2O}}}{\overset{\overrightarrow{k_{2O}}}{\leftrightarrow}} Pt - O_2H_{(ads)}$$

$$\text{(Rate-Determining Step)} \qquad (1.XV)$$

$$Pt - O_2H_{(ads)} + 3H^+ + 3e^- \underset{\overleftarrow{k_{3O}}}{\overset{\overrightarrow{k_{3O}}}{\leftrightarrow}} Pt + 2H_2O \qquad (1.XVI)$$

In all three reaction equations, $\overrightarrow{k}_{iO}$ and $\overleftarrow{k}_{iO}$ represent the reaction rate constants in the forward and backward directions, respectively. For electrochemical reactions (1.XV) and (1.XVI), the rate constants $\overrightarrow{k}_{2O}, \overleftarrow{k}_{2O}, \overrightarrow{k}_{3O}$, and $\overleftarrow{k}_{3O}$ are all dependent on the electrode potential, and can be expressed as follows:

$$\overrightarrow{k}_{2O} = \overrightarrow{k}_{2O}^{\,o} \exp\left(-\frac{(1 - \alpha_{2O})n_{\alpha2O}FE_c}{RT}\right) \qquad (1.39)$$

$$\overleftarrow{k}_{2O} = \overleftarrow{k}_{2O}^{\,o} \exp\left(\frac{\alpha_{2O}n_{\alpha2O}FE_c}{RT}\right) \qquad (1.40)$$

$$\overrightarrow{k}_{3O} = \overrightarrow{k}_{3O}^{\,o} \exp\left(-\frac{(1 - \alpha_{3O})n_{\alpha3O}FE_c}{RT}\right) \qquad (1.41)$$

$$\overleftarrow{k}_{3O} = \overleftarrow{k}_{3O}^{\,o} \exp\left(\frac{\alpha_{3O}n_{\alpha3O}FE_c}{RT}\right) \qquad (1.42)$$

where $\overrightarrow{k}_{2O}^{\,o}, \overleftarrow{k}_{2O}^{\,o}, \overrightarrow{k}_{3O}^{\,o}$, and $\overleftarrow{k}_{3O}^{\,o}$ are the forward and backward reaction rate constants for Reactions (1.XV) and (1.XVI), respectively, which are independent of the electrode potential; α_{20} and α_{30} are the electron transfer coefficients; $n_{\alpha 20}$ and $n_{\alpha 30}$ are the electron transfer numbers for Reactions (1.XV) and (1.XVI), respectively; and E_c is the cathode potential.

The concentrations of Pt-related surface species or surface reaction sites such as Pt–O_2, Pt–O_2H, and Pt in Reactions (1.XIV) to (1.XVI) may be expressed as the corresponding surface coverage, such as θ_{Pt-O_2}, θ_{Pt-O_2H}, and θ_{Pt}. They have the following relationship:

$$\theta_{Pt-O_2} + \theta_{Pt-O_2H} + \theta_{Pt} = 1 \tag{1.43}$$

Assuming Reaction (1.XIV) is fast and always in an equilibrium state, the following equation will apply:

$$\theta_{Pt-O_2} = \frac{\overrightarrow{k_{10}}}{\overleftarrow{k_{10}}} C_{O_2} \theta_{Pt} \tag{1.44}$$

where C_{O_2} is the concentration or the partial pressure of oxygen in the CL. By combining Eqns (1.43) and (1.44), we obtain

$$\theta_{Pt-O_2H} + \left(1 + \frac{\overrightarrow{k_{10}}}{\overleftarrow{k_{10}}} C_{O_2}\right) \theta_{Pt} = 1 \tag{1.45}$$

Because Reaction (1.XV) is the slowest and Reaction (1.XVI) is fast, it is reasonable to assume that the surface coverage value of the Pt–O_2H site during the whole reaction process remains constant (steady-state assumption)—that is, the amount of Pt–O_2H produced is equal to the amount consumed. Then, the following equation will apply:

$$\overrightarrow{k_{20}} C_{H^+} \theta_{Pt-O_2} + \overleftarrow{k_{30}} \theta_{Pt} C_{H_2O}^2 = \overleftarrow{k_{20}} \theta_{Pt-O_2H} + \overrightarrow{k_{30}} C_{H^+}^3 \theta_{Pt-O_2H} \tag{1.46}$$

where C_{H^+} is the concentration of protons in the aqueous phase. By combining Eqns (1.39)–(1.42),(1.45),and (1.46), one obtains

$$\theta_{Pt-O_2} = \frac{\overrightarrow{k_{10}}}{\overleftarrow{k_{10}}} C_{O_2}$$

$$\times \frac{\overleftarrow{k_{20}^o}\exp\left(\dfrac{\alpha_{20}n_{\alpha 20}FE_c}{RT}\right) + \overrightarrow{k_{30}^o}C_{H^+}^3\exp\left(-\dfrac{(1-\alpha_{30})n_{\alpha 30}FE_c}{RT}\right)}{\left\{\begin{array}{l}\overrightarrow{k_{20}^o}\dfrac{\overrightarrow{k_{10}}}{\overleftarrow{k_{10}}}C_{O_2}C_{H^+}\exp\left(-\dfrac{(1-\alpha_{20})n_{\alpha 20}FE_c}{RT}\right) + \overleftarrow{k_{20}^o}\left(1 + \dfrac{\overrightarrow{k_{10}}}{\overleftarrow{k_{10}}}C_{O_2}\right)\exp\left(\dfrac{\alpha_{20}n_{\alpha 20}FE_c}{RT}\right) \\[3mm] + \overleftarrow{k_{30}^o}C_{H_2O}^2\exp\left(\dfrac{\alpha_{30}n_{\alpha 30}FE_c}{RT}\right) + \overrightarrow{k_{30}^o}C_{H^+}^3\left(1 + \dfrac{\overrightarrow{k_{10}}}{\overleftarrow{k_{10}}}C_{O_2}\right)\exp\left(-\dfrac{(1-\alpha_{30})n_{\alpha 30}FE_c}{RT}\right)\end{array}\right\}} \tag{1.47}$$

$$\theta_{\text{Pt-O}_2\text{H}} = 1 - \left(1 + \frac{\overrightarrow{k}_{10}}{\overleftarrow{k}_{10}}C_{\text{O}_2}\right)$$

$$\times \frac{\overleftarrow{k}^{o}_{20}\exp\left(\dfrac{\alpha_{20}n_{\alpha20}FE_c}{RT}\right) + \overrightarrow{k}^{o}_{30}C^{3}_{\text{H}^+}\exp\left(-\dfrac{(1-\alpha_{30})n_{\alpha30}FE_c}{RT}\right)}{\left\{\begin{array}{l}\overrightarrow{k}^{o}_{20}\dfrac{\overrightarrow{k}_{10}}{\overleftarrow{k}_{10}}C_{\text{O}_2}C_{\text{H}^+}\exp\left(-\dfrac{(1-\alpha_{20})n_{\alpha20}FE_c}{RT}\right) + \overleftarrow{k}^{o}_{20}\left(1+\dfrac{\overrightarrow{k}_{10}}{\overleftarrow{k}_{10}}C_{\text{O}_2}\right)\exp\left(\dfrac{\alpha_{20}n_{\alpha20}FE_c}{RT}\right) \\[2ex] + \overleftarrow{k}^{o}_{30}C^{2}_{\text{H}_2\text{O}}\exp\left(\dfrac{\alpha_{30}n_{\alpha30}FE_c}{RT}\right) + \overrightarrow{k}^{o}_{30}C^{3}_{\text{H}^+}\left(1+\dfrac{\overrightarrow{k}_{10}}{\overleftarrow{k}_{10}}C_{\text{O}_2}\right)\exp\left(-\dfrac{(1-\alpha_{30})n_{\alpha30}FE_c}{RT}\right)\end{array}\right\}} \tag{1.48}$$

If the electrode potential E_c in Eqns (1.47) and (1.48) is replaced by the equilibrium electrode potential $E^r_{\text{O}_2/\text{H}_2\text{O}}$, θ_{Pt} and $\theta_{\text{Pt-O}_2\text{H}}$ will become θ^o_{Pt} and $\theta^o_{\text{Pt-O}_2\text{H}}$, which can expressed as in Eqns (1.49) and (1.50), respectively:

$$\theta^o_{\text{Pt-O}_2} = \frac{\overrightarrow{k}_{10}}{\overleftarrow{k}_{10}}C_{\text{O}_2}$$

$$\times \frac{\overleftarrow{k}^{o}_{20}\exp\left(\dfrac{\alpha_{20}n_{\alpha20}FE^r_{\text{O}_2/\text{H}_2\text{O}}}{RT}\right) + \overrightarrow{k}^{o}_{30}C^{3}_{\text{H}^+}\exp\left(-\dfrac{(1-\alpha_{30})n_{\alpha30}FE^r_{\text{O}_2/\text{H}_2\text{O}}}{RT}\right)}{\left\{\begin{array}{l}\overrightarrow{k}^{o}_{20}\dfrac{\overrightarrow{k}_{10}}{\overleftarrow{k}_{10}}C_{\text{O}_2}C_{\text{H}^+}\exp\left(-\dfrac{(1-\alpha_{20})n_{\alpha20}FE^r_{\text{O}_2/\text{H}_2\text{O}}}{RT}\right) + \overleftarrow{k}^{o}_{20}\left(1+\dfrac{\overrightarrow{k}_{10}}{\overleftarrow{k}_{10}}C_{\text{O}_2}\right)\exp\left(\dfrac{\alpha_{20}n_{\alpha20}FE^r_{\text{O}_2/\text{H}_2\text{O}}}{RT}\right) \\[2ex] + \overleftarrow{k}^{o}_{30}C^{2}_{\text{H}_2\text{O}}\exp\left(\dfrac{\alpha_{30}n_{\alpha30}FE^r_{\text{O}_2/\text{H}_2\text{O}}}{RT}\right) + \overrightarrow{k}^{o}_{30}C^{3}_{\text{H}^+}\left(1+\dfrac{\overrightarrow{k}_{10}}{\overleftarrow{k}_{10}}C_{\text{O}_2}\right)\exp\left(-\dfrac{(1-\alpha_{30})n_{\alpha30}FE^r_{\text{O}_2/\text{H}_2\text{O}}}{RT}\right)\end{array}\right\}} \tag{1.49}$$

$$\theta^o_{\text{Pt-O}_2\text{H}} = 1 - \left(1 + \frac{\overrightarrow{k}_{10}}{\overleftarrow{k}_{10}}C_{\text{O}_2}\right)$$

$$\times \frac{\overleftarrow{k}^{o}_{20}\exp\left(\dfrac{\alpha_{20}n_{\alpha20}FE^r_{\text{O}_2/\text{H}_2\text{O}}}{RT}\right) + \overrightarrow{k}^{o}_{30}C^{3}_{\text{H}^+}\exp\left(-\dfrac{(1-\alpha_{30})n_{\alpha30}FE^r_{\text{O}_2/\text{H}_2\text{O}}}{RT}\right)}{\left\{\begin{array}{l}\overrightarrow{k}^{o}_{20}\dfrac{\overrightarrow{k}_{10}}{\overleftarrow{k}_{10}}C_{\text{O}_2}C_{\text{H}^+}\exp\left(-\dfrac{(1-\alpha_{20})n_{\alpha20}FE^r_{\text{O}_2/\text{H}_2\text{O}}}{RT}\right) + \overleftarrow{k}^{o}_{20}\left(1+\dfrac{\overrightarrow{k}_{10}}{\overleftarrow{k}_{10}}C_{\text{O}_2}\right)\exp\left(\dfrac{\alpha_{20}n_{\alpha20}FE^r_{\text{O}_2/\text{H}_2\text{O}}}{RT}\right) \\[2ex] + \overleftarrow{k}^{o}_{30}C^{2}_{\text{H}_2\text{O}}\exp\left(\dfrac{\alpha_{30}n_{\alpha30}FE^r_{\text{O}_2/\text{H}_2\text{O}}}{RT}\right) + \overrightarrow{k}^{o}_{30}C^{3}_{\text{H}^+}\left(1+\dfrac{\overrightarrow{k}_{10}}{\overleftarrow{k}_{10}}C_{\text{O}_2}\right)\exp\left(-\dfrac{(1-\alpha_{30})n_{\alpha30}FE^r_{\text{O}_2/\text{H}_2\text{O}}}{RT}\right)\end{array}\right\}} \tag{1.50}$$

Again, Eqns (1.49) and (1.50) are fairly complicated and contain all the reaction constants, the concentrations of both oxygen and protons, and the cathode potential. However, these two equations indicate that both $\theta_{\text{Pt-O}_2\text{H}}$ and θ_{Pt} can be expressed as functions of the oxygen and proton concentrations, which are practically controllable and measurable.

The overall ORR rate can be expressed as in Eqn (1.51):

$$r_{\text{O}_2/\text{H}_2\text{O}} = \overrightarrow{r}_{\text{O}_2/\text{H}_2\text{O}} - \overleftarrow{r}_{\text{O}_2/\text{H}_2\text{O}}$$

$$= \overrightarrow{k}^{o}_{20}\theta_{\text{Pt-O}_2}C_{\text{H}^+}A\exp\left(-\frac{(1-\alpha_{20})n_{\alpha20}FE_c}{RT}\right) - \overleftarrow{k}^{o}_{20}\theta_{\text{Pt-O}_2\text{H}}A\exp\left(\frac{\alpha_{20}n_{\alpha20}FE_c}{RT}\right) \tag{1.51}$$

Then, the net ORR current density can be expressed as in Eqn (1.52):

$$
i_{O_2/H_2O} = \frac{r_{O_2/H_2O}}{A}
$$

$$
= n_O F \left[\overset{\rightarrow o}{k}_{20} \theta_{Pt-O_2} C_{H^+} \exp\left(-\frac{(1-\alpha_{20})n_{\alpha 20}FE_c}{RT} \right) \right.
$$

$$
\left. - \overset{\leftarrow o}{k}_{20} \theta_{Pt-O_2H} \exp\left(\frac{\alpha_{20}n_{\alpha 20}FE_c}{RT} \right) \right] \tag{1.52}
$$

where, in both equations, $\vec{r}_{O_2/H_2O}$ and $\overset{\leftarrow}{r}_{H_2/H^+}$ are the forward and backward ORR rates, respectively, and n_o is the overall electron transfer number for the ORR, which is 4. By using the method as for the anode HOR, the exchange current density $(i^0_{O_2/H_2O})$ for the cathode ORR can also be derived as Eqn (1.53):

$$
i^0_{O_2/H_2O} = n_O F \overset{\rightarrow o}{k}_{20} \theta^o_{Pt-O_2} C_{H^+} \exp\left(-\frac{(1-\alpha_{20})n_{\alpha 20}FE^r_{O_2/H_2O}}{RT} \right)
$$

$$
= n_O F \overset{\leftarrow o}{k}_{20} \theta^o_{Pt-O_2H} \exp\left(\frac{\alpha_{20}n_{\alpha 20}FE^r_{O_2/H_2O}}{RT} \right) \tag{1.53}
$$

where $\theta^o_{Pt-O_2H}$ and θ^o_{Pt} are the surface coverage of the $Pt-O_2H$ and Pt sites at the equilibrium cathode potential $(E^r_{O_2/H_2O})$. This exchange current density can also be expressed in another form. For example, from Eqn (1.53), the equilibrium cathode potential can be expressed as follows:

$$
E^r_{O_2/H_2O} = \frac{2.303RT}{n_{\alpha 20}F} \log\left(\frac{\overset{\rightarrow o}{k}_{20}}{\overset{\leftarrow o}{k}_{20}} \right) + \frac{2.303RT}{n_{\alpha 20}F} \log\left(\frac{\theta^o_{Pt-O_2} C_{H^+}}{\theta^o_{Pt-O_2H}} \right) \tag{1.54}
$$

If we combine Eqns (1.53) and (1.54), we get an alternative expression of exchange current density:

$$
i^0_{O_2/H_2O} = n_O F (\overset{\rightarrow o}{k}_{20})(\overset{\leftarrow o}{k}_{30})^{1-\alpha_{20}} (\theta^o_{Pt-O_2})^{\alpha_{20}} (C_{H^+})^{\alpha_{20}} (\theta^o_{Pt-O_2H})^{1-\alpha_{20}} \tag{1.55}
$$

The right-hand side of Eqn (1.55) contains seven parameters: n_O, $\overset{\rightarrow o}{k}_{20}$, $\overset{\leftarrow o}{k}_{20}$, $\theta^o_{Pt-O_2}$, $\theta^o_{Pt-O_2H}$, C_{H^+}, and α_{20}. Theoretically, if these seven parameters are known, one can calculate the exchange current density. However, both $\theta^o_{Pt-O_2}$ and $\theta^o_{Pt-O_2H}$, as shown in Eqns (1.49) and (1.50), contain too many parameters to be practical for calculation.

Fortunately, as discussed in the above section on the HOR, the exchange current density can be directly measured experimentally. The cathode overpotential, η_c, is the difference between the cathode potential and its equilibrium potential and can be expressed as in Eqn (1.56):

$$
\eta_c = E_c - E^r_{O_2/H_2O} \tag{1.56}
$$

On combining Eqns (1.52), (1.53), and (1.56), one obtains an alternative Butler–Volmer equation:

$$i_{O_2/H_2O} = i^o_{O_2/H_2O}\left[\exp\left(-\frac{(1-\alpha_{2O})n_{\alpha2O}F\eta_c}{RT}\right) - \exp\left(\frac{\alpha_{2O}n_{\alpha2O}F\eta_c}{RT}\right)\right]$$

(1.57)

If the overpotential is large enough, for example, $\eta_c > 60$ mV, the second term on the right-hand side of Eqn (1.57) can be omitted when compared to the first term; this equation will then become

$$i_{O_2/H_2O} = i^o_{O_2/H_2O}\exp\left(-\frac{(1-\alpha_{2O})n_{\alpha2O}F\eta_c}{RT}\right)$$

(1.58)

By taking the log for both the sides of Eqn (1.58), we get

$$E_c = E^r_{O_2/H_2O} + \frac{2.303RT}{(1-\alpha_{2O})n_{\alpha2O}F}\log(i^o_{O_2/H_2O}) - \frac{2.303RT}{(1-\alpha_{2O})n_{\alpha2O}F}\log(i_{O_2/H_2O})$$

(1.59)

Equation (1.59) are also known as the Tafel equation. In experiments, the electrode potential (E_c) at different current densities (i_{O_2/H_2O}) can be measured. Then, the plot of E_c vs. $\log(i_{O_2/H_2O})$ can be obtained, as shown in Figure 1.3. From the slope ($= \dfrac{2.303RT}{(1-\alpha_{2O})n_{\alpha2O}F}$, called the Tafel slope), the kinetic parameter $(1-\alpha_{2O})n_{\alpha2O}$ can be calculated according to Eqn (1.59), and from the intercept ($= E^r_{O_2/H_2O} + \dfrac{2.303RT}{(1-\alpha_{2O})n_{\alpha2O}F}\log(i^o_{O_2/H_2O})$) of this plot at $i_{O_2/H_2O} = 1.0$, the exchange current density ($i^o_{O_2/H_2O}$) can be calculated if $E^r_{O_2/H_2O}$ is known. Therefore, three important electrode kinetic

FIGURE 1.3 Tafel plot for the ORR on Pt/C electro-catalyst at 23 °C.

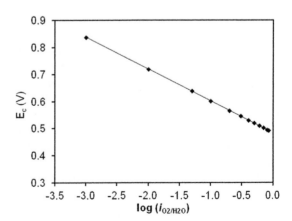

parameters—exchange current density $(i^o_{O_2/H_2O})$, electron transfer coefficient $(1 - \alpha_{2O})$, and electron transfer number $(n_{\alpha 2O})$—can be experimentally measured based on Tafel Eqn (1.59).

Same as the case for the HOR, in a very small η_c range near the open circuit potential, a simplified form of Eqn (1.57) can also be obtained:

$$
\begin{aligned}
i_{O_2/H_2O} &= i^o_{O_2/H_2O}\left[\left(1 - \frac{(1-\alpha_{2O})n_{\alpha 2O}F\eta_c}{RT}\right) - \left(1 + \frac{\alpha_{2O}n_{\alpha 2O}F\eta_c}{RT}\right)\right] \\
&= i^o_{O_2/H_2O}\left(-\frac{n_{\alpha 2O}F\eta_c}{RT}\right)
\end{aligned}
$$

$$(1.60)$$

By rearranging Eqn (1.60), we obtain an alternative expression:

$$
R^c_{ct} = \left|-\frac{\partial \eta_c}{\partial i_{O_2/H_2O}}\right| = \left|-\frac{\partial(E_c - E^r_{O_2/H_2O})}{\partial i_{O_2/H_2O}}\right| = \frac{\partial E_c}{\partial i_{O_2/H_2O}} = \frac{RT}{n_{\alpha 2O}i^o_{O_2/H_2O}}
$$

$$(1.61)$$

where R^c_{ct} is the charge transfer resistance for the ORR, which can be measured using EIS at the open circuit potential. From Eqn (1.61), if the R^c_{ct} and $n_{\alpha 2O}$ are known, the exchange current density can be calculated.

Similar to the case for the HOR, both the exchange current density $(i^o_{O_2/H_2O})$ and the electron transfer coefficient (α_{2O}) are temperature dependent. The temperature dependency of $i^o_{O_2/H_2O}$ can be expressed as in the Arrhenius form:

$$
\log(i^o_{O_2/H_2O}) = \log(I^o_{O_2/H_2O}) - \frac{E^o_{O_2}}{2.303R}\frac{1}{T}
$$

$$(1.62)$$

where $I^o_{O_2/H_2O}$ is the exchange current density at infinite temperature, and $E^o_{O_2}$ is the reaction activation energy for the ORR. The plot of $\log(i^o_{O_2/H_2O})$ vs. $1/T$ allows one to obtain both the reaction activation energy and the exchange current density at infinite temperature.

Similar to the situation for the HOR, the mass transfer of oxygen inside a porous fuel cell CL will also have an effect on the overall reaction rate. The limiting diffusion current densities can be expressed as in Eqns (1.63) and (1.64):

$$
\overrightarrow{I}^l_{O_2/H_2O} = n_O F \overrightarrow{m}_{O_2/H_2O} C_{O_2}
$$

$$(1.63)$$

$$
\overleftarrow{I}^l_{O_2/H_2O} = n_O F \overleftarrow{m}_{O_2/H_2O} C_{H_2O}
$$

$$(1.64)$$

where $\overrightarrow{I}^l_{O_2/H_2O}$ and $\overleftarrow{I}^l_{O_2/H_2O}$ are the mass transfer limiting current densities for the ORR (forward) and the water oxidation reaction (backward), respectively, and $\overrightarrow{m}_{O_2/H_2O}$ and $\overleftarrow{m}_{O_2/H_2O}$ are the corresponding mass transfer coefficients. Similar to the case for the HOR, Eqn (1.57) can be modified to Eqn (1.65):

$$i_{O_2/H_2O} = i^o_{O_2/H_2O}\left[\left(1 - \frac{i_{O_2/H_2O}}{\overrightarrow{I}^l_{O_2/H_2O}}\right)\exp\left(-\frac{(1-\alpha_{2O})n_{\alpha 2O}F\eta_c}{RT}\right)\right.$$

$$\left. -\left(1 - \frac{i_{O_2/H_2O}}{\overleftarrow{I}^l_{O_2/H_2O}}\right)\exp\left(\frac{\alpha_{2O}n_{\alpha 2O}F\eta_c}{RT}\right)\right] \tag{1.65}$$

This equation is the current density expression after the O_2 and H_2O mass transfer effects are taken into account.

If the overpotential is large enough and the mass transfer is taken into account, Eqn (1.65) can be simplified as Eqn (1.66):

$$E_c = E^r_{O_2/H_2O} + \frac{2.303RT}{(1-\alpha_{2O})n_{\alpha 2O}F}\log(i^o_{O_2/H_2O})$$

$$-\frac{2.303RT}{(1-\alpha_{2O})n_{\alpha 2O}F}\log\left(\frac{i_{O_2/H_2O}\,\overrightarrow{I}^l_{O_2/H_2O}}{\overrightarrow{I}^l_{O_2/H_2O} - i_{O_2/H_2O}}\right) \tag{1.66}$$

It is worth pointing out that the equations for both the HOR and the ORR in the above sections are for cases when the electrochemical reactions occur on a smooth planar electrode or on a catalyst surface rather than in a porous matrix CL. The situation in the CL may be more complicated than on the planar surface. However, on modification by using the apparent parameters and the real electrochemical active surface, one finds that the equations are still valid for quantitative treatment of experimental data.

1.4. PEM FUEL CELL CURRENT–VOLTAGE EXPRESSION

As shown in Fig. 1.1, the center of a single fuel cell contains the anode, membrane (or solid electrolyte), and cathode. When a current density (I_{cell}) passes though this cell, the cell voltage (V_{cell}) can be expressed as follows:

$$V_{cell} = E_c - E_a - I_{cell}R_m \tag{1.67}$$

where R_m is the membrane resistance (or electrolyte resistance), which will receive detailed attention in Chapter 5. By combining this equation with Eqns (1.38) and (1.59), and letting both the anode and cathode current densities be equal to the cell current density (i.e. $I_{cell} = i_{H_2/H^+} = i_{O_2/H_2O}$), the fuel cell voltage can be derived with Eqn (1.68):

$$V_{cell} = E^r_{O_2/H_2O} - E^r_{H_2/H^+} + \frac{2.303RT}{(1-\alpha_{2O})n_{\alpha2O}F}\log(i^o_{O_2/H_2O})$$

$$+ \frac{2.303RT}{\alpha_H n_{\alpha H}F}\log(i^o_{H_2/H^+}) - \frac{2.303RT}{(1-\alpha_{2O})n_{\alpha2O}F}\log\left(\frac{i_{O_2/H_2O}\,\overrightarrow{I}^l_{O_2/H_2O}}{\overrightarrow{I}^l_{O_2/H_2O} - i_{O_2/H_2O}}\right)$$

$$- \frac{2.303RT}{\alpha_H n_{\alpha H}F}\log\left(\frac{\overrightarrow{I}^l_{H_2/H^+}i_{H_2/H^+}}{\overrightarrow{I}^l_{H_2/H^+} - i_{H_2/H^+}}\right) - I_{cell}R_m$$

$$(1.68)$$

where the term $(E^r_{O_2/H_2O} - E^r_{H_2/H^+})$ is the fuel cell theoretical OCV (V^{OCV}_{cell}). Note that Eqn (1.68) are only valid when both the anode and the cathode overpotentials are >60 mV; otherwise, a large error will result. Based on Eqn (1.68), one can use experimental data to simulate the kinetic parameters shown in this equation. It is also worth noting that when using this equation, the effects of the CL structure and catalyst morphology should be considered before carrying out the simulation.

The relationship between cell voltage and current density can be measured using fuel cell hardware; a typical schematic polarization curve is shown in Fig. 1.4. It can be seen that there are four losses in the whole current density range: (1) the OCV loss caused by H_2 crossover and cathode mixed potential,

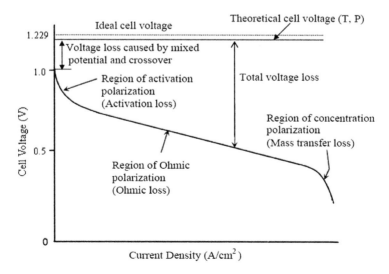

FIGURE 1.4   Typical polarization curve of a PEM fuel cell [12].

(2) the loss caused by slow reaction kinetics, (3) the loss caused by membrane resistance, and (4) the loss caused by mass transfer limitation. These four losses can also be reflected by Eqn (1.68). This will be discussed in detail in Chapter 4.

The fuel cell OCV (V_{cell}^{OCV}) in Eqn (1.68) can be expressed as follows:

$$V_{cell}^{OCV} = E_{O_2/H_2O}^r - E_{H_2/H^+}^r \tag{1.69}$$

At standard conditions (1.0 atm and 25 °C), if we use both $E_{O_2/H_2O}^r = 1.229$ V (vs. SHE) and $E_{H_2/H^+}^r = 0.000$ V (vs. SHE), the fuel cell OCV should be 1.229 V. However, the measured OCV at room temperature is around 1.0 V, which is much lower than the value calculated from thermodynamic electrode potentials. This difference can be explained by the losses caused by H_2 crossover and mixed cathode potential, which result from the oxidation of the Pt surface at the cathode side [11]. Thus, the measured fuel cell OCV could be expressed by using Eqn (1.70):

$$V_{measured}^{OCV} = V_{cell}^{OCV} - \Delta V_{H_2-xover} - \Delta V_{PtO-mixed} \tag{1.70}$$

where $V_{measured}^{OCV}$ is the measured OCV, $\Delta V_{H_2-xover}$ is the voltage loss caused by the hydrogen crossover from the anode to the cathode, and $\Delta V_{Pt-O-mixed}$ is the voltage loss caused by the cathode Pt surface mixed potential. The factors that affect fuel cell OCV will be discussed in detail in Chapter 7.

1.5. FUEL CELL COMPONENTS

Figure 1.5 shows a single PEM fuel cell. The key components include the membrane electrode assembly (MEA), flow field plate/bipolar plate, current collector, as well as other components such as sealing materials and end plate.

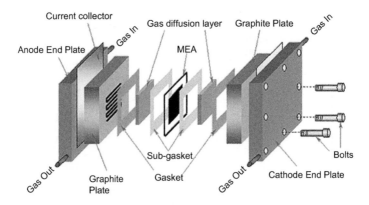

FIGURE 1.5   The structure of a typical PEM fuel cell [13]. (For color version of this figure, the reader is referred to the online version of this book.)

The MEA is a unit comprising the anode, membrane, and cathode and consists of the electrocatalyst, CL, GDL, and membrane. The following sections will discuss these key components separately.

1.5.1. Fuel Cell Electrocatalysts

In a H_2/air PEM fuel cell, the HOR and the ORR take place in their respective CLs. Thus, the anode and cathode electrocatalysts both play critical roles in fuel cell performance. To date, the most active and widely employed catalysts in PEM fuel cells are highly dispersed Pt-based catalysts. Although they pose several challenges, such as costliness, sensitivity to impurities/contaminants, and insufficient stability/durability under fuel cell operating conditions [14], Pt-based catalysts are recognized as the most practical choice in current PEM fuel cell technology.

The sluggish kinetics of the cathode ORR means that more active electrocatalysts are definitely necessary to overcome the large overpotential on the cathode. Although the current carbon-supported Pt catalysts have the most active catalytic activity toward the ORR, they are still not fully satisfactory in terms of activity and stability. In recent years, one focus of catalyst studies has been to further develop more active and durable catalysts.

During the previous few decades of PEM fuel cell development, many kinds of Pt-based electrocatalysts have been developed for the ORR. These include Pt and Pt-based binary alloy catalysts, such as PtFe [15–18], PtCo [17,19–22], and PtNi [17,21]. In particular, the Pt-based binary alloy catalysts show enhanced catalytic activity for the ORR due to their electronic and configuration effects. Normally, these catalyst particles (size ~2–6 nm) are dispersed on supports with particle sizes of 20–50 nm to increase the active surface area. The most commonly used supports are carbon materials, which have large surface areas and good electronic conductivity as well as relative chemical and thermal stabilities. The widely used carbon support materials are carbon blacks, carbon nanotubes, and carbon nanofibers. However, one drawback of such carbon supports is their electrochemical instability (called carbon corrosion) under PEM fuel cell operating conditions, in particular in high-temperature operations. Carbon support corrosion causes Pt catalyst particle isolation and CL deactivation, which results in fuel cell performance degradation. In recent years, to overcome this issue, much effort has been put into exploring noncarbon support materials. These new support materials include metal oxides, carbides, and nitrides. Although some performance has been achieved with these supports, they are still not as good as carbon supports in terms of ORR activity and fuel cell performance. To lower the cost of Pt-based catalysts, intensified efforts have been made in recent years to reduce Pt loading in the cathode CL and/or replace Pt using non-noble metal catalysts. The latter approach in particular has become a hot area in the last several years. The non-noble metal catalysts include carbides [23–25], nitrides [26,27], transition metal chalcogenides [28–30], metal macrocycles (e.g. porphyrin and phthalocyanine) [31,32], and others. The most promising non-noble metal catalysts for the ORR seem to be Fe- and Co-based nitrogen-containing catalysts obtained by heat treatment at 700–1000 °C [31,33–39]. Although these catalysts exhibit high electrocatalytic activity toward the ORR, in practical terms, there is still a long way to go before they can be compared with Pt-based catalysts. It is believed that non-noble metal catalysts are the necessary choice for sustainable fuel cell commercialization in terms of cost and global abundance. Regarding anode catalysts, Pt-based catalysts seem to be good enough for catalyzing the HOR process. Since the amount of Pt catalyst required at the anode is much less than at the cathode, little attention is being paid to reducing anode catalyst loading or replacing Pt catalysts by using non-noble metal catalysts. However, the anode Pt-based catalysts are very sensitive to impurities, such as carbon monoxide (CO) and hydrogen sulfide (H_2S) in the fuel. Therefore, various kinds of CO-tolerant catalysts have been developed, including carbon-supported PtRu [40–42], PtSn [40,43,44], PtMo [45–48], and PtAu [49–51]. All of these showed improved tolerance to impurities (e.g. CO) and thus enhanced fuel cell performance.

In the development of fuel cell catalysts, catalyst synthesis plays a critical role in improving catalyst activity and stability. Over the last several decades, many syntheses methods have been developed, including the impregnation–reduction, colloid, sol–gel, and microwave-assisted methods. Experimental results showed all these methods to be effective in synthesizing catalysts for PEM fuel cells. The most important progress in recent years has been in the synthesis of nanostructured catalysts for fuel cell applications [52]. Nanostructured Pt-based catalysts are claimed to be much more active than the commercially available Pt/C catalysts.

1.5.2. Catalyst Layers

The anode and cathode CLs are the key components in PEM fuel cells because both the ORR and the HOR take place within them to yield fuel cell performance. They are thin layers (~10–100 μm, usually <50 μm), mainly composed of catalyst powders, proton conductive ionomer (normally Nafion® ionomer), and polytetrafluoroethylene (PTFE).

An ideal CL should provide passages for proton transport, electron transport, and reactant gas transport and should also be able to remove product water. In general, there are two kinds of CLs: hydrophobic CL and hydrophilic CL. The composition ratio in the CL greatly influences fuel cell performance and will be discussed in detail in Chapter 2.

1.5.3. Gas Diffusion Layer

In a single fuel cell, there are two pieces of GDLs, located on either side of the MEA. The GDL can provide support for the CL as well as passages for the transport of reactant gases and water. It can also provide electronic contact with the flow field plates. Usually, a GDL includes two layers: the backing layer or gas-diffusion media (GDM) and the microporous layer (MPL).

The commonly used GDMs are carbon-fiber-based materials such as carbon paper and carbon cloth, which have a high porosity ($\geq 70\%$) and good electrical conductivity [53]. Figure 1.6 shows the scanning electron microscope (SEM) micrographs of carbon paper and carbon cloth. GDMs are usually pretreated to change their hydrophilicity/hydrophobicity. For example, PTFE resin is often used to impregnate carbon paper to increase its hydrophobicity.

The MPL also plays an important role in improving GDL performance. It can be made as follows: a certain amount of carbon powder, solvent (e.g. isopropanol), and bonder (e.g. PTFE) is mixed thoroughly to form a slurry, which when coated onto the GDM forms a thin MPL layer. The MPL can not only change the porosity of the GDM, thus improving both water transport and gas transport in the GDL, but it can also support the CL to prevent the catalyst from penetrating the GDM. The composition of the MPL, including that of carbon powder and PTFE, can significantly affect the GDL performance and thereby the fuel cell performance [54–58]. Studies have shown that an MPL containing 15–20% PTFE exhibits the best performance [59,60]. The properties and loadings of the particular carbon powder can also affect the PTFE content in an MPL [57–59]. To enhance the transportation of both liquid water and reactant gases, a bifunctional MPL was also suggested recently; this employed a composite carbon powder consisting of 80 wt.% acetylene black and 20 wt.% Black Pearls®2000 carbon [61].

1.5.4. Membrane (or Solid Electrolyte)

The membrane is one of the key components of PEM fuel cells, as it not only acts as a solid electrolyte to conduct protons but also serves to separate the

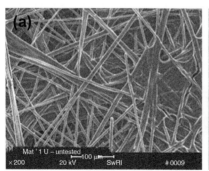

FIGURE 1.6   SEM micrographs of two gas-diffusion-medium substrates, both with approximately 7-μm diameter fibers. (a) Carbon-fiber paper, Spectracorp 2050 A, with no PTFE. The reference bar indicates 100 μm. (b) Carbon cloth, Textron Avcarb 1071 HCB. The reference bar indicates 600 μm [53].

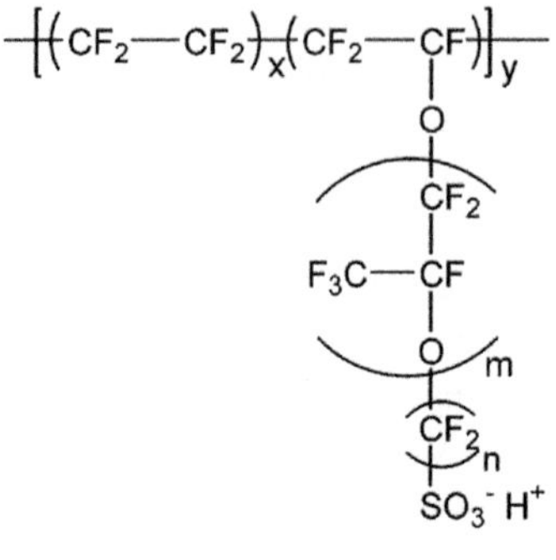

$$-\left[\left(CF_2-CF_2\right)_x\left(CF_2-CF\right)_y\right]-$$

Flemion: m=0 or 1, n=1-5

Nafion: m≥1, n=2

x=5-13.5, y=1000

$SO_3^-\ H^+$

FIGURE 1.7   Molecular structure of Nafion® [62].

anode and the cathode. Those commonly used are perfluorosulfonic acid (PFSA) membranes, such as Nafion® and Flemion®. Nafion® is a sulfonated tetrafluoroethylene-based polymer developed by DuPont in the late 1960s. There are three kinds of groups in the structure of Nafion®: the tetrafluoroethylene (Teflon®)-like backbone, the $-O-CF_2-CF-O-CF_2-CF_2-$ side chains that connect the backbone and the third group, and the clusters consisting of sulfonic acid ions, as shown in Fig. 1.7.

Nafion® membrane is a nonreinforced film based on Nafion® ionomers in acid (H^+) form. Under PEM fuel cell operating conditions, the Nafion® membrane is the most practical solid electrolyte due to its unique structure, excellent thermal and mechanical stability, and high proton conductivity (but it is also an isolator for electronic conduction). According to the cluster network model [63–66] of Nafion® membrane, Nafion® contains some sulfonic ion clusters with a diameter of approximately 4 nm. The clusters are equally distributed within a continuous fluorocarbon lattice, and are interconnected by narrow channels with a diameter of about 1 nm; this provides passages for the transport of protons. Figure 1.8 shows this cluster network model. Proton

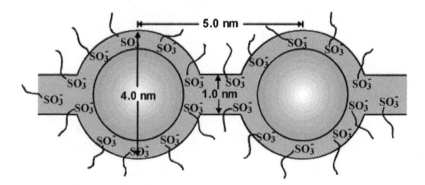

FIGURE 1.8   Cluster network model for the Nafion® membrane [63–66]. (For color version of this figure, the reader is referred to the online version of this book.)

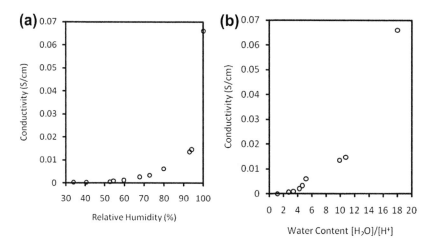

FIGURE 1.9 Conductivity of Nafion® as a function of RH (a) and water content (b) [68].

transport will occur through these channels and bring 3–5 water molecules together from the anode to the cathode.

The proton conductivity of PFSA depends on the membrane's relative humidity (RH) or the water content. Figure 1.9 shows that Nafion® conductivity increases with increasing values of RH or water content. PEM fuel cell testing results have also indicated that membrane conductivity increases with increasing RH [67]. The temperature can also affect proton conductivity.

1.5.5. Membrane Electrode Assembly

In a PEM fuel cell, a membrane having both anode and cathode sides coated with CLs is called an MEA. The MEA is the core of a PEM fuel cell, as it provides the location for the electrochemical reactions. A typical five-layer MEA includes the anode GDL, anode CL, membrane, cathode CL, and cathode GDL. As the reaction in a CL is "three phased," an MEA should provide passages for the transport of electrons, protons, and reactant gases. Furthermore, it should also have the ability to remove product water. During the recent decades of fuel cell development, many MEA structures have been developed, and they will be discussed in detail in Chapter 2.

1.5.6. Flow Field Plate/Bipolar Plate

The flow field plate, an electronic conductive metal or nonmetal plate with flow channels on one side, also plays an important role in determining fuel cell performance. In a fuel cell stack, the flow channels occur on both sides of the plate, and it is then called a bipolar plate. The flow field plate or bipolar plate provides structural support for the mechanically weak MEA, supplies reactant gases to the electrodes, and removes product water from its flow channels [69]. In a single cell, the flow field provides the electronic connection with the MEA and current collector, and in a stack, the bipolar plate provides the connection between two adjacent cells. The requirements for bipolar plate materials include high electronic conductivity, good chemical and mechanical stability, impermeability to reactant gases, low cost, light weight, and easy fabrication. The most commonly used bipolar plates in PEM fuel cells are graphite plates and metal plates (a metal plate is usually coated with Ni or Au). A graphite plate has a high electronic conductivity and chemical stability. However, its cost and weight are not satisfactory because it is thicker than a metal plate, which can be made to be very thin and has good conductivity. However, the disadvantage of a metal plate is that it corrodes in a PEM fuel cell operating environment [70].

The configuration of the flow field and the dimensions of the flow channels (such as the width and depth of the channel and the width of the ribs) can significantly affect the distribution of reactant gases in the flow channels and the removal of product water [71–74]. In the quest to improve fuel cell performance, various flow field patterns have been employed, such as the (a) pin-type flow field, (b) straight parallel flow field, (c) interdigitated flow field, (d) single serpentine flow field, and (e) multiple serpentine flow field [12,69]. In practice, the most typical patterns used in PEM fuel cells are the straight parallel and serpentine flow fields. Figure 1.10 shows a straight parallel flow field, which includes many straight parallel flow channels connected to the gas inlet and outlet. Because the flow channels are short and there are no channel direction changes inside the plate, the gas pressure drop is small, which may result in nonuniform gas distribution over the whole flow field. Another problem associated with this small gas pressure drop is low water removal ability. This results in the accumulation of water in the flow channels and forms water droplets and blocks the channels for gas transport, which causes the performance to drop dramatically.

To overcome the drawbacks of the straight parallel flow field design, a serpentine flow field pattern has been developed. As shown in Fig. 1.11, in the serpentine flow field, the reactant gas flows mainly along the flow channel, which leads to a uniform gas distribution inside the plate. In addition, the gas pressure drop from the inlet to the outlet is large, which favors water removal.

1.5.7. Current Collectors

A current collector is a plate attached to a flow field plate or bipolar plate to collect the current generated by fuel cell reactions. Sometimes, metal flow field plates/bipolar plates can also serve as current collectors in a single PEM fuel cell or a stack. The electrons generated by the HOR at the anode must be conducted through the anodic electrode and current collector and must then

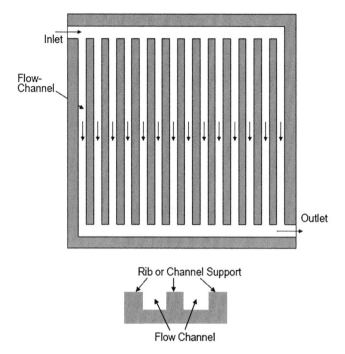

FIGURE 1.10 Straight parallel flow field, and flow channel cross-section [69].

travel through an external circuit and enter the cell via the cathode current collector and electrode. The material used for the current collector should have a good electronic conductivity, strong electrochemical and mechanical stability, low cost, and a light weight to reduce the weight of the fuel cell stack. Typical materials include copper, stainless steel, titanium, and aluminum. To increase their electronic conductivity, their surfaces are usually coated by another metal

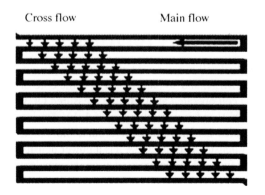

FIGURE 1.11 Serpentine flow field design [75,76].

(e.g. gold-coated copper and gold-coated aluminum). Copper and stainless steel are inexpensive, robust, and easy to fabricate. However, they are also dense and heavy, which decreases the mass specific power density of a fuel cell stack. Exfoliated graphite has been explored as a nonmetallic current collector in PEM fuel cells [77]; this material exhibited good performance, but it is not mechanically robust, so a stack constructed from exfoliated graphite must be protected.

1.5.8. Other Components

Sealing poses a challenge in PEM fuel cells. Sealing materials are placed between the MEA and flow field plates/bipolar plates to prevent gas and coolant leakage; hence, sealing failure will result in reactant gas leakage and resultant safety problems. A good sealing material should meet the following require-ments: good chemical and electrochemical stability in PEM fuel cell envi-ronments, good thermal stability, suitable compressibility, low cost, low gas permeability, and easy fabrication. The commonly used materials are PTFE films and silicon elastomers [78]. For PTFE-based sealing, compressibility is a challenge, so a larger blade pressure is required for fuel cell assembly. For silicon-based sealing, gasket degradation is a concern during long-term fuel cell operation. It has been recognized that decomposition products from silicon materials can deposit on or into the CLs and can poison the catalyst as well as change the hydrophilic/hydrophobic properties of the CL, resulting in fuel cell degradation [79].

The end plate is another important component in a PEM fuel cell or fuel cell stack. End plates are placed at each end of the fuel cell anode and cathode sides. Sometimes, the end plates also serve as flow field plates with flow channels on one side. For example, when a metal plate is used as the flow field plate, the end plates at the anode and cathode sides also serve as the flow field plates on their respective sides. However, a separate plate is often used as the end plate to assemble a fuel cell if graphite plates are used as the flow fields. To reduce the weight of fuel cell stacks, robust but light materials, such as aluminum and polymers, are often used for the end plates.

1.6. SINGLE CELL AND FUEL CELL STACK OPERATION

As shown in Fig. 1.5, a single PEM fuel cell is composed of end plates, current collectors, flow field plates, gaskets, MEA, and so on. The MEA is the core of the single fuel cell. For a H_2/air fuel cell operated at 80 °C, the measured OCV is much lower than the theoretical OCV and is normally around 1.0 V. During operation, a single fuel cell generates a cell voltage of around 0.6–0.8 V, depending on the controlled current density, and the MEA has a power density of <1.0 W cm^{-2}. Therefore, a single fuel cell cannot be used as an independent power unit in applications.

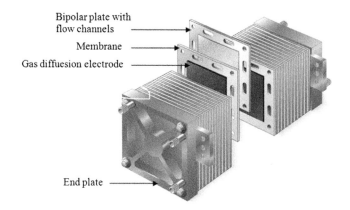

Bipolar plate with
flow channels

Membrane

Gas diffuesion electrode

End plate

FIGURE 1.12   Typical structure of a PEM fuel cell stack [80]. (For color version of this figure, the reader is referred to the online version of this book.)

To achieve higher voltage and power density, several cells have to be connected in series to form a fuel cell stack. As shown in Fig. 1.12, the bipolar plate and MEA are the repeat units in the stack. From the number of single cells in a stack, the maximum voltage and power can be determined. In a fuel cell stack, the fuel passage and oxidant passage are shared by all the single cells. Due to the heat generated during operation, a cooling system with coolant circulation is required as an accessory to the stack to control the temperature. For operating a PEM fuel cell stack, one also requires several other accessories, such as an air supply system (air compressor), a fuel pump, humidifiers, temperature controllers, and electronic systems for DC to AC conversion, as well as fuel cell load and voltage control. Normally, these accessories consume about 30% of the energy generated by the stack. This 30% energy loss is called parasitic power loss. The PEM fuel cell stack needs to have a service life time of 5000 h for automobile applications, and of 40,000–50,000 h for stationary applications.

1.7. FUEL CELL PERFORMANCE

1.7.1. Fuel Cell Power Density

In a PEM fuel cell, the MEA power density (P_{MEA}) is a product of the generated voltage and current density and is expressed in watts per square centimeter:

$$P_{MEA} = I_{cell} V_{cell} \qquad (1.71)$$

Figure 1.13 shows the power density curve obtained using a H_2/air PEM fuel cell at 80 °C and 3.0 atm with 100% RH. It can be seen that the power

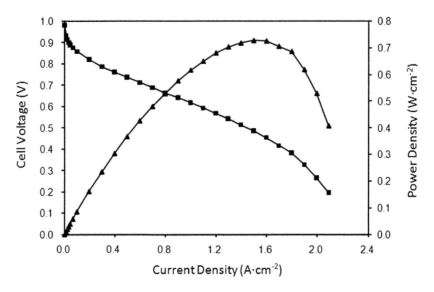

FIGURE 1.13 Polarization and power density curves of a H_2/air PEM fuel cell at 80 °C and 3.0 atm with 100% RH. Active area: 4.4 cm^2; stoichiometries of H_2 and O_2 are 1.5 and 2.0, respectively.

density initially increases with increasing current density, reaching a peak value of 0.74 W cm^{-2} at a current density of approximately 1.6 A cm^{-2}. The peak power density is usually called the maximum power density and is often used to evaluate the MEA performance of a fuel cell. From Fig. 1.13, it is obvious that the maximum power density is not achieved at maximum current density.

For a fuel cell stack, mass- and volume-specific power densities are more useful parameters. The mass power density is the ratio of power to stack weight, measured in kilowatts per kilogram. Clearly, light-weight stack materials are required to increase the mass power density of a stack. Volume power density is the ratio of power to stack volume, measured in kilowatts per liter or kilowatts per cubic meter. Thus, for increasing the volume power density, one requires to reduce the stack size and simplify the stack system.

1.7.2. Fuel Crossover

In general, the membranes used in PEM fuel cells, such as Nafion® membranes, are thin (typically 25–50 μm). The permeation of reactant gases (e.g. hydrogen and oxygen) across the membrane is not negligible in H_2/air PEM fuel cells. This gas permeation is also called gas crossover. Compared to oxygen crossover, H_2 crossover is much more pronounced in PEM fuel cells because H_2 is smaller than O_2. The hydrogen crossover rate is related to fuel cell operating conditions that include temperature, backpressure, RH, and current density

[81,82]. In general, hydrogen crossover has a negative impact on fuel cells; this includes reduced fuel efficiency, decreased fuel cell OCV [11], and accelerated degradation of the membrane and the CLs [81,83].

Factors affecting hydrogen crossover rate and the impact of crossover on fuel cell performance will receive detailed attention in Chapter 6.

1.7.3. Practical Electrical Energy Efficiency of Fuel Cells

In Section 1.2.3, we discussed the thermodynamic efficiency of a PEM fuel cell as the ratio of theoretical electrical energy (ΔG_{cell}) to the overall reaction heat energy (ΔH). This is the maximum efficiency that can be obtained from a fuel cell. However, in reality, fuel cell efficiency is less than this ideal due to several losses in fuel cell performance. Therefore, the actual electrical energy efficiency (η_e) should be written as in Eqn (1.72):

$$\eta_e = \frac{nFV_{cell}}{\Delta H} \times 100\% \qquad (1.72)$$

In addition, the amount of supplied hydrogen is usually more than what is consumed. For example, the hydrogen stoichiometry (λ_H) is usually controlled at 1.2–2.0. Thus, the hydrogen utilization (σ_H) is <100%. Hydrogen utilization can be expressed as follows:

$$\sigma_H = \frac{1}{\lambda_H} \qquad (1.73)$$

Therefore, the fuel cell electrical energy efficiency expressed in Eqn (1.72) should be modified by Eqn (1.73) to give Eqn (1.74):

$$\eta_e = \frac{nFV_{cell}}{\Delta H} \frac{1}{\lambda} \times 100\% \qquad (1.74)$$

In addition, if some hydrogen is lost due to hydrogen crossover or diffusion out of the fuel cell, the fuel cell efficiency will be further reduced. If we assume that the current generated by hydrogen loss is i_{loss} and the actual current of the fuel cell is I_{cell}, then the electrical energy efficiency of a fuel cell can be further modified as in Eqn (1.75):

$$\eta_e = \frac{nFV_{cell}}{\Delta H} \frac{1}{\lambda} \frac{I_{cell}}{I_{cell} + i_{loss}} \times 100\% \qquad (1.75)$$

For example, at room temperature, if a H_2/air fuel cell is operated at a current of 100 A ($i = 100$ A) and a cell voltage of 0.6 V, and the hydrogen stoichiometry is 1.2, and the hydrogen loss current is 5 A ($i_{loss} = 5$A), then the calculated fuel cell efficiency according to Eqn (1.75) is only about 35.5% (with $\Delta H = 285.8$ kJ mol^{-1} and $n = 2$), which is much smaller than the theoretical electrical energy efficiency of 83%.

1.8. FUEL CELL OPERATING CONDITIONS

1.8.1. Operating Temperature

Temperature is one of the critical operating conditions. Conventional PEM fuel cells are usually operated at temperatures <90 °C because the proton conductivity of the Nafion® membrane is strongly dependent on the membrane's water content. Above 90 °C, the membrane will become dehydrated, which leads to lower proton conductivity and thus decreased fuel cell performance. In recent years, high-temperature PEM fuel cells (operated at >90 °C) have received much attention due to their advantages over conventional PEM fuel cells, such as enhanced electrode kinetics, improved impurity tolerance, and simplified water and thermal management [84]. Thus, membranes that can operate at >90 °C are highly desirable. One kind of high-temperature membrane, based on polybenzimidazole [85–87], has been extensively explored in recent years. Although an increase in the operating temperature can have several benefits, as mentioned above [88], there are also several disadvantages, such as decreased OCV and accelerated degradation of fuel cell components. The effects of operating temperature on fuel cell performance will be discussed in detail in Chapter 4.

1.8.2. Operating Pressure

Pressure is another important operating condition. PEM fuel cells can be operated under a wide pressure range, from ambient to 5 atm. An increase in the operating pressure can effectively enhance the electrode kinetics and improve the mass transport process and can result in better fuel cell performance [89]; it can also increase fuel cell OCV. However, higher operating pressures require more power to supply reactant gases to the fuel cell, causing more parasitic power loss. Normally, PEM fuel cells are operated between 1 and 3 atm. The effects of operating pressure on fuel cell performance will be discussed in detail in Chapter 9.

1.8.3. Relative Humidity

The RH is defined as the ratio of water vapor pressure (P_{H_2O}) to saturated water vapor pressure $(P^\circ_{H_2O}(T))$ at a temperature T, and can be expressed as in Eqn (1.76) [67]:

$$\text{RH} = \frac{P_{H_2O}}{P^\circ_{H_2O}(T)} \times 100\% \tag{1.76}$$

where $P^\circ_{H_2O}(T)$ is the saturated water vapor pressure at temperature T, which defines the maximum amount of water that can be present in the gas, and it is

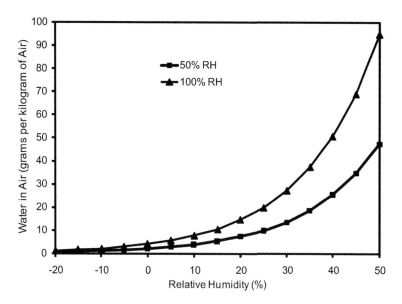

FIGURE 1.14 Amount of water in air at 50% and 100% RHs across a range of temperatures [91].

a function of temperature only [90]. For a mixture of gas and water, the RH can be expressed as in Eqn (1.77):

$$\text{RH} = \frac{P_{\text{total}} - P_{\text{g}}}{P_{\text{H}_2\text{O}}^{\circ}(T)} \times 100\% \tag{1.77}$$

where P_{total} and P_{g} are the total pressure of the system and the gas partial pressure, respectively. As shown in Fig. 1.14, the water content increases with increasing temperature and RH.

In a PEM fuel cell, the RH is a very important operating condition because the proton conductivity of Nafion® is proportional to the water content of the membrane. During fuel cell operation, the reactant gases H_2 and air pass through their respective humidifiers to increase their RHs before they are fed into the fuel cell. In general, an increase in the RH can improve fuel cell performance. However, an increase in the RH can also cause water management difficulty inside the fuel cell, which results in electrode "water flooding." The effects of RH on fuel cell performance will be discussed in detail in Chapter 8.

1.8.4. Gas Flow Rates and Stoichiometries

During operation of a PEM fuel cell, it is necessary to control gas flow rates or stoichiometries because they can significantly affect performance. According to Faraday's Law, the needed amounts of H_2 and O_2 in a H_2/air PEM fuel cell operated at a current density (I_{cell}) can be expressed as follows:

$$M_{H_2}^{needed} = \frac{I_{cell}A_{MEA}}{2F} \tag{1.78}$$

$$M_{O_2}^{needed} = \frac{I_{cell}A_{MEA}}{4F} \tag{1.79}$$

where $M_{H_2}^{needed}$ and $M_{O_2}^{needed}$ are needed hydrogen (mol s^{-1}) and needed oxygen (mol s^{-1}), respectively, and A_{MEA} is the active MEA area. Here the units for M_{H_2} and M_{O_2} are mol s^{-1}, and the units for I_{cell}, A_{MEA}, and F are A cm^{-2} (or C s^{-1}), cm^2, and 96,487 C mol^{-1}, respectively. Because $M_{H_2}^{needed}$ and $M_{O_2}^{needed}$ have a unit of mol s^{-1}, they are also called the needed mass flow rates for H_2 and O_2.

However, in fuel cell operation, the amounts of supplied reactant gases are normally more than those of actually consumed gases. The ratio between the supplied and needed (consumed) gases is called the stoichiometry (λ). Obviously, this stoichiometry is directly related to the gas utilization efficiency. For fuel cell operation, the stoichiometries of H_2 (λ_{H_2}) and O_2 (λ_{O_2}) are normally controlled at 1.2–2.0 and 1.5–2.5, respectively. According to the stoichiometries, the corresponding gas flow rates can be controlled. Normally, the fuel cell gas supply system only controls the gas flow rate at the fuel cell inlet. The flow rate at the outlet is the difference between the inlet flow rate and the gas consumption rate.

Considering the stoichiometry, the controlled mass flow rates of H_2 and O_2 at the inlets of a fuel cell can be expressed as in Eqns (1.80) and (1.81):

$$M_{H_2}^{controlled} = \frac{I_{cell}A_{MEA}\lambda_{H_2}}{2F} \tag{1.80}$$

$$M_{O_2}^{controlled} = \frac{I_{cell}A_{MEA}\lambda_{O_2}}{4F} \tag{1.81}$$

where $M_{H_2}^{controlled}$ and $M_{O_2}^{controlled}$ are the controlled mass flow rates (mol s^{-1}) at the fuel cell inlets for H_2 and O_2, respectively. If H_2 and O_2 are treated as ideal gases, then the volume flow rates for H_2 and O_2 at standard conditions (1 atm, 25 °C) can be expressed as in Eqns (1.82) and (1.83):

$$V_{H_2}^{controlled} = M_{H_2}^{controlled} = \frac{I_{cell}A_{MEA}\lambda_{H_2}}{2F} \times V_m \times 60 \tag{1.82}$$

$$V_{O_2}^{controlled} = M_{O_2}^{controlled} = \frac{I_{cell}A_{MEA}\lambda_{O_2}}{2F} \times V_m \times 60 \tag{1.83}$$

where $V_{H_2}^{controlled}$ and $V_{O_2}^{controlled}$ are the controlled volume flow rates (L min^{-1}) at the fuel cell inlets for H_2 and O_2 at standard conditions, respectively, and V_m is the volume of one molar ideal gas at standard conditions, with a value of 22.4 L mol^{-1}. Of course, Eqns (1.82) and (1.83) should be calibrated according to the actual operating pressure and temperature. For a H_2/air PEM fuel cell, the controlled volume flow rate for air ($V_{air}^{controlled}$) can be expressed as in Eqn (1.84), by calibrating Eqn (1.83) with the molar ratio of O_2 in air (~0.21):

$$V_{air}^{controlled} = \frac{1}{0.21} \frac{I_{cell}A_{MEA}\lambda_{O_2}}{4F} \times V_m \times 60 \qquad (1.84)$$

Equations (1.83) and (1.84) can also be expressed as weight flow rates:

$$W_{H_2}^{controlled} = \frac{I_{cell}A_{MEA}\lambda_{H_2}}{2F} \times m_{H_2} \times 60 \qquad (1.85)$$

$$W_{O_2}^{M} = \frac{I_{cell}A_{MEA}\lambda_{O_2}}{4F} \times m_{O_2} \times 60 \qquad (1.86)$$

where $W_{H_2}^{controlled}$ and $W_{O_2}^{controlled}$ are the weight flow rates of H_2 and O_2 at the fuel cells inlets in g min^{-1}, and m_{H_2} and m_{O_2} are the molar weights of H_2 and O_2, with respective values of 2 g mol^{-1} and 32 g mol^{-1}. Similarly, the weight flow rate of air at the cathode inlet can be expressed as in Eqn (1.87):

$$W_{air}^{controlled} = \frac{1}{0.21} \frac{I_{cell}A_{MEA}\lambda_{O_2}}{4F} \times m_{O_2} \times 60 \qquad (1.87)$$

where $W_{air}^{controlled}$ is the weight flow rate of air at the cathode inlet (g min^{-1}).

1.9. CHAPTER SUMMARY

This chapter mainly deals with the fundamentals of H_2/air PEM fuel cells, including fuel cell reaction thermodynamics and kinetics, as well as a brief introduction to the single fuel cell and the fuel cell stack. The electrochemistry and reaction mechanisms of H_2/air fuel cell reactions, including the anode HOR and the cathode ORR, are discussed in depth. Several concepts related to PEM fuel cell performance, such as fuel cell polarization curves, OCV, hydrogen crossover, and fuel cell efficiencies, are also introduced. With respect to fuel cell structures and components, the material properties and effects on fuel cell performance are also discussed. In addition, several important conditions for fuel cell operation, including temperature, pressure, RH, and gas stoichiometries and flow rates, and their effects on fuel cell operation, are also briefly presented. This chapter provides the requisite baseline knowledge for the remaining chapters.

REFERENCES

[1] http://www1.eere.energy.gov/hydrogenandfuelcells/fuelcells/fc_types.html.
[2] Larminie J, Dicks A, editors. Fuel cell systems explained. John Wiley & Sons, Ltd; 2003.
[3] Stonehart P, Kohlmayr G. Electrochim Acta 1972;17:369–82.
[4] Vogel W, Lundquist L, Ross P, Stonehart P. Electrochim Acta 1975;20:79–93.
[5] Watanabe M, Igarashi H, Yosioka K. Electrochim Acta 1995;40:329–34.
[6] Springer TE, Raistrick ID. J Electrochem Soc 1989;136:1594–603.
[7] Kinoshita K, editor. Electrochiemical oxygen technology. New York: John Wiley & Sons; 1992.
[8] Services EG, Parsons I, editors. Fuel cell handbook. 5th ed. Morgantown, West Virginia: DOE; 2000.
[9] Bard AJ, Faulkner LR, editors. Electrochemical methods, fundamentals and applications. 2nd ed. New York: Wiley & Sons Inc; 2001.
[10] Liu H, Zhang J. Electrocatalysis of direct methanol fuel cells: from fundamentals to applications. In: Claude Lamy CC, Alonso-Vante Nicolas, editors. Weiheim: Wiley-VCH; 2009.
[11] Zhang J, Tang Y, Song C, Zhang J, Wang H. J Power Sources 2006;163:532–7.
[12] Zhang J, Zhang J. In: Zhang J, editor. PEM fuel cell electrocatalysts and catalyst layers - fundamentals and applications. London: Springer; 2008. p. 965–1002.
[13] http://www.scientific-computing.com/features/feature.php?feature_id=126.
[14] Siroma Z, Ishii K, Yasuda K, Inaba M, Tasaka A. J Power Sources 2007;171:524–9.
[15] Tae Hwang J, Shik Chung J. Electrochim Acta 1993;38:2715–23.
[16] Li W, Zhou W, Li H, Zhou Z, Zhou B, Sun G, et al. Electrochim Acta 2004;49:1045–55.
[17] Xiong L, Kannan AM, Manthiram A. Electrochem Commun 2002;4:898–903.
[18] Shukla AK, Raman RK, Choudhury NA, Priolkar KR, Sarode PR, Emura S, et al. J Electroanal Chem. 2004;563:181–90.
[19] Yu P, Pemberton M, Plasse P. J Power Sources 2005;144:11–20.
[20] Xu Q, Kreidler E. J He T. Electrochim Acta 2010;55:7551–7.
[21] Travitsky N, Ripenbein T, Golodnitsky D, Rosenberg Y, Burshtein L, Peled E. J Power Sources 2006;161:782–9.
[22] He Q, Mukerjee S. Electrochim Acta 2010;55:1709–19.
[23] Yang XG, Wang CY. Appl Phys Lett 2005;86:224104–224103.
[24] Lee K, Ishihara A, Mitsushima S, Kamiya N, Ota K-I. Electrochim Acta 2004;49:3479–85.
[25] Meng H, Shen PK. Electrochem Commun 2006;8:588–94.
[26] Zhong H, Zhang H, Liu G, Liang Y, Hu J, Yi B. Electrochem Commun 2006;8:707–12.
[27] Zhong H, Zhang H, Liang Y, Zhang J, Wang M, Wang X. J Power Sources 2007;164:572–7.
[28] SarI Ozenler S, KadIrgan F. J Power Sources 2006;154:364–9.
[29] Papageorgopoulos DC, Liu F, Conrad O. Electrochim Acta 2007;52:4982–6.
[30] Alonso-Vante N, Tributsch H, Solorza-Feria O. Electrochim Acta 1995;40:567–76.
[31] Zhang L, Lee K, Bezerra CWB, Zhang J, Zhang J. Electrochim Acta 2009;54:6631–6.
[32] Baker R, Wilkinson DP, Zhang J. Electrochim Acta 2008;53:6906–19.
[33] Rajesh B, Piotr Z. Nature 2006;443:63.
[34] Lefevre M, Proietti E, Jaouen F, Dodelet J-P. Science 2009;324:71–4.
[35] Bezerra CWB, Zhang L, Lee K, Liu H, Marques ALB, Marques EP, et al. Electrochim Acta 2008;53:4937–51.
[36] Li S, Zhang L, Liu H, Pan M, Zan L. Zhang J. Electrochim Acta 2010;55:4403–11.
[37] Lalande G, Faubert G, Côté R, Guay D, Dodelet JP, Weng LT. Bertrand P. J Power Sources 1996;61:227–37.

[38] Faubert G, Côté R, Guay D, Dodelet JP, Dénès G, Bertrandc P. Electrochim Acta 1998;43: 341–53.

[39] Charreteur F, Jaouen F, Dodelet J-P. Electrochim Acta 2009;54:6622–30.

[40] Lee D, Hwang S, Lee I. J Power Sources 2005;145:147–53.

[41] Ioroi T, Akita T, Yamazaki S-i, Siroma Z, Fujiwara N, Yasuda K. Electrochim Acta 2006;52: 491–8.

[42] Pitois A, Pilenga A, Tsotridis G. Appl Catal A 2010;374:95–102.

[43] Lim D-H, Choi D-H, Lee W-D, Lee H-I. Appl Catal B 2009;89:484–93.

[44] Lee SJ, Mukerjee S, Ticianelli EA, McBreen J. Electrochim Acta 1999;44:3283–93.

[45] Santiago EI, Camara GA, Ticianelli EA. Electrochim Acta 2003;48:3527–34.

[46] Papageorgopoulos DC, Keijzer M, de Bruijn FA. Electrochim Acta 2002;48:197–204.

[47] Ordóñez LC, Roquero P, Sebastian PJ, Ramírez J. Int J Hydrogen Energy 2007;32:3147–53.

[48] Lebedeva NP, Janssen GJM. Electrochim Acta 2005;51:29–40.

[49] Ma L, Zhang H, Liang Y, Xu D, Ye W, Zhang J, et al. Catal Commun 2007;8:921–5.

[50] Wang J, Yin G, Liu H, Li R, Flemming RL, Sun X. J Power Sources 2009;194:668–73.

[51] Hernández-Fernández P, Rojas S, Ocón P, de Frutos A, Figueroa JM, Terreros P, et al. J Power Sources 2008;177:9–16.

[52] Bing Y, Liu H, Zhang L, Ghosh D, Zhang J. Chem Soc Rev 2010.

[53] Mathias M, Roth J, Fleming J, Lehnert W. In: Vielstich W, Gasteiger HA, Lamm A, editors. Handbook of fuel cells-fundamentals, technology and applications, vol. 3. London: John Wiley & Sons, Ltd; 2003.

[54] Tseng C-J, Lo S-K. Energy Convers Manage 2010;51:677–84.

[55] Park G-G, Sohn Y-J, Yim S-D, Yang T-H, Yoon Y-G, Lee W-Y, et al. J Power Sources 2006;163:113–8.

[56] Antolini E, Passos RR, Ticianelli EA. J Power Sources 2002;109:477–82.

[57] Jordan LR, Shukla AK, Behrsing T, Avery NR, Muddle BC, Forsyth M. J Power Sources 2000;86:250–4.

[58] Passalacqua E, Squadrito G, Lufrano F, Patti A, Giorgi L. J Appl Electrochem 2001;31: 449–54.

[59] Paganin VA, Ticianelli EA, Gonzalez ER. J Appl Electrochem 1996;26:297–304.

[60] Lufrano F, Passalacqua E, Squadrito G, Patti A, Giorgi L. J Appl Electrochem 1999;29: 445–8.

[61] Wang X, Zhang H, Zhang J, Xu H, Zhu X, Chen J, et al. J Power Sources 2006;162:474–9.

[62] Saito M, Hayamizu K, Okada T. J Phys Chem B 2005;109:3112–9.

[63] Heitner-Wirguin C. J Memb Sci 1996;120:1–33.

[64] Mauritz KA, Moore RB. Chem Rev 2004;104:4535–86.

[65] Hsu WY, Gierke TD. J Memb Sci 1983;13:307–26.

[66] Gierke TD, Munn GE, Wilson FC. J Polym Sci B Polym Phys 1981;19:1687–704.

[67] Zhang J, Tang Y, Song C, Xia Z, Li H, Wang H, et al. Electrochim Acta 2008;53:5315–21.

[68] Anantaraman AV, Gardner CL. J Electroanal Chem 1996;414:115–20.

[69] Li X, Sabir I. Int J Hydrogen Energy 2005;30:359–71.

[70] Antunes RA, Oliveira MCL, Ett G. Ett V. Int J Hydrogen Energy 2010;35:3632–47.

[71] Kumar A, Reddy RG. J Power Sources 2003;113:11–8.

[72] Kumar A, Reddy RG. J Power Sources 2006;155:264–71.

[73] Yoon Y-G, Lee W-Y, Park G-G, Yang T-H, Kim C-S. Int J Hydrogen Energy 2005;30:1363–6.

[74] Yoon Y-G, Lee W-Y, Park G-G, Yang T-H, Kim C-S. Electrochim Acta 2004;50:709–12.

[75] Nguyen TV. ECS Trans 2006;3:1171–80.

[76] Li H, Tang Y, Wang Z, Shi Z, Wu S, Song D, et al. J Power Sources 2008;178:103–17.

[77] Hentall PL, Lakeman JB, Mepsted GO, Adcock PL, Moore JM. J Power Sources 1999;80: 235–41.

[78] Frisch L. Sealing Technol 2001;2001:7–9.

[79] Schulze M, Knöri T, Schneider A, Gülzow E. J Power Sources 2004;127:222–9.

[80] G.T.P. Database, https://www.ticona-photos.com/sites/PhotoDB/PL/Forms/DispForm.aspx? ID=1420.

[81] Inaba M, Kinumoto T, Kiriake M, Umebayashi R, Tasaka A, Ogumi Z. Electrochim Acta 2006;51:5746–53.

[82] Cheng X, Zhang J, Tang Y, Song C, Shen J, Song D, et al. J Power Sources 2007;167:25–31.

[83] Collier A, Wang H, Zi Yuan X, Zhang J, Wilkinson DP. Int J Hydrogen Energy 2006;31: 1838–54.

[84] Zhang J, Xie Z, Zhang J, Tang Y, Song C, Navessin T, et al. J Power Sources 2006;160: 872–91.

[85] Li QF, Rudbeck HC, Chromik A, Jensen JO, Pan C, Steenberg T, et al. J Memb Sci 2010;347: 260–70.

[86] Kongstein OE, Berning T, Børresen B, Seland F, Tunold R. Energy 2007;32:418–22.

[87] Ainla A, Brandell D. Solid State Ionics 2007;178:581–5.

[88] Zhang J, Tang Y, Song C, Cheng X, Zhang J, Wang H. Electrochim Acta 2007;52:5095–101.

[89] Zhang J, Li H, Zhang J. ECS Trans 2009;19:65–76.

[90] http://hyperphysics.phy-astr.gsu.edu/hbase/kinetic/watvap.html.

[91] http://en.wikipedia.org/wiki/Relative_humidity.

Chapter 2

Design and Fabrication of PEM Fuel Cell MEA, Single Cell, and Stack

PEM Fuel Cell Testing and Diagnosis. http://dx.doi.org/10.1016/B978-0-444-53688-4.00002-4

2.1. INTRODUCTION

The design and fabrication of proton exchange membrane (PEM) fuel cell components, single cells, and stacks are two of the most important processes in fuel cell technology development. In general, the design and assembly of a fuel cell can have a strong effect on its performance. Given the materials and components used in the fuel cell, design and fabrication have to be optimized with respect to the corresponding fuel cell power output to achieve the best performance. To date, designs and assembly methods have been optimized and validated using fuel cell testing as well as real operation in various application systems, such as portable power devices, stationary power generators, and automobiles. The major challenges still hindering their commercialization are high cost and insufficient durability. The basic components of H_2/air (O_2) PEM fuel cells/stack have been briefly introduced in Chapter 1. In this chapter, the designs of the key components of PEM fuel cells and the resultant effects on cell performance will be discussed in detail, including the fabrication of the PEM fuel cell membrane electrode assembly (MEA), single cell, and stack.

2.2. MEA DESIGN AND ASSEMBLY

The MEA consists of an anodic electrode, PEM, and a cathodic electrode. Because the electrode reactions take place in the MEA, it is the heart of a H_2/air PEM fuel cell. The components of an MEA include the anode gas diffusion medium (A-GDM), anode microporous layer (A-MPL), anode catalyst layer (A-CL), PEM, cathode catalyst layer (C-CL), cathode microporous layer (C-MPL), and cathode gas diffusion medium (C-GDM), as shown in Fig. 2.1. A commonly used term is "gas diffusion layer" (GDL), which actually contains the gas-diffusion-medium layer and the microporous layer. Each component shown in Fig. 2.1 has specific characteristics and functions in fuel cell operation and performance. Therefore, they differ significantly in their design and fabrication. These topics will be addressed in the following sections.

2.2.1. Gas Diffusion Layer Design

The GDL is a key component in H_2/air PEM fuel cells. As shown in Fig. 2.1, the GDL has a gas diffusion medium as a backing layer and a microporous layer as a sublayer. The A-CL or C-CL is coated on this MPL sublayer. The GDL performs the following functions: (1) providing passages for gas diffusion through the GDL from the flow channel to the CL, (2) providing electron pathways from the CL to the flow field or vice versa, (3) retaining some water on its surface to maintain proton conductivity through the PEM, (4) removing excess water to prevent the CL and GDL from flooding, and (5) serving as a physical microporous support for the CL when it is applied onto the GDL. A GDL design must include the design of both the GDM and the MPL.

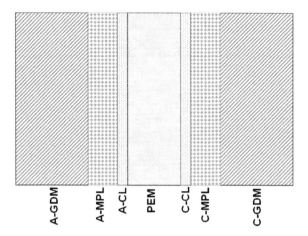

FIGURE 2.1 Schematic structure of the MEA, including A-GDM, A-MPL, A-CL, PEM, C-CL, C-MPL, and C-GDM. (For color version of this figure, the reader is referred to the online version of this book.)

2.2.1.1. Gas Diffusion Medium

Carbon-fiber paper [1–4] and carbon-fiber cloth [5–10] with a thickness of 100–300 μm are widely used as the GDM for H_2/air PEM fuel cells due to their high porosity (>70%) and good electronic conductivity. In Chapter 1, Fig. 1.6 showed scanning electron micrographs of carbon fiber paper and cloth. The typical properties of these GDMs are listed in Table 2.1. Recently, metallic porous materials have also been explored as GDM candidates. For example, Zhang et al. [11] studied a 12.5-μm thick copper foil as a GDM, and the results showed that the thinness and straight-pore feature of this material improved water management.

The hydrophobicity/hydrophilicity of a GDM is an important property for gas transportation. Although commercially available carbon paper and/or carbon cloth are hydrophobic, they are usually pretreated using hydrophobic materials (such as fluorinated ethylene propylene (FEP) [13] and polytetrafluoroethylene (PTFE) [14,15]), which seems to be a necessary step to further increase their hydrophobicity and thereby prevent "water flooding." The details of GDM pretreatment will be covered in Section 2.3.1.

2.2.1.2. Microporous Layer

The MPL is a substrate layer applied onto the GDM, and it consists of carbon or graphite powders with a hydrophobic material (such as PTFE) as a binder. In the industry, the MPL is also known as a carbon sublayer. The MPL thickness is typically optimized to suit the fuel cell operating conditions [16,17], while the MPL's average pore diameter is <20 μm [18,19]. The primary purpose of the

TABLE 2.1 Typical Properties of Carbon Fiber Paper and Carbon Fiber Cloth [12]

	Method	Carbon-Fiber Paper*	Woven Fabric**
Thickness (mm)	Calipers at 7 kPa	0.19	0.38
Areal weight (g m^{-2})	Gravimetric	85	118
Density (g cm^{-3})	At 7 kPa calculated	0.45	0.31
Resistance (through-plane, Ω cm^2)	Two flat graphite blocks at 1.3 MPa	0.009[†]	0.005[†]
Bulk resistivity (through-plane. Ω cm)	Mercury contacts	0.08	Not available
Bulk resistivity (in-plane, Ω cm)	Four-point probe	0.0055[‡]	0.009[‡]
Gas permeability (through-plane, Darcys)	Gurley 4301 permeometer	8[$]	55[$]
Material description		Toray TGP-H-060	Avcarb 1071 HCB

* Reported by Toray (unless indicated otherwise).
** Reported by Ballard Material Systems (unless indicated otherwise).
[†]Measured at General Motors (GM), includes diffusion-medium bulk resistance and two contact resistances (plate to diffusion media).
[‡]Measured at GM. uncompressed, average of resistivity in machine, and crossmachine direction.
[$]Measured at GM, uncompressed, see Eqn (2.12), 1 Darcy = 10^{-12} m^2.

MPL is to improve water management [20,21] and redistribute the reactants, as (1) it can wick liquid water away from the CL to the GDL and thus facilitate gas transportation in the opposite direction [19]; (2) it changes the porosity of the GDM, preventing the catalyst ink from penetrating the GDM; and (3) it reduces the contact resistance between the GDL and the adjacent CL and thus decreases the internal resistance of a PEM fuel cell.

Normally, the MPL is fabricated by thoroughly mixing carbon or graphite powders, solvents (such as ethanol or isopropanol), and binders (such as PTFE) to form a paste that is then spread onto the GDM using the doctor blade technique [22], spraying [23,24], painting [25], rolling [24], or screen printing [24]. The GDM with MPL is afterward put into an oven and heated slowly to 350 °C to evaporate the solvent and other organics, and to sinter the binder. The properties of an MPL thus formed can be changed by adjusting its composition (e.g. the loadings of binder and carbon) [13,18,26–28] and the properties of the carbon or graphite powders [16,17,22,23,29,30].

2.2.1.3. Effect of PTFE Loading in GDL on Fuel Cell Performance

As a hydrophobic material, PTFE serves as a binder to maintain the integrity of the carbon particles in the MPL and provides suitable hydrophobicity to avoid water flooding. The amount of PTFE in the GDL has a significant influence on fuel cell performance. Depending on the MEA structure and the fuel cell operating conditions, the optimized PTFE content reported in the literature usually varies from 10 to 40% by weight [27,28,31–33]. For example, Lufrano et al. [32] reported that the best fuel cell performance could be achieved with a PTFE loading of 20 wt.% in a system operated at 70 °C, with 2.5- and 3.0-bar operating pressures for H_2 and air, respectively. The same optimal PTFE loading was reported by Park et al. [27]. Giorgi et al. [28] investigated electrodes with different PTFE contents in the diffusion layer using Hg-instrusion porosimetry, scanning electron microscopy (SEM), and electrochemical techniques such as cyclic voltammetry, galvanostatic polarization, and electrochemical impedance spectroscopy. Their results indicated that the total porosity and macroporosity of the GDL decreased with increasing PTFE content, as shown in Fig. 2.2 [28], resulting in decreased active catalyst surface area and gas permeability. At a high current density (in the mass transfer control region), both the mass transport rate and cell performance increased with decreasing PTFE content, due to the increased total porosity. The best fuel cell performance was obtained with the lowest PTFE content (i.e. 10 wt.%).

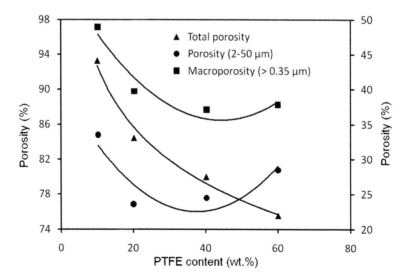

FIGURE 2.2   Total porosity, porosity (2–50 μm), and macroporosity of a GDL as a function of the PTFE content [28].

2.2.1.4. Effects of MPL Carbon Loading on Fuel Cell Performance

Carbon loading in the MPL is directly related to the thickness, porosity, and electrical conductivity of a layer. A thinner MPL may not properly perform its water/gas management function. However, a thicker MPL means a longer pathway, which will increase the mass transport resistance. Therefore, cell operating conditions dictate the optimal carbon loading to obtain the best cell performance. For example, the maximum fuel cell performance was achieved at 75 °C ambient pressure and 100% relative humidity (RH) with a carbon (acetylene black) loading of 0.5 mg cm^{-2} in the MPL [18]. A higher carbon loading creates a thicker MPL and consequently a longer path for gas diffusion. But if the carbon loading is too small, the MPL may be too thin to improve water management. The optimal carbon loading depends on the properties of carbon. For example, Jordon et al. [16] reported that the optimal loading for acetylene black carbon with 10 wt.% PTFE is 1.9 mg m^{-2}. Generally, the optimal amount in the MPL increases with decreasing carbon surface area [23].

Different carbon blacks have different characteristics that affect the MPL porosity, pore size distribution, and electrical conductivity. The typical carbon blacks used in MPLs are ketjen black, Vulcan XC-72$^{®}$, acetylene black, and their composites [16,22,23,29,30,34]. Passalacqua et al. [23] compared the performance of MPLs made with Vulcan XC-72$^{®}$, Shawinigan acetylene black (SAB), Mogul L, and Asbury 850 graphite, each carbon having the different specific surface areas listed in Table 2.2. The porosimetric characteristics of the GDLs obtained using these MPLs are listed in Table 2.3. The best performance was achieved using SAB, which has a high pore volume and a small average pore size. This result is attributable to improved mass transport and water management. Jordan et al. [16] also reported that a GDL with acetylene black as the MPL carbon powder showed a better cell performance than one with Vulcan XC-72$^{®}$. The same conclusion was reported by Antolini et al. [29]. One study [33] used carbon cloth rather than carbon paper as the GDM, and investigated the effect of MPL carbon powders on fuel cell performance.

TABLE 2.2 Materials Used for MPL Preparation [23]

Material	Surface Area (m^2 g^{-1})
Asbury graphite 850	13
SAB	70
Mogul L	140
Vulcan XC-72$^{®}$	250

TABLE 2.3 Porosimetric Characteristics of GDLs with Different Carbon Powders in Their MPLs. APR Is the Average Pore Radius; APR_p and APR_s Are the Average Pore Radius for the Primary and Secondary Pores, Respectively; and V_p and V_s Are the Specific Pore Volume for the Primary and Secondary Pores, Respectively [23]

Carbon in MPL	Pore Volume $(cm^3\,g^{-1})$	APR (μm)	V_p $(cm^3\,g^{-1})$	V_s $(cm^3\,g^{-1})$	APR_p (μm)	APR_s (μm)
Asbury 850	0.346	3.5	0.212	0.134	0.29	8.6
SAB	0.594	1.7	0.368	0.226	0.27	4.3
Mogul L	0.276	6.0	0.157	0.119	0.20	13.6
Vulcan XC-72®	0.489	1.8	0.319	0.17	0.24	4.9

Unlike with carbon paper, both sides of the carbon cloth were coated with a carbon/PTFE mixture to form a gas diffusion half-layer (GDHL) on each side, then the CL was applied onto one of the GDHLs. The results suggested that fuel cell performance can be improved under high pressure by using cathodes with Vulcan XC-72® carbon powder on the catalyst side and acetylene black on the gas side.

Wang et al. [22,30] investigated acetylene black (AB), Black Pearls® 2000 (BP), and their composite in MPLs and proposed a bifunctional MPL with an AB–BP composite. The best cell performance, with a peak power density of 0.91 W cm^{-2}, was achieved at 80 °C and 0.2 MPa using an MPL with 10 wt.% BP and 90 wt.% AB and a total carbon loading of 0.5 mg m^{-2} on each side of the gas diffusion backing (GDB; TGPH-030 Toray carbon paper). This result is attributable to the pore size distributions and the wettability of the pore walls. BP has a larger surface area and the ability to adsorb water internally [35], so the MPL using BP was more hydrophilic. Moreover, BP has the most micropores and macropores and the least mesopores, as shown in Fig. 2.3, which can lead to low gas transport and easy occupation by liquid water, thus limiting the mass transport in the MPL with BP. Compared to BP, AB has fewer macropores but more mesopores, and these can provide more passages for gas transport in the MPL. In addition, it has been reported [30] that the surface of AB is more hydrophobic than that of BP, so an MPL with AB provides more hydrophobic pores for gas transport. However, the hydrophobic pores in AB might hinder liquid water removal. On combining the advantages of AB and BP, it is found that a composite of the two carbon powders in a suitable ratio will retain the hydophobicity of AB and simultaneously provide more passages for gas transport. Experimental results [22,30] showed that the micropores for water

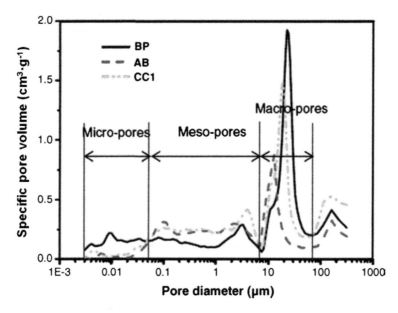

FIGURE 2.3   Pore size distributions in GDLs (including MPL on GDB (TGPH-030 Toray carbon paper)) with three carbon powders in the MPL: BP, AB, and AB–BP composite (CC1, 80 wt.% AB and 20 wt% BP with a loading of 0.5 mg cm^{-2} on each side of the GDB) [22]. (For color version of this figure, the reader is referred to the online version of this book.)

flow were increased by adding a small number of BP in AB, while retaining the latter's hydrophobicity.

2.2.2. Catalyst Layer Design

The CL is where the electrochemical reactions occur, which makes it another key component inside the MEA of PEM fuel cells. The CL is a uniform layer with a thickness of 10–100 μm (usually <50 μm), composed of electrocatalyst powders, proton-conducting ionomer (e.g. Nafion®), and/or binder (e.g. PTFE). Almost all the important challenges in PEM fuel cell development, such as high cost and low durability, arise from the CLs because they are complex, heterogeneous, contain expensive Pt-based catalysts, and have low stability. The reactions in PEM fuel cells have three phases, involving the reactant gases (e.g. H_2 or O_2), proton conductive ionomer (e.g. Nafion®), and electron conductor (e.g. carbon-supported Pt catalyst). Therefore, when designing a CL, it is desirable to extend and maximize the three-phase reaction zone to optimize fuel cell performance.

The three-phase reaction boundary inside the CL is depicted in Fig. 2.4. It can be seen that every active reaction site must simultaneously possess a reactant gas, proton conductive ionomer, and electron conductor. The

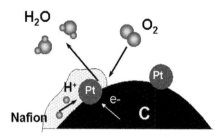

FIGURE 2.4 Schematic of a three-phase reaction boundary [36]. (For color version of this figure, the reader is referred to the online version of this book.)

passages for the transportation of the reactant gas, electrons, and protons must be tailored to the reaction zones. In addition, water is required to maintain the proton conductivity of the ionomer. However, water produced by the electro-chemical reactions in the CL must be removed, so passages for water transport in the reaction zones are also necessary.

Anode and cathode CLs must be designed to generate high rates for the desired reactions and minimize the amount of expensive catalyst required to achieve the target performance. In addressing these requirements, the ideal CLs should (i) maximize the active surface area per unit mass of the electrocatalysts, (ii) minimize the obstacles for reactant transport to the catalyst, (iii) enable proton transport to the exact required position, and (iv) facilitate water removal. These are also the main requirements in extending the three-phase reaction boundary. To meet these requirements, each material's property specifications should be considered during designing. Some compromise between conflicting requirements is also necessary. In addition, the CL structure should be carefully tailored by using materials that permit the proper interactions between components.

The important properties of a CL, such as electron and proton conductivi-ties, porosity, surface area, and catalytic activity, are determined by its struc-ture, fabrication method, and component properties. In the development of PEM fuel cells to date, many kinds of CL fabrication methods have been used, which include the doctor blade technique, painting, printing, spraying, rolling, screening, and others. Some CL structures, such as the PTFE-bonded electrode, the Nafion®-bonded electrode, and the catalyst-coated membrane, have been well developed. The first generation of CLs, using PTFE-bonded Pt black electrocatalysts, exhibited excellent long-term performance but at a prohibi-tively high cost [37]. These conventional electrodes generally featured high platinum loading, that is, 4 mg cm^{-2}. One of the most significant improve-ments has been made by Raistrick [38], who fabricated a CL with dispersed Pt/C, which was followed by painting/spraying a solubilized ionomer on its surface. These electrodes used 0.4 mg cm^{-2} and demonstrated the same performance as did the first-generation electrodes with 4 mg cm^{-2} [38]. Further research has led to the lowering of Pt loading to $0.1-0.3 \text{ mg cm}^{-2}$ by using thin-film methods [39–42], and even to $0.01-0.02 \text{ mg cm}^{-2}$ with the sputtering-deposition method [43].

During PEM fuel cell development, two typical classes of CLs have been explored: hydrophobic and hydrophilic. These will be addressed in the following sections.

2.2.2.1. Hydrophobic CL

PTFE-bonded hydrophobic CLs are commonly developed for H_2/air PEM fuel cells. In this type of CLs, the catalyst particles (e.g. Pt/C) are thoroughly mixed with a certain amount of hydrophobic binder (such as PTFE, poly-vinylidene difluoride) to form a catalyst ink, then the ink is cast onto the GDL. To provide ionic transport to the catalyst site, PTFE-bonded CLs are generally impregnated with an ionomer (commonly Nafion®) by brushing or spraying, forming a typical gas diffusion electrode (GDE) with a hydrophobic CL. The PTFE-bonded hydrophobic CL was a remarkable advance for PEM fuel cell development. First, the substitution of platinum black with carbon-supported platinum decreased the platinum loading >10-fold while still achieving a similar performance. More importantly, proton transport was enhanced significantly by impregnating the CL with a proton-conducting material. With experimental progress, a PTFE content of 20–40 wt.% and a ratio of Nafion® to carbon (in Pt/C catalyst) of 0.8–1.0 have proven to be the optimal parameters for creating an efficient electrode. To date, the performance of such electrodes has been significantly improved, and mass manufacturing has been achieved.

Figure 2.5 shows a schematic of the structure of a PTFE-bonded GDE. The unique virtue of this electrode is that the gas transport limitation is significantly reduced because the PTFE forms passages for gas transport. However, its disadvantages are obvious as well. The PTFE (especially with a high loading) may wrap around the catalyst particles and decrease both the electron

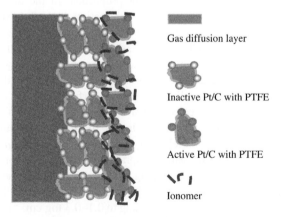

Gas diffusion layer

Inactive Pt/C with PTFE

Active Pt/C with PTFE

Ionomer

FIGURE 2.5 Schematic of a PTFE-bonded hydrophobic electrode [44]. (For color version of this figure, the reader is referred to the online version of this book.)

conductivity and the catalyst utilization. In addition, application of Nafion® to the electrode surface leads to asymmetric distribution because the sprayed Nafion® cannot penetrate deeply into the CL. As a result, the catalyst particles inside the CL may be inaccessible to the Nafion® ionomer, leading to a higher proton transport resistance and leaving some Pt inactive, as denoted by the blank circles in Fig. 2.5. According to estimates, the platinum utilization in such electrodes is only 10–20% [45]. Finally, the MEAs assembled with such GDEs are prone to delamination because the electrode and membrane swell to different degrees, which creates a discontinuity in the ion path and decreases the durability of the PEM fuel cells.

2.2.2.2. Hydrophilic CL

Unlike hydrophobic CLs, hydrophilic CLs use a hydrophilic perfluorosulfonate ionomer (PFSI) such as Nafion® as a binder instead of PTFE. Hence, this kind of CL can be called an ionomer-bonded hydrophilic CL. During preparation, the catalyst powder (e.g. Pt/C), PFSI (e.g. Nafion®), and solvent (e.g. ethanol or isopropanol) are mixed thoroughly to form a uniform hydrophilic catalyst ink/paste that is then transferred to a GDL or a membrane. Hydrophilic CLs can be classified into two groups, according to the transfer method: GDL-based hydrophilic CL and catalyst coated membrane (CCM).

2.2.2.2.1. GDL-Based Hydrophilic CL

In the GDL-based hydrophilic CL, the hydrophilic catalyst ink/paste is coated onto the GDL [31,46–48] with the same methods used in hydrophobic CL fabrication, such as brushing, spraying, and the doctor blade technique. After the catalyst ink is spread, the electrode is first dried slowly at room temperature and then dried at 80–135 °C for about 30 min. For example, Qi and Kaufman [47] produced a low Pt loading, high-performance electrode for PEM fuel cells by casting catalyst ink made of E-TEK 20 wt.% Pt/C, Nafion®, and solvent water onto the GDL to form the catalyzed electrode. The best performance, with a peak power density of 0.72 W cm^{-2} under ambient pressure, was achieved with a Nafion® loading of 30 wt.% and a Pt loading of 0.12 mg cm^{-2}. To improve the fuel cell performance and catalyst utilization further, Qi and Kaufman proposed various activation methods, such as steaming or boiling [49], high temperature and pressure operation [50,51], as well as H$_2$ evolution on the electrode [52]. In this CL structure, the catalyst particles came in contact with the proton conductor (i.e. Nafion®). In this way, both electron and proton transfer were ensured, and Pt catalyst utilization was improved. However, the passages for gas transport and water transfer can be limited by the lack of hydrophobic agent (such as PTFE), so mass transfer could be an issue in the CL if "water flooding" occurs. To overcome this challenge, a thinner CL, such as a CCM electrode, may be required.

2.2.2.2.2. Catalyst Coated Membrane

To further increase the ionic connection between the membrane and the CL, and improve the mass transfer in a hydrophilic CL, a thin-film CL (TFCL) was proposed by Wilson and Gottesfeld [39,41]. This TFCL involves casting the catalyst ink onto the membrane directly rather than onto the GDL to form a CCM. Wilson and Gottesfeld [41,42] suggested a decal transfer method for fabricating the ionomer-bonded hydrophilic CL. This process includes two key steps: coating the catalyst ink onto a blank substrate film and then transferring the coat onto the proton conductive membrane (e.g. Nafion® membrane).

Besides Nafion® ionomer, cation (such as tetrabutylammonium ion (TBA$^+$) or Na$^+$) exchange ionomers and membranes, which have a higher glass transition temperature, are often alternatively adopted in catalyst ink preparation and decal transfer [37,42,53,54]. By using this ionomer, one can increase the transferring temperature to as high as 160–210 °C without any structural damage. The high temperature facilitates effective contact between the ionomer and the catalyst particles and forms a more intimate membrane/CL interface. It can also introduce a robust, pseudocrystalline structure to the ionomer in the CL. After the decal transfer process, the catalyzed membrane assembly is converted to the H$^+$ form by lightly boiling it in diluted H$_2$SO$_4$ and rinsing it in deionized water before incorporating it into the MEA for testing. This procedure is depicted in Fig. 2.6 as the "conventional decal method."

Recently, a modified decal transfer technique for CCM fabrication was reported [55]. In this method, a colloidal catalyst ink was used, as described by Uchida [56]. First, the ink was coated onto a Teflon® substrate. After drying, the CL was transferred to a H$^+$ form membrane (e.g. Nafion® 112 membrane) by hot pressing at 120–135 °C. Finally, the Teflon® substrate was peeled off the CCM.

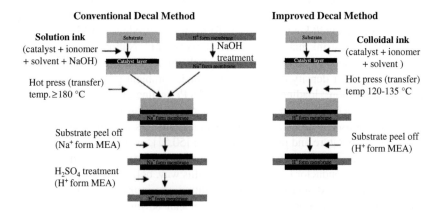

FIGURE 2.6 Schematic flowcharts of the conventional decal method and the improved decal method for making a CCM [55]. (For color version of this figure, the reader is referred to the online version of this book.)

This procedure is depicted in Fig. 2.6 as the "improved decal method." This is the simpler of the two methods. In addition, the fuel cell testing results [55] indicated that the MEA made by the improved decal method yielded a better fuel cell performance. The authors attributed the superior cell performance to the higher porosity of the agglomerates in the MEA, which facilitated mass transport.

The CCM technology has been well developed in recent years [57–60] due to the advantages of having (i) a tight contact between the CL and the membrane to achieve low interfacial resistance, (ii) a thin CL with low mass transfer resistance, and (iii) good contact among the electrode components. Note that during the preparation of CCM catalyst ink, alcohol or isopropanol is generally used as the solvent rather than glycerol, which is used in the conventional decal method. In a CCM, catalyst utilization can also be improved, with the Pt loading reduced to levels as low as 0.07 mg cm^{-2} [39] and even 0.02 mg cm^{-2} [59], yet yielding a highly satisfactory fuel cell performance.

Regarding the ionomer (PFSI) content in a hydrophilic CL, the optimal amount and distribution of the ionomer in the CL is a tradeoff among three requirements: (i) maximum contact between the ionomer and the Pt particles to guarantee proton transport, (ii) minimal electron resistance, and (iii) minimal gas transport resistance. Normally, gas transport can be affected by both decreased porosity due to the presence of a solid ionomer and liquid water accumulation due to the hydrophilicity of the CL. When carbon-supported platinum (Pt/C) is used as the catalyst, the carbon particles have a much larger surface area than the Pt particles, so only if the carbon surface is covered by the ionomer can contact between the ionomer and the Pt particles be ensured. This indicates that the ratio between the ionomer and the carbon in the CL is quite important for achieving high performance. The suggested ratio of ionomer to carbon ($I:C$) is about 0.8:1.0, which is calculated based on the assumption that the ionomer forms a thin layer (~1 nm) on the carbon surface.

2.2.2.3. Partially Pyrolyzed Nafion®-Ionomer-Bonded Electrode

As discussed above, in the structure of a conventional hydrophobic CL, the hydrophobic agent (e.g. PTFE) is normally used during the catalyst ink prep-aration, and a certain amount of proton conductor (ionomer, e.g. Nafion®) is sprayed onto the CL surface. However, because the sprayed Nafion® solution may not effectively penetrate the interior of the CL, the three-phase reaction zone will not be extended sufficiently. As a result, the contact between the catalyst particles and the proton conductor (e.g. Nafion®) might not be very tight, which leads to low Pt catalyst use. Conversely, in the structure of a Nafion®-bonded hydrophilic CL, the catalyst particles and Nafion® ionomer have good contact because they are mixed during the catalyst ink preparation. The passages for both electron and proton transfer are guaranteed, and Pt utilization is improved as well. However, there are too few passages in the CL for the reactant gas and water, due to the lack of a hydrophobic agent.

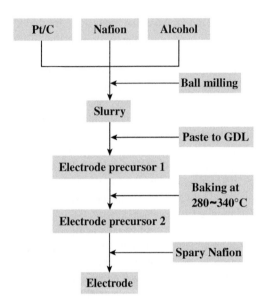

FIGURE 2.7 Schematic flowchart for the preparation of partially pyrolyzed Nafion®-ionomer-bonded electrode. (For color version of this figure, the reader is referred to the online version of this book.)

To utilize the advantages of PTFE-bonded hydrophobic and Nafion®-bonded hydrophilic CLs, a novel electrode structure was proposed [56]. In this new structure, the catalyst slurry was made by mixing catalyst powders, Nafion® ionomer as the binder, and alcohol as the solvent. The prepared catalyst slurry was then coated onto the GDL to obtain the electrode precursor, which was baked at 280–340 °C under nitrogen to pyrolyze some of the Nafion® ionomer. Part of the ionomer in the CL was partially pyrolyzed during baking by controlling the temperature and time. Nafion® has two functional groups in its molecular structure: a sulfonic acid group and a fluorinated carbon chain. During pyrolysis at approximately 280 °C, part of the Nafion® ionomer will lose its sulfonic acid group, leaving the fluorinated carbon chain. The left carbon chain has properties similar to PTFE, and can serve as the hydrophobic agent and bonder in a CL, as PTFE does in a PTFE-bonded hydrophobic CL. However, the unpyrolyzed Nafion® ionomer in the CL retains its original structure, with the sulfonic acid group attached to the fluorinated carbon chain, and can therefore be the bonder and proton conductor, as Nafion® ionomer is in a Nafion®-bonded hydrophilic CL. Consequently, after baking, a certain amount of Nafion® was sprayed onto the electrode surface to further extend the three-phase zone. The processes for this electrode preparation are depicted in Fig. 2.7. In this partially pyrolyzed Nafion®-ionomer-bonded electrode, the hydrophobic and

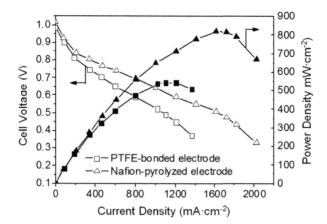

FIGURE 2.8   PEM fuel cell performances of MEAs with different electrodes. Nafion® 1035 membrane; cell temperature: 80°; backpressure: 0.2 MPa; humidity temperatures for H_2 and air: 90° and 85°; stochiometries of H_2 and air: 1.5 and 2.5 [56].

hydrophilic structures can be distributed uniformly in the CL. The transport of electrons, protons, reactant gases, and water can be facilitated. In addition, the three-phase reaction zone can be effectively extended. Figure 2.8 presents a typical result, showing improved fuel cell performance and Pt utilization [56]. It is worth pointing out that the Nafion ionomer loading, baking temperature, and baking time are three key parameters in achieving an electrode with high fuel cell performance [56].

2.2.3. Proton Exchange Membrane Design

As shown in Fig. 2.1, the PEM is another key component in the MEAs of H_2/air PEM fuel cells, serving not only as a solid electrolyte but also as a separator to prevent direct contact between the anode and cathode compartments. The most practically and extensively used PEM in H_2/air PEM fuel cells is the perfluorosulfonic acid (PFSA) membrane, such as Nafion® membrane. As shown in Chapter 1, Fig. 1.7, the molecular structure of Nafion® consists of three groups: the tetrafluoroethylene (Teflon)-like backbone; the sulfonate acid group; and the $-O-CF_2-CF-O-CF_2-CF_2-$ side chain, which connects the backbone and the sulfonate group. So far, PFSA membranes are considered to be the best and have been widely used in H_2/air PEM fuel cells because of their high proton conductivities and good stabilities in both oxidative and reductive environments. However, they also have disadvantages, such as high cost, high degradation rate at high temperatures (>80 °C), and dependence of proton conductivity on the membrane's water content. Thus, the development of new membrane materials continues to be a hot topic for research and development in PEM fuel cell technology. The requirements for PEM materials are (1) low cost; (2) strong mechanical stability; (3) strong chemical stability; (4) strong electrochemical stability in both oxidative and reductive environments; (5) high proton conductivity in the operating temperature range of fuel cells; (6) insensitivity of proton conductivity to water content; and (7) low gas permeability.

In recent years, many kinds of materials have been developed to synthesize proton-conducting membranes for H_2/air PEM fuel cells, and some have exhibited promising performance as potential candidates to replace PFSA membranes. The major membranes are (1) fluorinated membrane, (2) partially fluorinated membrane, (3) nonfluorinated (including hydrocarbon) membrane, and (4) nonfluorinated composite membrane. Among these, the hydrocarbon membrane is considered a promising alternative due to its low cost compared with PFSA membranes [61].

Several literature reviews [62–69] provide more detailed information about fuel cell membranes and their design and fabrication.

2.2.4. MEA Assembly

As a key component in PEM fuel cells, the MEA consists of the anode, PEM, and cathode. Its assembly mainly involves a hot-pressing process to make a "sandwich" with the membrane in the middle and the anode and cathode on either side. Both the anode and the cathode should contain a GDM, MPL (carbon sublayer), and CL. When the MEA is being hot pressed, the CL must face one side of the membrane. Before MEA assembly, pretreatment of the membrane is usually required to remove possible impurities and to completely protonate the membrane.

Hot pressing is an effective and simple way for assembling electrodes and PEMs to achieve good interfacial contacts between them. Hot-pressing conditions, such as temperature, pressure, and pressing duration, influence the performance and durability of the resulting MEA [70–74]. A study [72] showed that the combination of temperature, pressure, and time should be optimized to achieve a high-performance MEA.

Apparently, the temperature plays a major role in this optimization. For a Nafion®-based membrane, the hot-pressing temperature is normally limited by its glass transition temperature (T_g, ~128 C). At a temperature lower than T_g, the Nafion® resin in both the CL and the membrane will not melt and can result in poor ionomeric contact between the CL and the membrane, which leads to low catalyst utilization and higher ionic resistance. In contrast, a temperature much higher than T_g may lead to a loss in the water retention properties of Nafion®, and acidic group degradation in the ionomer. Therefore, there is an optimal temperature for the hot-pressing process [70,73,74]. Under this optimal temperature, the CL and membrane combine most effectively to provide a maximum electrochemical area at the interface, leading to the highest catalyst utilization and fuel cell performance [70].

The hot-pressing pressure is related to the mechanical strength, porosity, and thickness of the electrode. Normally, this porosity decreases with increasing pressure, which can restrict the mass transport of the gas. Moreover, the carbon fibers are prone to be crushed under high pressure. However, the electrode thickness can be decreased under a high hot-pressing pressure, which shortens the mass transportation pathway. A study [72] showed that a lower hot-pressing pressure could result in a better fuel cell performance than a higher hot-pressing pressure.

Hot-pressing time is another important parameter that affects the contact between the membrane and the electrode, as well as the electrode porosity. It is recognized that with an increase in hot-pressing time, the ionic conductivity and the three-phase reaction area in the CL can be first increased and then decreased, and the electrode porosity can also be decreased. Liang et al. [71] used direct methanol fuel cells to investigate the durability of MEAs assembled under different hot-pressing conditions, and found that a longer hot-pressing time could induce significantly improved MEA durability without sacrificing cell performance, because of a stronger interfacial binding between the CL and the membrane, which suppresses their delamination. Although the optimal hot-pressing conditions for PEM fuel cells are slightly different due to the differences in the materials and structures of the electrode and membrane, hot-pressing of PEM fuel cell MEAs is usually conducted at 120–160 °C and 2000–35,000 kPa pressure for 30–300 s.

Typically, the anode CL and cathode CL are applied onto their respective GDLs to form the anode and cathode GDEs, then the anode and cathode GDEs are hot pressed with the membrane in the middle to form a sandwich-like structure. In recent years, there have been significant developments in the CCM, which is a typical three-layer MEA. For the CCM, however, the hot-pressing process joining the membrane and the anode and cathode GDLs is unnecessary. Usually, the GDLs are simply pressed together with the CCM using blade pressure, during the fuel cell assembly process.

2.3. TYPICAL EXAMPLES FOR MEA FABRICATION

To provide a better understand of MEA fabrication, this section offers some typical examples, including GDM, MPL, and CL preparation, membrane pretreatment, and MEA assembly.

2.3.1. GDM Preparation

As described in Section 2.2.1.1, the GDM is the backing layer that supports the MPL and CL. The most commonly used materials for the GDM are carbon-based sheets, such as carbon paper and carbon cloth. Carbon paper is presently more widely used because it is relatively inexpensive, and MPLs and CLs are easier to make on it than on carbon cloth. Section 2.3 on MEA fabrication thus describes the techniques that use carbon paper.

Although as-received carbon paper is to some extent hydrophobic, it is still necessary to increase its hydrophobicity by pretreating it before use in MEA fabrication, to prevent water flooding. This pretreatment method is known as "carbon paper wet proofing," and the procedure is as follows: (1) dip the GDM into an aqueous solution or suspension of a hydrophobic agent such as PTFE or FEP; assuming PTFE is used, the concentration can be varied from 1 to 10 wt.%; (2) remove the GDM and eliminate the excess solution or suspension; (3) dry the PTFE-impregnated GDM in an oven; (4) repeat the above three steps until the expected PTFE content in the GDM is achieved (usually 5–30 wt.%); (5) bake the impregnated GDM in an oven at 350 °C to remove the solvent and surfactants contained in the PTFE suspension, to sinter the PTFE particles, and to fix the PTFE to the GDM surface. The desired PTFE content can easily be reached by adjusting the concentration of the PTFE suspension. But a lower PTFE concentration and more dipping times are helpful in achieving a uniform PTFE distribution in the GDM. Figure 2.9 shows an SEM image of carbon paper with 20% PTFE [2], wherein it is evident that the PTFE uniformly covers the carbon paper surface.

Figure 2.10 shows SEM images of (a) carbon paper wet proofed with 20% FEP and (b) original carbon paper. Evidently, the FEP dispersed evenly on the surface of the carbon paper. Figure 2.10 also shows that some pores on the carbon paper surface were covered by FEP after wet proofing. These results suggest that this method can increase the hydrophobicity of the carbon paper, which can be characterized by measuring its contact angle.

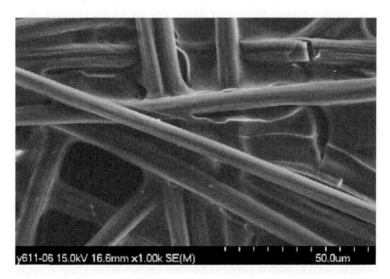

y611-06 15.0kV 16.6mm x1.00k SE(M)   50.0um

FIGURE 2.9   SEM image of carbon paper (Toray, TGPH-090) pretreated with 20% PTFE [2].

20 wt.% FEP carbon paper

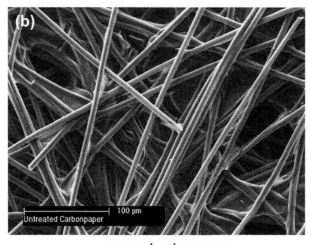

untreated carbon paper

FIGURE 2.10   Comparison of surface SEM images of (a) Carbon paper (Toray, TGPH-090, E-TEK) impregnated with 20 wt.% FEP hydrophobic polymer, to (b) Untreated carbon paper [13].

For the wet-proofing treatment of carbon papers, the PTFE distribution through the thickness of the GDM is sensitive to the drying method. As shown in Fig. 2.11 [12], a fast drying method results in more concentrated PTFE on the exposed GDM surfaces. However, a slow drying method tends to form a more uniform distribution of PTFE through the bulk of the GDM.

Note that the PTFE content or loading in the GDM must be optimized with respect to the fuel cell performance. Too much PTFE in the GDL can decrease

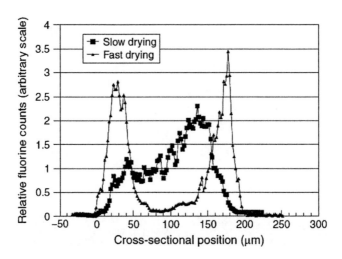

FIGURE 2.11 Cross-sectional fluorine maps across carbon-fiber paper (Toray TGP-H-060). PTFE distribution through the paper depends heavily on drying conditions [12].

its conductivity, and too little PTFE will lead to insufficient hydrophobicity for smooth gas transportation.

2.3.2. MPL Preparation

As described in Section 2.2.1.2, the MPL is a sublayer of carbon powders made on the GDB that can change the GDB porosity and support the CL. The typical procedure for MPL preparation is as follows:

(1) In a suitable ratio, ultrasonically mix the carbon powder (e.g. AB, BP, or Vulcan X-72® carbon), the hydrophobic agent (e.g. PTFE), and the solvent to create a uniform ink/paste.

(2) Coat the above ink/paste onto one or both sides of the GDB by brushing, spraying, spreading, doctor blade, screen printing, or by using other techniques.

(3) Bake the carbon-coated GDM in an oven with nitrogen flow at 240 °C for 30 min, followed by another 40 min at 350 °C; the MPL will then be formed on the GDB.

It must be noted that the MPL can be applied on one or both sides of the GDB, depending on the GDB material. For example, the MPL is usually created on both sides of an SGL carbon paper but on only one side of a Toray TGPH carbon paper. The loadings of carbon powder and PTFE can significantly affect the GDL performance. Hence, these loadings need to be optimized according to the MEA structure and operating conditions [18,27], as discussed in Section 2.2.1.2. The category of carbon powder used in the MPL also significantly affects the performance of the GDL and, consequently, that of the fuel cell [22,30,75].

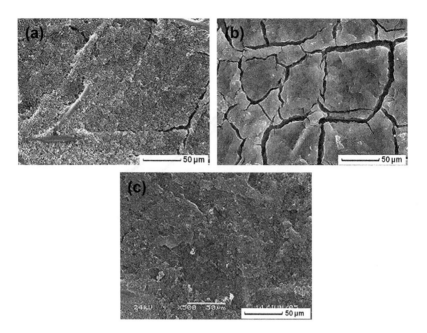

FIGURE 2.12   SEM micrographs of cathode GDL surfaces with MPLs prepared using different carbon powders: (a) Acetylene black (AB), (b) Black Pearls 2000 (BP), and (c) Composite carbon with 90% AB and 10% BP [30].

Figure 2.12 shows SEM micrographs of the GDL surfaces with MPLs prepared using different carbon powders. It can be seen that the MPL with AB possesses rich pores and a uniform surface, whereas the MPL with BP is dense and has only large cracks. These features can be attributed to the properties of AB and BP carbon powders, which have been addressed in Section 2.2.1.2. As shown in Fig. 2.12, an MPL with composite carbon powders (CC) presents a more uniform surface with finer pores. The fuel cell testing results indicate that the best performance is achieved using an MPL with CC, as shown in Fig. 2.13. This can be explained by the more functional pore structure formed on the MPL when CC facilitates the mass transport of gas and water [22,30].

2.3.3. CL Fabrication

2.3.3.1. PTFE-Bonded Hydrophobic CLs

The PTFE-bonded hydrophobic CL is one of the classic CLs. The preparation process for this kind of CL is similar to the MPL preparation process and can be described as follows [76–78]:

(1) Mix the catalyst powder (e.g. Pt/C), hydrophobic bonder (PTFE emulsion), and solvent (e.g. ethanol or isopropanol) ultrasonically to form a catalyst ink/paste;

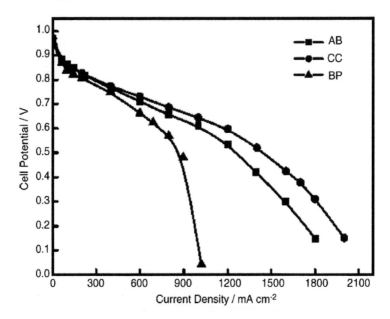

FIGURE 2.13 H_2/air PEM fuel cell performance of electrodes with MPLs prepared using different carbon powders. Cell temperature: 80 °C; humidifier temperatures for anode and cathode: 90° and 85°, respectively; operating pressure: 0.2 MPa [30].

(2) Spread the catalyst ink/paste onto a GDL (with the MPL on wet-proofed carbon paper) using the doctor blade technique or by spraying, painting, screen printing, etc.;

(3) Dry the precursor prepared in the previous step at 240 °C for 30 min under an inert gas atmosphere to remove the solvent and the surfactants contained in the PTFE emulsion;

(4) Bake the precursor at 350 °C for 30–40 min to sinter the PTFE and thereby hydrophobilize the CL;

(5) Spray a certain amount (e.g. 0.5 mg cm^{-2}) of ionomer (e.g. Nafion) solution onto the CL surface to form ionic pathways and increase the three-phase reaction zone;

(6) Dry the as-prepared electrode at room temperature to evaporate the solvent contained in the ionomer, and finally obtain the PTFE-bonded hydrophobic electrode.

In this PTFE-bonded hydrophobic electrode, the PTFE content significantly affects the fuel cell performance. The passages for gas and water transport are tailored by introducing the PTFE during the catalyst ink/paste preparation stage. However, these proton conductors are not enough because the impregnating Nafion, located on the CL surface, cannot deeply penetrate the electrode. Thus, the three-phase reaction zone is not extended efficiently, leading to low Pt utilization.

2.3.3.2. Ionomer-Bonded Hydrophilic CLs

Unlike a PTFE-bonded hydrophobic CL, in an ionomer-bonded hydrophilic CL, the proton-conducting ionomer (e.g. Nafion) is used as the bonder. The classic preparation process for this CL is as follows [46,47,79,80]:

(1) Prepare the catalyst ink by thoroughly mixing the catalyst (e.g. Pt/C), ionomer (e.g. Nafion), and solvent (e.g. alcohol, usually isopropanol);

(2) Spray the catalyst ink onto the GDL (with the MPL on wet-proofed carbon paper);

(3) Dry the electrode first at room temperature, then at 70–135 °C for 30 min, to obtain an electrode with an ionomer-bonded hydrophilic CL on the GDL.

Figure 2.17 shows the degradation of an MEA made by applying a hydrophilic catalyst ink to a GDL (i.e. using the conventional method). This MEA exhibited good durability.

Aside from being sprayed on the GDL, the catalyst ink can also be applied to the membrane to make a Nafion-bonded hydrophilic CL. To efficiently extend the three-phase reaction zone and reduce the Pt loading, Wilson et al. [37,39,41] developed a thin-film electrode using Nafion ionomer as the bonder. Their preparation process uses a decal method, the details of which are as follows [39]:

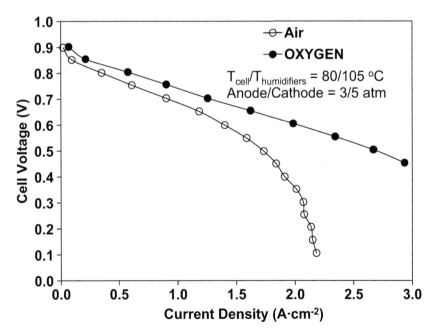

FIGURE 2.14   Air and oxygen fuel cell polarization curves for directly catalyzed developmental Dow membrane with a catalyst loading of 0.13 mg Pt cm^{-2} [39].

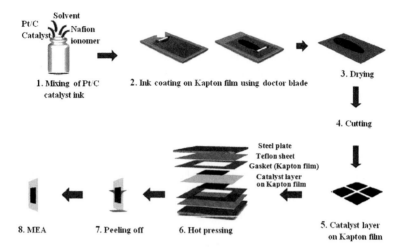

FIGURE 2.15 A schematic representation of the procedure for MEA fabrication using the decal transfer process [83]. (For color version of this figure, the reader is referred to the online version of this book.)

(1) Prepare a uniform hydrophilic ink composed of Nafion solution (in Na^+ form), Pt/C catalyst, and solvent (e.g. glycerol). The weight ratio of Pt/C catalyst to Nafion is typically between 5:2 and 3:1.

(2) Paint this catalyst ink onto a dry Nafion membrane (in Na+ form) to form a CL, then bake the CL at approximately 160–190 °C to dry the ink.

(3) Cast the catalyst ink onto the reverse side of the Nafion membrane using the same process.

(4) Rehydrate and ion exchange the membrane into the H^+ form by immersing the catalyzed membrane into a slightly boiling 0.1 M sulfuric acid solution for 2 h.

(5) Rinse the catalyzed membrane sufficiently and air dry it.

The result is a thin-film, Nafion®-bonded hydrophilic electrode in which the catalyst and ionomer are thoroughly mixed, but which lacks the passage for gas and water transport because it has no hydrophobic agent. So, this electrode is usually made very thin (5–10 μm) to avoid "water flooding" Fig. 2.14 shows the fuel cell performance of a thin-film electrode prepared by the above method. It can seen that the electrode exhibited a good performance with a Pt loading of just 0.13 mg cm^{-2}.

Aside from the above process, a decal transfer method has also been developed to make a thin-film electrode [81–83]. The catalyst ink is first coated onto a decal substrate (such as PTFE film or Kapton® film) by spraying or by using the doctor blade technique. The catalyst ink is then transferred to a Nafion® membrane by a hot-pressing process to form a catalyzed membrane. This decal transfer method is presented schematically in Fig. 2.15 [83].

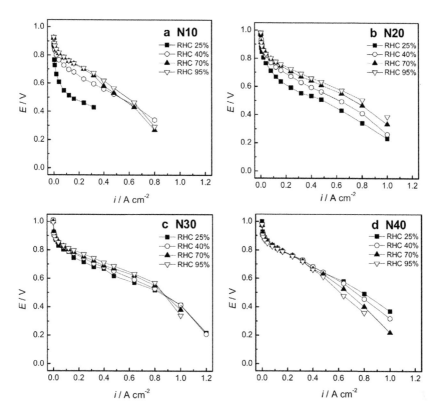

FIGURE 2.16   Polarization curves of MEAs made by the decal transfer method, with different Nafion contents (N10, N20, N30, and N40), and at various RHs on the cathode, from 25 to 95%. Cell temperature: 65 °C; stoichiometries of H_2 and air: 1.5 and 2.0, respectively; anode RH was fixed at 80% [82].

Figure 2.16 shows the polarization curves of MEAs made by this decal transfer method with different Nafion® contents (N10, N20, N30, and N40), at RHs from 25 to 95% on the cathode. It can be seen from Fig. 2.16 that the Nafion® content plays an important role in determining the fuel cell performance, and the optimal Nafion® content is related to the operating conditions.

The CCM has been well developed in recent years [79,84,85]. During CCM preparation, the catalyst, Nafion® ionomer, and solvent are mixed to form a uniform catalyst ink, which is then directly sprayed onto the membrane. In the CCM, the contact between the CL and the membrane is tight, so the CL does not tend to be delaminated during long-term fuel cell operation, suggesting good MEA durability. As shown in Fig. 2.17, the CCM in study [79] exhibited very good durability, with a low fuel cell degradation rate during 1000 h of fuel cell testing.

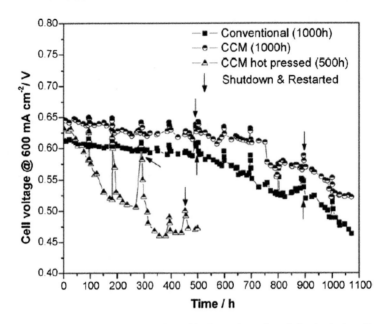

FIGURE 2.17 Effects of MEA fabrication method on voltage degradation in single cells operated at 600 mA cm^{-2}. Cell temperature: 80 °C; stoichiometries of H$_2$ and air: 1.5 and 3.0, with RHs of 100% and 55%, respectively [79].

2.3.4. Membrane Pretreatment

The PEM is the heart of an MEA, and it not only separates the anode from the other compartments but it also conducts the protons produced at the CL/membrane interface. The as-received membrane may contain impurities or may not be fully protonated; either factor will affect the membrane's performance and thus eventually influence the fuel cell's performance. It is therefore necessary to pretreat membranes before they are used in MEAs.

The typical membrane pretreatment process includes the following [72]: (1) boiling the membrane at 60–80 °C in a dilute H$_2$O$_2$ solution (H$_2$O$_2$ concentration of ~3–5%) to remove the organic and inorganic impurities contained in the membrane; (2) boiling the membrane at 60–80 °C in an H$_2$SO$_4$ aqueous solution (a 0.5 M H$_2$SO$_4$ solution is usually employed) to protonate the membrane; and finally, (3) twice rinsing the membrane at a high temperature, such as 70 °C, with deionized water. The resulting pretreated membrane can be stored for future use.

2.3.5. MEA Fabrication

Hot pressing is a widely used method to prepare conventional hydrophobic and hydrophilic electrodes for MEAs, and descriptions of the process can be found elsewhere [71–73,76]. Hot-pressing parameters, such as temperature, pressure, and duration, significantly affect MEA performance, as shown in Fig. 2.18. If a CCM is used, it is usually put between the anode and cathode GDLs, which are then pressed together by blade pressure during fuel cell assembly. Hot pressing of a CCM and GDLs is not necessary because the CL has good contact with the membrane. Indeed, hot pressing reduces the porosity of GDLs and leads to decreased durability, as shown in Fig. 2.17.

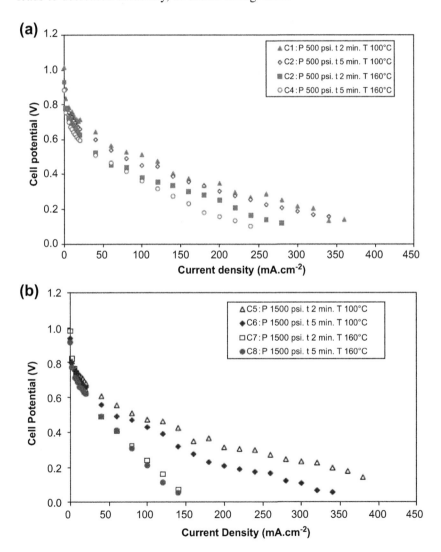

FIGURE 2.18   The effects of hot-pressing temperature and time on the polarization curves of MEAs prepared at (a) 500 psi and (b) 1500 psi. [72]. (For color version of this figure, the reader is referred to the online version of this book.)

2.4. FLOW FIELD DESIGN

Flow field plates or bipolar plates are key components of PEM fuel cells and stacks. In a PEM fuel cell, the functions of a flow field plate are as follows: (1) it provides flow channels for the fuel and oxidant gases to their respective anodic and cathodic electrode surfaces, (2) it provides flow channels for the removal of the water coming from the humidifier and generated by electrochemical reactions, (3) it provides mechanical support for the anodic and cathodic electrodes, (4) it serves as a current collector, although a separate current collector is often used in a single cell, (5) it electronically connects one cell to another in a stack, and (6) it acts as a physical barrier to prevent the fuel, oxidant, and coolant fluids from mixing. In addition, a flow field plate is helpful for heat management. Flow field plates must perform the above functions simultaneously to achieve a good fuel cell performance. However, the requirements sometimes conflict, necessitating an optimized flow field design. In the past several decades of PEM fuel cell development, many flow field plate designs have been tested. Li et al. [86] extensively reviewed the subject. Figure 2.19 shows several typical flow field designs.

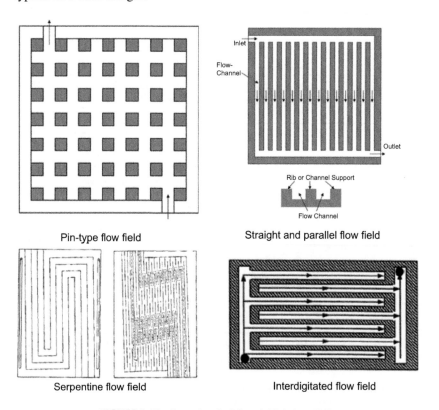

FIGURE 2.19   Several typical flow field designs [86].

2.4.1. Materials for a Flow Field Plate

Material selection is very important in designing a flow field plate, given the functions it performs in a PEM fuel cell. The materials must have a good electronic conductivity, low gas permeability, good mechanical stability to support the electrodes, good chemical/electrochemical stability, good machinability for making the flow field, light weight, and low cost. The most commonly used materials for flow field plates in PEM fuel cells are metal plates [87,88] and graphite plates [89], although other composite materials are also used [90–92]. Metal plates have good electronic conductivity and can be manufactured very thin to achieve light weights [88]. But their disadvantage is that they get corroded in the PEM fuel cell operating environment, which leads to flow field plate failure and thereby shortens the fuel cell's lifetime. Gold coatings are often applied to metal flow field plates to solve this problem. At the present stage of PEM fuel cell technology, graphite plates are widely used because they have good electronic conductivity and excellent chemical/electrochemical stability [93,94]. The disadvantages of graphite plates are obvious, though: high cost, brittleness, and greater weight compared to metal plates.

2.4.2. Flow Field Layout (Channel Pattern)

Flow field layout plays an important role in flow field design, as it affects both reactant gas distribution and water removal. Several flow field layouts have been developed, according to their flow channels: the pin-type flow field [95,96], the straight parallel flow field [97,98], the serpentine flow field [98,99], and the interdigitated flow field [100–103], as shown in Fig. 2.19. All these patterns have their own characteristics in terms of reactant gas distribution, water, and heat management, and their advantages and disadvantages have been extensively reviewed by Li and Sabir [86].

Of all the patterns, the straight parallel flow field and serpentine flow field are presently the mostly widely used; these patterns are shown in Figs 1.10 and 1.11 of Chapter 1. In the straight parallel flow field, the gas distributions and the pressure along the flow channels are nonuniform due to the lack of change in the flow channel. In addition, the nonuniform pressure drops caused by the short flow channel can result in low water removal ability, which has been addressed in Section 1.5.6 of Chapter 1. The serpentine flow field design can be classified as either single-channel or multichannel (the latter having two or more serpentine channels). The single-channel serpentine flow field has a long flow channel, leading to a large pressure drop along the flow channel between the gas inlet and outlet, which is good for water removal but results in nonuniform gas distribution. The multichannel serpentine flow field has more flow channels, and more reactant gas flows along the channels, leading to a uniform gas distribution inside the flow field plate. However, the drop in gas pressure between the inlet and outlet is lower compared to the drop in the single-channel flow field. Thus, the water removal ability is relatively low [104].

2.4.3. Flow Channel Parameters

The channel geometry can significantly affect cell performance [105–107] because of its impact on the reactant gas flow and distribution, as well as the water management inside the flow field. Figure 2.20 shows a cross-section of flow field channels. The main geometric parameters of flow channels are the channel width w, channel depth d, rib/land width l, and wall angle θ. Channel length L is another geometric parameter that is dependent on the size constraint of the particular application. Given a certain size and layout for a flow field, optimization of the geometric parameters of the flow channels can yield a more uniform reactant gas distribution as well as better water and heat management, leading to better fuel cell performance. The typical parameters for the flow channels are 0.5–2.5 mm for the channel width, 0.2–2.5 mm for the channel depth, 0.2–2.5 mm for the rib width, and 0–15° for the wall angle [99].

2.4.4. Flow Field Plate Fabrication

Flow channels can be made on either or both sides of a graphite, metal, or composite material plate. In a fuel cell stack, a plate with flow channels on both sides is called a bipolar plate. The fabrication process of a flow field plate depends on the materials used. Carbon-based plates (using materials such as graphite felt, flexible graphite, and carbon resin) are usually brittle; consequently, the plate thickness and the cross-sectional flow channel area should be made larger during fabrication. Metal plates usually have excellent mechanical

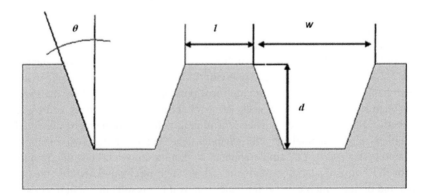

FIGURE 2.20 Cross-section of flow field channels: rib width l, channel width w, channel depth d, and wall angle θ [99,107]. (For color version of this figure, the reader is referred to the online version of this book.)

stability and can be made into thinner flow field plates with a smaller cross-sectional area. Although metal plates usually have good electronic conductivity, they are subject to corrosion in a PEM fuel cell operating environment, resulting in increased surface resistance and decreased PEM fuel cell performance. Therefore, protective coatings are usually applied to metal plates to prevent flow field plate corrosion [87,88].

Flow field fabrication is a purely mechanical process; therefore, any workshop with appropriately designed machines can accomplish it.

2.5. SEALING DESIGN

The sealing gasket is yet another important component in both single PEM fuel cells (as shown in Fig. 1.5 in Chapter 1) and stacks (as shown in Fig. 2.12). Two sealing gaskets are required for one MEA unit, placed between each side of the MEA and the flow field plates. Generally, the sealing gaskets perform three important functions in a PEM fuel cell or stack: (1) sealing off gases to prevent gas crossover and leakage, (2) insulating to prevent the fuel cell from shorting, and (3) sealing the coolant. Sealing gaskets must be designed carefully because a poor-quality gasket can cause fuel cell leakage, leading to low performance and safety issues, especially when the fuel is H_2 gas.

2.5.1. Sealing Material Selection

To perform their functions, sealing materials in PEM fuel cells must meet stringent criteria: (1) electronic insulation, (2) high chemical/electrochemical stability, (3) suitable compressibility, and (4) good compatibility with the reactant gases and coolant fluids. The most widely used sealing materials are PTFE, fluoroelastomer, and silicon-based materials such as silicon rubber and silicon elastomers [108–110]. If the sealing materials are not stable enough after long-term PEM fuel cell operation, the decomposed impurities or other products may get into the fuel cell components [108–111], leading to contamination. For example, some decomposition products of silicon-based materials may get adsorbed on the membrane and/or CL, causing decreased membrane conductivity and/or catalyst activity, and certain decomposition fragments may get into the GDL, changing its hydrophilicity/hydrophobicity.

2.5.2. Sealing Design and Fabrication

In accordance with MEA and flow field designs, many sealing designs have been developed during the past several decades [108,112]. The major materials used are silicon rubbers. Normally, two methods are used to fabricate these sealings: die cutting or molding, such as screen printing. In some fabrication methods, the sealings are directly molded onto the GDL [100], bipolar plate [101,102], and MEA, as shown in Fig. 2.21 [103–105]. In recent years, the

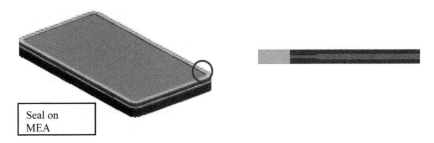

Seal on
MEA

FIGURE 2.21 Schematic representation of a seal integrated with an MEA [112]. (For color version of this figure, the reader is referred to the online version of this book.)

design shown in Fig. 2.21 has become more popular with CCM development. The CCM is usually laminated together with one sealing gasket sheet on each side to form a sealed MEA. Another popular sealing arrangement is simply to put a gasket sheet between the flow field plate and the MEA, and then press them all together during fuel cell assembly, as shown in Fig. 1.5 in Chapter 1.

2.6. SINGLE CELL DESIGN AND ASSEMBLY

2.6.1. Single Cell Hardware

Single fuel cell hardware, shown in Fig. 1.5 of Chapter 1, consists of end plates, current collectors, sealing gaskets, flow field plates, and bolts. Depending on the materials used, single cell hardware design should consider several factors, including but not limited to material selection, stability, flow field layout, and flow channel pattern. However, the cell hardware components must be designed to achieve optimal combined performance, including compatibility with the MEA.

2.6.2. Single Cell Assembly

Single cell assembly is the process of putting together all the requisite components to form a single fuel cell. As shown in Fig. 1.5 in Chapter 1, all the components are pressed together with the MEA in the middle, then tightened using bolts. In some tests, to exert a uniform and constant pressure along the MEA surface, a gas bladder is used with a piston along one side of the single cell. The pressure used to hold the components together is important in achieving a high-performance single cell. For example, if the pressure is too high, the porosity of the GDE may be reduced; in the worst-case scenario, the MEA may be damaged, leading to a large mass transfer resistance and consequently poor fuel cell performance. However, if the pressure is too low, the contacts between the fuel cell components will be an issue, again resulting in large internal resistance and consequently low cell performance. In our

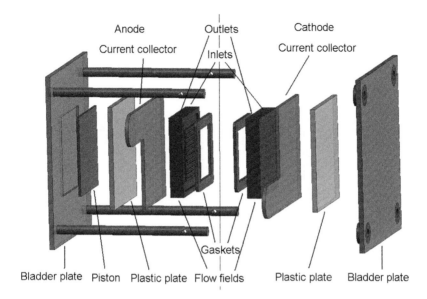

FIGURE 2.22   A designed single PEM fuel cell with an active area of 4.4 cm^2 [113]. (For color version of this figure, the reader is referred to the online version of this book.)

laboratory, the designed fuel cell hardware includes bladder plates, piston, plastic plates, current collectors, flow field plates, and gaskets, as shown in Fig. 2.22 [113]. Note that on the end plates there are the gas inlet and outlet mouths to connect to the gas manifolds, for fuel and oxidant feeding, which are not shown in this figure. The plastic plate is used to isolate the current collector/ flow field plate from the metal bladder. Gaskets are attached to the flow field plates to seal the MEA on both sides. The cell is pressed and sealed using the bladder plates and piston, powered by gas (e.g. nitrogen or air). The tightness of the fuel cell can easily be controlled by adjusting the bladder pressure.

2.7. STACK DESIGN AND ASSEMBLY

2.7.1. Hardware of a Fuel Cell Stack

As shown in Fig. 2.23, a fuel cell stack consists of many single cells connected in series. It also contains end plates, flow field plates (current collector), gaskets, and bolts. The difference is that there is a cooling plate between each two adjacent single cells. Due to the heat generated during fuel cell operation, the extra heat must be removed to maintain the operating temperature. Therefore, this cooling plate is necessary. Another method is to integrate the anode (or cathode) flow field plate of one cell, the cooling plate, and the cathode (or anode) flow field plate of the adjacent cell into one shared plate. On

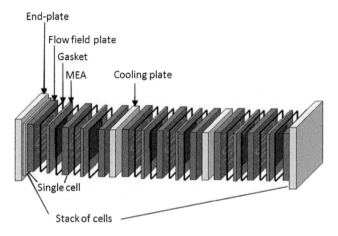

FIGURE 2.23   Hardware of an H$_2$/air PEM fuel cell stack [61].

this shared plate, both sides are fabricated with flow field channels, then one side serves as the anode (or cathode) flow field of the one cell and the other as the cathode (or anode) flow field of the other single cell. Inside this bipolar plate are some channels for coolant flow.

2.7.2. Heat Management in a PEM Fuel Cell Stack

Normally, PEM fuel cells are operated between 60 and 80 °C. At the startup of a fuel cell stack, the electrochemical reactions inside the stack produce energy that will rapidly heat the whole stack to this temperature range within 1–2 min. Due to the limitations of the PEM (e.g. Nafion® membrane), the stack temperature must be kept <95 °C, otherwise the fuel cell performance will rapidly decline. Of course, if another high-temperature membrane is used, the stack operating temperature can be elevated. To maintain the stack temperature within a desired range, such as 60–80 °C, a stack cooling system must be added to the fuel cell system to remove the extra heat. It is also well recognized that the temperature distribution in a stack affects the fuel cell's performance. The cooling system is therefore helpful in homogenizing the temperature distribution within the stack.

The cooling system in a fuel cell system can be designed as either internal cooling [114] or external cooling [115]. Within these categories, internal cooling can be either air or liquid cooling (liquid being water or a water ethylene glycol mixture). Normally, an air or liquid cooling plate can be designed to be integrated with the bipolar plate, to simplify the fuel cell stack [114,116]. The coolant plates in a fuel cell stack must be sealed to prevent fluid from leaking into the reactant gas channels or outside of the fuel cell. For the external cooling system, there is no cooling liquid present in the cell active area, which eliminates any sealing problems with respective to the electrode [115] and yields a simplified fuel cell system.

2.7.3. Fuel Cell Stack Assembly

A fuel cell stack is assembled by packing many single cells in series, as shown in Fig. 2.23. The electronic series connection of all these single cells is realized by the electronic conducting bipolar plates. The number of single cells depends on the desired stack power and size, and the performance of single cells. In other words, the power that can be generated by a fuel cell stack is determined by the number of its cells, the total active area of the MEA, and the single cell performance.

2.8. CHAPTER SUMMARY

This chapter has presented the design and assembly of single PEM fuel cells and stacks, with a focus on the MEA and the fabrication of its components. This is because the MEA is the most important component of a PEM fuel cell, where electrochemical reactions occur for power generation. The chapter discusses MEA design and assembly/fabrication, including GDLs (gas diffusion medium and microporous sublayer), CLs, and PEMs, as well as analyzes various factors that affect MEA design and performance. The chapter also provides a typical example of step-by-step MEA fabrication from various component materials to the whole MEA. Other components, including flow field plates, sealing gaskets, and their corresponding designs and fabrication, are also presented in this chapter. Finally, the design and fabrication of both single fuel cells and stacks are introduced, the intent being to provide readers with the basic information and procedures for achieving workable PEM fuel cell and stack hardware. We believe that this chapter forms a solid foundation for the PEM fuel cell testing and diagnosis described in the following chapters.

REFERENCES

[1] Cindrella L, Kannan AM, Lin JF, Saminathan K, Ho Y, Lin CW, et al. J Power Sources 2009;194:146–60.

[2] Du C, Wang B, Cheng X. J Power Sources 2009;187:505–8.

[3] Kannan AM, Munukutla L. J Power Sources 2007;167:330–5.

[4] Lin JF, Wertz J, Ahmad R, Thommes M, Kannan AM. Electrochim Acta 2010;55:2746–51.

[5] Fuel Cells Bull 2001;4. 15–15.

[6] Ko T-H, Liao Y-K, Liu C-H. New Carbon Mater 2007;22:97–101.

[7] Ko T-H, Liao Y-K, Liu C-H. Carbon 2007;45. 2321–2321.

[8] Liu C-H, Ko T-H, Kuo W-S, Chou H-K, Chang H-W, Liao Y-K. J Power Sources 2009;186:450–4.

[9] Liu C-H, Ko T-H, Shen J-W, Chang S-I, Chang S-I, Liao Y-K. J Power Sources 2009;191:489–94.

[10] Yang H, Tu HC. Chiang IL. Int J Hydrogen Energy 2010;35:2791–5.
[11] Zhang F-Y, Advani SG, Prasad AK. J Power Sources 2008;176:293–8.
[12] Mathias M, Roth J, Fleming J, Lehner W. In: Vielstich W, Gasteiger HA, Lamm A, editors. Handbook of fuel cells-fundamentals, technology and applications, vol. 3. John Wiley & Sons, Ltd.; 2003.
[13] Lim C, Wang CY. Electrochim Acta 2004;49:4149–56.
[14] Pai Y-H, Ke J-H, Huang H-F, Lee C-M, Zen J-M, Shieu F-S. J Power Sources 2006;161:275–81.
[15] Park G-G, Sohn Y-J, Yang T-H, Yoon Y-G, Lee W-Y, Kim C-S. J Power Sources 2004;131:182–7.
[16] Jordan LR, Shukla AK, Behrsing T, Avery NR, Muddle BC, Forsyth M. J Power Sources 2000;86:250–4.
[17] Tseng C-J, Lo S-K. Energy Convers Manage 2010;51:677–84.
[18] Park S, Lee J-W, Popov BN. J Power Sources 2006;163:357–63.
[19] Kong CS, Kim D-Y, Lee H-K, Shul Y-G, Lee T-H. J Power Sources 2002;108:185–91.
[20] Qi Z, Kaufman A. J Power Sources 2002;109:38–46.
[21] Nam JH, Kaviany M. Int J Heat Mass Transf 2003;46:4595–611.
[22] Wang X, Zhang H, Zhang J, Xu H, Zhu X, Chen J, et al. J Power Sources 2006;162:474–9.
[23] Passalacqua E, Squadrito G, Lufrano F, Patti A, Giorgi L. J Appl Electrochem 2001;31:449–54.
[24] Lee H-K, Park J-H, Kim D-Y, Lee T-H. J Power Sources 2004;131:200–6.
[25] Ambrosio E, Francia C, Gerbaldi C, Penazzi N, Spinelli P, Manzoli M, et al. J Appl Electrochem 2008;38:1019–27.
[26] Moreira J, Ocampo AL, Sebastian PJ, Smit MA, Salazar MD, del Angel P, et al. Int J Hydrogen Energy 2003;28:625–7.
[27] Park S, Lee J-W, Popov BN. J Power Sources 2008;177:457–63.
[28] Giorgi L, Antolini E, Pozio A, Passalacqua E. Electrochim Acta 1998;43:3675–80.
[29] Antolini E, Passos RR, Ticianelli EA. J Power Sources 2002;109:477–82.
[30] Wang XL, Zhang HM, Zhang JL, Xu HF, Tian ZQ, Chen J, et al. Electrochim Acta 2006;51:4909–15.
[31] Paganin VA, Ticianelli EA, Gonzalez ER. J Appl Electrochem 1996;26:297–304.
[32] Lufrano F, Passalacqua E, Squadrito G, Patti A, Giorgi L. J Appl Electrochem 1999;29:445–8.
[33] Antolini E, Passos RR, Ticianelli EA. J Appl Electrochem 2002;32:383–8.
[34] Neergat M, Shukla AK. J Power Sources 2002;104:289–94.
[35] Wang X, Hsing IM, Yue PL. J Power Sources 2001;96:282–7.
[36] Zhang H, Wang X, Zhang J, Zhang J. In: Zhang J, editor. PEM fuel cell electrocatalysts and catalyst layers—fundamentals and applications. London: Springer; 2008. p. 889–916.
[37] Wilson MS, Valerio JA, Gottesfeld S. Electrochim Acta 1995;40:355–63.
[38] Raistrick ID. US Patent 4876115, 1989.
[39] Wilson MS, Gottesfeld S. J Electrochem Soc 1992;139:L28–30.
[40] Springer TE, Wilson MS, Gottesfeld S. J Electrochem Soc 1993;140:3513–26.
[41] Wilson MS, Gottesfeld S. J Appl Electrochem 1992;22:1–7.
[42] Wilson MS. Membrane catalyst layer for fuel cells. In: US Patent 5234777, 1993.
[43] O'Hayre R, Lee S-J, Cha S-W, Prinz FB. J Power Sources 2002;109:483–93.
[44] Zhang J, Zhang J. In: Zhang J, editor. PEM fuel cell electrocatalysts and catalyst layers—fundamentals and applications. London: Springer; 2008. p. 965–1002.
[45] Srinivasan S, Velev OA, Parthasarathy A, Manko DJ, Appleby AJ. J Power Sources 1991;36:299–320.

[46] Shin SJ, Lee JK, Ha HY, Hong SA, Chun HS, Oh IH. J Power Sources 2002;106:146–52.
[47] Qi Z, Kaufman A. J Power Sources 2003;113:37–43.
[48] Dhar HP. Method for catalyzing a gas diffusion electrode. In: US Patent 5521020, 1996.
[49] Qi Z, Kaufman A. J Power Sources 2002;109:227–9.
[50] Qi Z, Kaufman A. J Power Sources 2002;111:181–4.
[51] Qi Z, Kaufman A. J Power Sources 2003;114:21–31.
[52] He C, Qi Z, Hollett M, Kaufman A. Electrochem Solid State Lett 2002;5:A181–3.
[53] Song SQ, Liang ZX, Zhou WJ, Sun GQ, Xin Q, Stergiopoulos V, et al. J Power Sources 2005;145:495–501.
[54] Xie J, More KL, Zawodzinski TA, Smith WH. J Electrochem Soc 2004;151:A1841–6.
[55] Saha MS, Paul DK, Peppley BA, Karan K. Electrochem Commun 2010;12:410–3.
[56] Zhang J, Wang X, Hu J, Yi B, Zhang H. Bull Chem Soc Jpn 2004;77:2289–90.
[57] Hsu CH, Wan CC. J Power Sources 2003;115:268–73.
[58] Mussell RD. Active layer for membrane electrode assembly. In: US Patent 5882810 1999.
[59] Debe MK. In: Vielstich W, Gastieger HA, Lamm A, editors. Handbook of fuel cells-fundamentals, technology and applications. John Wiley & Sons; 2003. p. 576–89.
[60] Song Y, Fenton JM, Kunz HR, Bonville LJ, Williams MV. J Electrochem Soc 2005;152:A539–44.
[61] Mehta V, Cooper JS. J Power Sources 2003;114:32–53.
[62] Alberti G, Casciola M. Ann Rev Mater Res 2003;33:129–54.
[63] Brandon NP, Skinner S, Steele BCH. Ann Rev Mater Res. 2003;33:183–213.
[64] Schuster MFH, Meyer WH. Ann Rev Mater Res 2003;33:233–61.
[65] Paddison SJ. Ann Rev Mater Res 2003;33:289–319.
[66] Rozière J, Jones DJ. Ann Rev Mater Res 2003;33:503–55.
[67] Nakao M, Yoshitake M. In: Vielstich W, Gasteiger HA, Lamm A, editors. Handbook of fuel cells-fundamentals, technology and applications. John Wiley & Sons, Ltd; 2003. p. 412–9.
[68] Jones DJ, Roziere J. In: Vielstich W, Gasteiger HA, Lamm A, editors. Handbook of fuel cells-fundamentals, technology and applications. John Wiley & Sons, Ltd; 2003.
[69] Hickner MA, Ghassemi H, Kim YS, Einsla BR, McGrath JE. Chem Rev 2004;104: 4587–612.
[70] Lin J-C, Lai C-M, Ting F-P, Chyou S-D, Hsueh K-L. J Appl Electrochem 2009;39:1067–73.
[71] Liang ZX, Zhao TS, Xu C, Xu JB. Electrochim Acta 2007;53:894–902.
[72] Therdthianwong A, Manomayidthikarn P, Therdthianwong S. Energy 2007;32:2401–11.
[73] Zhang J, Yin G-P, Wang Z-B, Lai Q-Z, Cai K-D. J Power Sources 2007;165:73–81.
[74] Liu P, Yin G-P, Wang E-D, Zhang J, Wang Z-B. J Appl Electrochem 2009;39:859–66.
[75] Yan W-M, Wu D-K, Wang X-D, Ong A-L, Lee D-J, Su A. J Power Sources 2010;195:5731–4.
[76] Ticianelli EA, Derouin CR, Redondo A, Srinivasan S. J Electrochem Soc 1988;135: 2209–14.
[77] Srinivasan S, Ticianelli EA, Derouin CR, Redondo A. J Power Sources 1988;22:359–75.
[78] Kumar GS, Raja M, Parthasarathy S. Electrochim Acta 1995;40:285–90.
[79] Prasanna M, Cho EA, Lim TH, Oh IH. Electrochim Acta 2008;53:5434–41.
[80] Cha SY, Lee WM. J Electrochem Soc 1999;146:4055–60.
[81] Suzuki T, Tsushima S, Hirai S. Int J Hydrogen Energy 2011;36:12361–9.
[82] Jeon S, Lee J, Rios GM, Kim H-J, Lee S-Y, Cho E, et al. Int J Hydrogen Energy 2010;35:9678–86.
[83] Cho HJ, Jang H, Lim S, Cho E, Lim T-H, Oh I-H, et al. Int J Hydrogen Energy 2011;36:12465–73.

[84] Kim KH, Kim HJ, Lee KY, Jang JH, Lee SY, Cho E, et al. Int J Hydrogen Energy 2008;33:2783–9.

[85] Kim K-H, Lee K-Y, Kim H-J, Cho E, Lee S-Y, Lim T-H, et al. Int J Hydrogen Energy 2010;35:2119–26.

[86] Li X, Sabir I. Int J Hydrogen Energy 2005;30:359–71.

[87] Wind J, LaCroix A, Braeuninger S, Hedrich P, Heller C, Schudy M. In: Vielstich W, Gasteiger HA, Lamm A, editors. Handbook of fuel cells-fundamentals, technology and applications. John Wiley & Sons, Ltd; 2003. p. 294–307.

[88] Tawfik H, Hung Y, Mahajan D. J Power Sources 2007;163:755–67.

[89] Robberg K, Trapp V. In: Vielstich W, Gasteiger HA, Lamm A, editors. Handbook of fuel cells-fundamentals, technology and applications. John Wiley & Sons, Ltd; 2003. p. 308–14.

[90] Dweiri R, Sahari J. J Power Sources 2007;171:424–32.

[91] Maheshwari PH, Mathur RB, Dhami TL. J Power Sources 2007;173:394–403.

[92] Mercuri RA, Gough JJ. US Patent 6037074, 2000.

[93] Qian P, Zhang H, Chen J, Wen Y, Luo Q, Liu Z, et al. J Power Sources 2008;175:613–20.

[94] Martin J, Oshkai P, Djilali N. J Fuel Cell Sci Technol 2005;2:70–80.

[95] Reiser CA, Sawyer RD. US Patent 4769297, 1988.

[96] Reiser CA. US Patent 4826742, 1989.

[97] Voss HH, Chow CY. US Patent 5230966A, 1993.

[98] Su A, Chiu YC, Weng FB. Int J Energy Res 2005;29:409–25.

[99] Wilkinson DP, Vanderleeden O. In: Vielstich W, Gasteiger HA, Lamm A, editors. Handbook of fuel cell-fundamentals, technology and applications. John Wiley & Sons, Ltd; 2003.

[100] Nguyen TV, He E. In: Vielstich W, Gasteiger HA, Lamm A, editors. Handbook of fuel cell-fundamentals, technology and applications. John Wiley & Sons, Ltd; 2003.

[101] Guo SM. In: Sunderam V, Albada G, Sloot P, Dongarra J, editors. Computational science—ICCS 2005. Springer Berlin/Heidelberg; 2005. p. 104–11.

[102] Yamada H, Hatanaka T, Murata H, Morimoto Y. J Electrochem Soc 2006;153:A1748–54.

[103] Wilson MS. US Patent 005641586A, 1997.

[104] Zhang J, Li H, Shi Z, Zhang J. Int J Green Energy 2010;7:461–74.

[105] Wang X-D, Yan W-M, Duan Y-Y, Weng F-B, Jung G-B, Lee C-Y. Energy Convers Manage 2010;51:959–68.

[106] Shimpalee S, Van Zee JW. Int J Hydrogen Energy 2007;32:842–56.

[107] Akhtar N, Qureshi A, Scholta J, Hartnig C, Messerschmidt M, Lehnert W. Int J Hydrogen Energy 2009;34:3104–11.

[108] Frisch L. Sealing Technol 2001;2001:7–9.

[109] Schulze M, Knöri T, Schneider A, Gülzow E. J Power Sources 2004;127:222–9.

[110] Tan J, Chao YJ, Yang M, Lee W-K. Van Zee JW. Int J Hydrogen Energy 2011;36:1846–51.

[111] Du B, Guo Q, Pollard R, Rodriguez D, Smith C, Elter J. J Oral Microbiol 2006;58:45–9.

[112] Koch S. PEM fuel cell stack sealing, report 2005 (Powerpoint slides; available online http://www.slideserve.com/Mercy/pem-fuel-cell-stack-sealing).

[113] Tang Y, Zhang J, Song C, Liu H, Zhang J, Wang H, et al. J Electrochem Soc 2006;153:A2036–43.

[114] Wu J, Galli S, Lagana I, Pozio A, Monteleone G, Yuan XZ, et al. J Power Sources 2009;188:199–204.

[115] Scholta J, Messerschmidt M, Jörissen L, Hartnig C. J Power Sources 2009;190:83–5.

[116] Peng J, Shin J-Y, Lee S-J. US Patent 20090162713A1, 2009.

Chapter 3

Techniques for PEM Fuel Cell Testing and Diagnosis

3.1. INTRODUCTION

The structures and components of proton exchange membrane (PEM) fuel cells and the designs of all the components, including the membrane electrode assembly (MEA), single cell, and stack, have been described and discussed in Chapter 2. For a practical PEM fuel cell, every feature of the components, key materials, and cell assembly should be achievable and optimized to achieve high performance. Because a fuel cell is a very complicated device, all the

PEM Fuel Cell Testing and Diagnosis. http://dx.doi.org/10.1016/B978-0-444-53688-4.00003-6

components should fully perform their individual roles and simultaneously function together synergistically. To investigate the individual functioning of each component and the synergistic effect, fuel cell testing and diagnosis have been recognized as the most popular and reliable ways to validate the designs of these components and of the fuel cell itself.

It is an ongoing challenge to fully understand the processes occurring inside fuel cells, because fuel cell science and technology spans multiple disciplines, including materials science, engineering design, chemistry, electrochemistry, interface science, mass transport phenomena, and electrocatalysis. During the development of PEM fuel cells, researchers and engineers have focused considerable attention on experimental and theoretical modeling to understand PEM fuel cell system processes, such as kinetics, thermodynamics, fluid dynamics, and chemical reactions. Rapid developments in physics and electrochemistry have yielded many testing and diagnostic tools that can be used to understand the processes inside a PEM fuel cell and accordingly improve its performance. This chapter will introduce these tools and the techniques for using them and will focus on their applications in the testing and diagnosis of PEM fuel cells.

3.2. TECHNIQUES FOR PEM FUEL CELL TESTING

3.2.1. Half-Cell Testing

To investigate individual components such as the catalyst and catalyst layer, and their effects on fuel cell performance, electrochemical half-cells are normally used as ex situ tools. The major advantage of using a half-cell is believed to be that it allows one to study how a specific component or experimental condition contributes to the overall cell performance, with little or no interference from other components or conditions. A half-cell test is usually conducted in a three-electrode system containing working, counter, and reference electrodes. Cyclic voltammetry (CV), rotating disk electrode, and rotating ring-disk electrode are the typical half-cell testing techniques for investigating the catalyst's electrochemical characteristics toward the hydrogen oxidation reaction (HOR) and the oxygen reduction reaction (ORR). Some special half-cell designs also allow one to study the effects of other operating conditions, such as catalyst layer/MEA design, temperature, pressure, humidity, as well as fuel and air flow rates. For more details, please see Chapter 12 of this book.

3.2.2. Fuel Cell Testing

Fuel cell testing is the most reliable and commonly used method to evaluate a PEM fuel cell. Its purpose is to assess the aspects of fuel cell performance, such as how the cell voltage changes when the fuel cell load (current or current density) is altered, or how the load changes when the cell voltage is altered, or how much power can be drawn from a fuel cell. PEM fuel cell testing can be classified into three modes: current control, voltage control, and power control. In general, the current and voltage control modes are the most commonly used.

In the current control mode, one controls the current (or current density) at a series of constant values and then records the corresponding cell voltages. Figure 3.1 shows a typical polarization curve, measured using the current control mode; Section 1.4 of Chapter 1 provides a detailed analysis of this polarization curve. The MEA power density can also be calculated using this curve, according to $P = IV$ (where P is the power density, I is the current density, and V is the cell voltage). Figure 3.1 presents power density as a function of current density, showing the maximum power density to be around 0.72 W cm^{-2}.

In the voltage control mode, one controls the cell voltage at a series of constant values and records the corresponding current or current density. By adjusting the cell voltage, a series of current or current density values can be recorded to obtain a fuel cell I–V curve, from which the corresponding power density can be estimated. Note that the power density here is for the MEA rather than for the fuel cell stack. Normally, there are three fuel cell power densities: (1) the MEA power density, expressed as the single cell's power divided by the MEA area, with a unit of watts per square centimeter of MEA (W cm^{-2}); (2) the stack mass power density, which is the stack power divided by the weight of the stack, with a unit of watts per kilogram of stack (W kg^{-1}); and (3) stack volume power density, which is the stack power divided by the

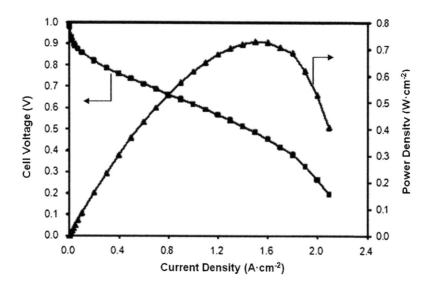

FIGURE 3.1   Polarization and power density curves of a Nafion®-112-membrane-based H$_2$/air PEM fuel cell at 80 °C and 3.0 atm with a 100% relative humidity (RH). Active area: 4.4 cm^2; stoichiometries of H$_2$ and O$_2$: 1.5 and 2.0, respectively.

stack volume, with a unit of watts per liter of stack ($W L^{-1}$). The fuel cell operating conditions, such as temperature, pressure, RH, and gas flow rate, can significantly affect cell performance. Thus, the operating conditions should be optimized for maximum performance. In general, fuel cell performance increases with increasing temperature, pressure, RH, and gas flow rate (or gas stoichiometry number).

3.2.3. Lifetime/Durability Testing

Lifetime/durability is very important for fuel cells in all applications, as it has a bearing on operating time and cost. Usually, lifetime/durability testing is the last step in fuel cell testing. But it is often a time-consuming process—for example, two years is required to conduct 17,520 h of lifetime testing.

One of the most popular lifetime testing modes is to control the current or current density under the desired operating conditions and to record the change in the cell voltage over time. At the beginning of lifetime (BOL) test, an I–V polarization curve is recorded, and at the end of lifetime (EOL) test, another curve is recorded. A comparison of the EOL and BOL I–V curves can yield the degradation rate (r_{d,I_i}), which is calculated using the measured voltage drop (ΔV) at a certain current density for the tested time period (Δt) and has a unit of microvolts per hour ($\mu V\ h^{-1}$):

$$r_{d,I_i} = \frac{\Delta V}{\Delta t} \tag{3.1}$$

This lifetime test mode is normally used for a fuel cell in stationary or portable applications where the load demand is relatively constant and stable. However, for automobile applications, the load demand is normally dynamic, so a dynamic lifetime test is needed. The usual approach is to control the current density change over time using an alternating pattern—that is, the current density is alternated between two magnitudes during the testing time, and the voltage changes are recorded. After the lifetime testing is completed, the EOL and BOL results are compared to obtain the degradation rates under different conditions.

3.2.4. Accelerated Testing

As indicated above, each component and operating condition requires lifetime testing, making this an extremely time-consuming process. To save time, some accelerated testing methods have been attempted, but with limited success.

Accelerated testing is often used to rapidly evaluate a PEM fuel cell design and screen the various component materials. Such testing is often conducted under very stressful conditions, such as open circuit voltage (OCV), high temperature, high current density, and the like. Accelerated testing of PEM fuel cells is covered in Chapter 11.

3.3. TECHNIQUES FOR PEM FUEL CELL DIAGNOSIS

During the research and development of PEM fuel cells, many techniques have been used to diagnose or characterize cells in situ or ex situ. The following sections will discuss the techniques widely used for PEM fuel cell diagnosis. This chapter will not provide in-depth coverage of the principles and instrumentation of well-known techniques, but it will mainly focus on their applications in PEM fuel cell testing and diagnosis.

3.3.1. Cyclic Voltammetry

3.3.1.1. CV Technique and Instrument

CV is a widely employed potentiodynamic electrochemical technique that can be used to acquire qualitative and quantitative information about electrochemical reactions, including electrochemical kinetics, the reversibility of reactions, reaction mechanisms, electrocatalytic processes, and other features. This technique uses an instrument called a potentiostat, an example of which is the Solartron 1287. The measurement is normally conducted in a three-electrode configuration or electrochemical cell containing a working electrode, counter electrode, and reference electrode. The electrolyte in this electrochemical cell is a liquid solution or solid membrane. During CV measurement, the potential of the working electrode in the studied system is measured with respect to the reference electrode, and the potential is scanned back and forth between specific upper and lower limits. At the same time, the current passing between the working electrode and the counter electrode is recorded. The potential is normally linear over time, with a slope, as shown in Fig. 3.2; this

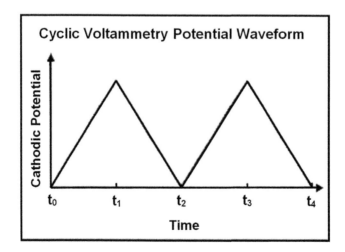

FIGURE 3.2 Typical CV potential waveform [1].

slope is the scan rate of the potential. The current passing though the working electrode is strongly dependent on the potential scan rate, the status of the working electrode, and the electrolyte composition. The plot of current vs. potential is called the cyclic voltammogram, as shown in Fig. 3.3. In this figure, the CV represents a reversible oxidation–reduction reaction. The upward peak indicates the oxidation of the active species in the solution, and the downward peak indicates the species' reduction. This technique can also be used to study the surface redox reaction on the working electrode.

The CV technique has been widely used in fuel cell research. The following sections will address in detail its applications in PEM fuel cell diagnosis.

3.3.1.2. In Situ Characterization of Electrocatalysts in PEM Fuel Cells

Electrocatalyst and catalyst layer measurements are of importance in the development of PEM fuel cells. This chapter only addresses in situ CV measurements in PEM fuel cells. For ex situ measurements, please refer to Chapter 12 of this book. In situ CV measurement is often the first experiment performed to evaluate the catalyst activity. To conduct the measurement, the anode side of the fuel cell, which is catalyzed by a Pt-based catalyst, is flushed with humidified hydrogen gas and serves as both the counter and reference electrodes; the cathode side, which is also catalyzed by a Pt-based catalyst (that may or may not be the same as at the anode), is flushed with humidified nitrogen (or nitrogen/carbon monoxide (CO) mixed gas in the case of a CO stripping experiment), and it serves as the working electrode. Because the kinetics of the HOR is relatively fast, even if some current passes through the anode, the anode potential may not be significantly changed. Therefore, the

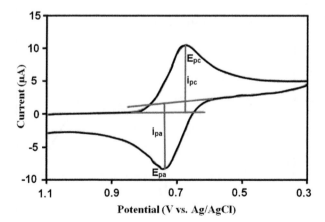

FIGURE 3.3 Typical cyclic voltammogram. E_{pc} and E_{pa} are the cathodic and anodic peak potentials, and i_{pc} and i_{pa} are the cathodic and anodic peak currents, respectively [1].

anode can serve as a reference electrode and a counter electrode simultaneously. In this way, the catalyst and its catalyst layer can be measured to obtain the electrochemical Pt surface area (EPSA or electrochemical active surface area). The EPSA normally represents how many active sites within the catalyst layer are available for the fuel cell reaction (i.e. the ORR) [2–6]. In general, the higher the EPSA, the more active the catalyst or catalyst layer will be.

Figure 3.4 shows a typical CV curve recorded from a fuel cell catalyst layer catalyzed by a carbon-supported Pt catalyst. A total of six peaks are apparent on the CV curve. Peaks 1 and 2 correspond to the hydrogen electroadsorption induced by the reduction of H^+ on the Pt(100) and Pt(111) crystal surfaces, respectively. The process can be expressed as follows:

$$Pt + H^+ + e^- \rightarrow Pt - H_{ads} \qquad (3.I)$$

Peaks 3 and 4 correspond to H_2 electrodesorption induced by the oxidation of H to produce H^+ on the Pt(111) and Pt(100) crystal surfaces, respectively, and can be expressed as follows:

$$Pt - H_{ads} \rightarrow Pt + H^+ + e^- \qquad (3.II)$$

Peak 5 represents oxidation of the Pt surface to form a PtO surface; this process can be expressed using the following reactions:

$$Pt + H - O - H \rightarrow Pt - O - H + H^+ + e^- \qquad (3.III)$$

$$Pt - O - H + H - O - H \rightarrow Pt - (O - H)_2 + H^+ + e^- \qquad (3.IV)$$

$$Pt - (O - H)_2 \rightarrow Pt - O + H - O - H \qquad (3.V)$$

Peak 6 represents the reduction of surface PtO to release the surface Pt, which can be expressed as follows:

$$Pt - O + 2H^+ + 2e^- \rightarrow Pt + H - O - H \qquad (3.VI)$$

The Q' and Q" shown in Fig. 3.4 represent the amounts of charge exchanged during the electroadsorption (Q') and desorption (Q") of atomic hydrogen on the active Pt sites. The shadows in Fig. 3.4 represent the contributions of the "double-layer" charge during electroadsorption and desorption. The EPSA ($m^2_{catalyst\ layer}\ g^{-1}$ Pt) can be calculated according to Eqn (3.2):

$$EPSA = \frac{Q_H}{L_{Pt}Q_f} \times 10^{-4} \qquad (3.2)$$

where Q_H is the Coulombic charge, with a unit of mC, obtained by averaging Q' and Q"; L_{Pt} is the Pt loading on the working electrode, with a unit of g; and Q_f is the Coulombic charge required to desorb the hydrogen when a clean Pt

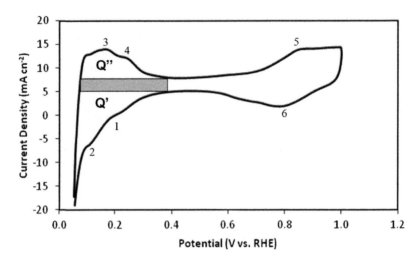

FIGURE 3.4 Typical CV curve recorded on a PEM fuel cell at 80 °C, 3.0 atm, 100% RH. Gore MEA with an active area of 46 cm^2.

surface is covered by a monolayer of hydrogen, its value being 0.21 mC cm^{-2} [2,4,5,7].

Similarly, to determine the anode EPSA, the anode compartment is purged with humidified N$_2$, and the cathode compartment is flushed with humidified H$_2$. In this case, the anode serves as the working electrode, and the cathode acts as both the reference and counter electrodes.

To achieve comparable results for different catalyst layers, EPSA measurements must be performed identically in terms of the measurement method, operating conditions, and catalyst layer state, otherwise large variability in the results can be expected. Of course, this measurement can also be used to study the effects of different operating conditions on the EPSA.

For a Pt/C-based catalyst layer in PEM fuel cells, according to the three-phase boundary theory, Pt catalysts that are not in the three-phase reaction zone are useless in the PEM fuel cell reaction as they are not accessible for reactants, electrons, or protons; these Pt catalysts are thus "inactive." To compare different catalyst layer designs, the Pt utilization ($u_{pt}(\%)$) can be calculated according to the following equation:

$$u_{pt}(\%) = \frac{EPSA}{TPSA} \qquad (3.3)$$

where TPSA is the total Pt surface area of the Pt catalyst powder used in the catalyst layer, and EPSA is measured using the CV technique described above and calculated according to Eqns (3.2). The TPSA can be calculated from the average Pt particle size obtained by X-ray diffraction (XRD) experiments or by high-resolution transmission electron microscopy (TEM) images, based on the assumption that the Pt particle is spherical with a surface area of $4\pi r_{Pt}^2$ (r_{Pt} being the radius of the Pt sphere) [8]. For example, if the Pt catalyst loading in the catalyst layer is L_{Pt} (mg cm^{-2}catalyst layer), the Pt density is d_{Pt} (mg cm^{-3}), and the diameter of each Pt sphere is r_{Pt} (cm), the weight of each Pt particle should be $\frac{4}{3}\pi r_{Pt}^3 d_{Pt}$ (mg). If the Pt catalyst loading in the catalyst layer is L_{Pt}, the Pt particle number density in the catalyst layer should be $\frac{3L_{Pt}}{4\pi r_{Pt}^3 d_{Pt}}$ (cm^{-2}catalyst layer), which corresponds to a Pt surface area of $\frac{3L_{Pt}}{r_{Pt}d_{Pt}}$ (cm^2 cm^{-2}catalyst layer); this value is then divided by the Pt loading in the catalyst layer (L_{Pt}) to get the TPSA, which is $\frac{3}{r_{Pt}d_{Pt}}$. Substituting this TPSA into Eqn (3.3), the Pt utilization can be alternatively expressed as follows:

$$u_{Pt}(\%) = \frac{\text{EPSA}}{3} r_{Pt}d_{Pt} \tag{3.4}$$

By using Eqn (3.4), if the Pt catalyst particle size and the EPSA are known, one can estimate the Pt utilization of the catalyst layer. For example, if EPSA = 400 cm^2 mg^{-1}, L_{Pt} = 0.2 mg cm^{-2}, r_{Pt} = 2 × 10^{-7} cm (2 nm), and d_{Pt} = 21,450 mg cm^{-3}, the Pt utilization is 57.2%. In actuality, there seems to be no consistency among different reports in the literature, due to differences in catalyst preparation, fuel cell testing conditions, and measuring methods.

Similar to the hydrogen desorption method, the CO stripping method has also been used to determine the EPSA in a single fuel cell [6,9,10]. The process is similar to that described above for EPSA measurement, the only difference being that the working electrode compartment is flushed with N_2/CO mixed gas instead of pure N_2 gas. The procedure is as follows: diluted CO (e.g. 0.1%) in N_2 gas is supplied to the working electrode compartment (anode or cathode side) at a constant flow rate while the working electrode potential is held at a constant value for a certain time (~5–30 min); note that the mixed gas needs to be humidified before flowing into the working electrode compartment. After that, the CO/N_2 mixed gas is switched off, and the compartment is purged with pure N_2 gas to remove all CO traces from the gas phase. After 5–30 min of purging, the electrode potential is scanned from a holding potential to the highest potential limit, then back to the lower limit, at a scan rate of $\geq$5 mV s^{-1}. Normally, two CV cycles are collected. The CO stripping value is determined from the first cycle, by using the second cycle as the baseline. Figure 3.5 shows two CV curves with (first cycle) and without (second cycle) a CO-adsorbed layer, obtained on Pt/C in the cathode of an MEA [10]. The figure shows a peak at a potential of about 0.78 V in the first cycle, which represents the electro-oxidation of the CO that was adsorbed on Pt sites. The disappearance of the hydrogen desorption peak in the low potential range indicates that all the Pt

FIGURE 3.5 Typical CO stripping voltammogram of the cathode in a MEA. Cell temperature: 30 °C (fully humidified); CO adsorption potential: 0.3 V; sweep rate: 20 mV s^{-1}. The filled area represents the charge related to the CO oxidation reaction (Q_{CO}) [10]. (For color version of this figure, the reader is referred to the online version of this book.)

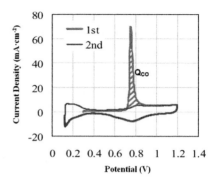

surfaces were covered by adsorbed CO. However, the hydrogen desorption peak can be observed in the second cycle because the Pt surface recovered after the adsorbed CO was removed by electrooxidation in the first cycle. The CO removal process in the first cycle can be expressed as follows:

$$Pt - CO + H_2O \rightarrow Pt + CO_2 + 2e^- + 2H^+ \qquad (3.VII)$$

The shadow of the peak in Fig. 3.5 represents the amount of charge (Q_{CO}) required to remove the adsorbed CO, which can be obtained by integrating the peak. When measuring this charge, the charges caused by the double layer and by Pt oxide formation have to be subtracted, which can be done using the second cycle as a baseline for subtraction. The EPSA can be calculated thus:

$$EPSA = \frac{Q_{CO}}{0.484 L_{Pt}} \qquad (3.5)$$

where L_{Pt} is the Pt loading and 0.484 (with a unit of mC cm^{-2}) represents the charge required to oxidize a monolayer of CO on a smooth Pt surface [2,5,11].

It should be noted that the EPSA calculated from CO stripping is often not consistent with that obtained from hydrogen desorption [5,6], due to the complexity of CO adsorption on a Pt surface, the effect of impurities during CO adsorption, and differences in the state of the Pt surface. For example, Vidakovic, et al. [6] characterized a fuel cell catalyst in situ using CO stripping and calculated the EPSA. CO adsorption was performed by flowing 0.1% CO in Ar at a flow rate of 140 ml min^{-1} through the working electrode compartment at 0.0 V vs. Ag/AgCl. Different adsorption times were used to confirm the achievement of saturated CO coverage. After CO adsorption, the gas was switched to N$_2$ to purge the system for 30 min, removing all CO traces from the gas phase. Then, the potential was scanned between 0 V and 1.2 V vs. Ag/AgC. Figure 3.6 shows the CO stripping voltammograms obtained on a Pt-based MEA with different CO adsorption times.

As shown in Fig. 3.6, the CO stripping peak potential occurs at 0.536 V vs. Ag/AgCl. High CO coverage was obtained with 1 min of CO adsorption, and

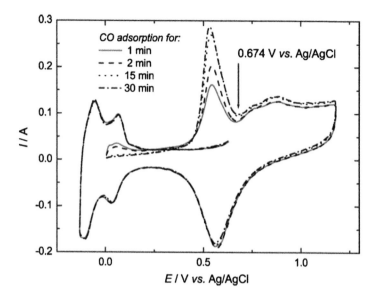

FIGURE 3.6   CO stripping voltammograms of unsupported Pt-based MEA after CO adsorption at 0.0 V vs. Ag/AgCl for different adsorption times. During CO stripping, N_2 was flushed through the working electrode compartment at a flow rate of 120 ml min^{-1}. Potential scan rate: 50 mV s^{-1}; room temperature [6].

a saturated CO monolayer was achieved after 15 min of CO adsorption. The charge amount, Q_{CO}, was obtained by integrating the CO stripping peak from 0.325 to 1.2 V vs. Ag/AgCl, and then the Pt surface area was obtained, which was larger than what had been obtained from hydrogen desorption. On considering the different modes of CO bonding on the surface of Pt—linear bonding and bridge bonding—one finds that an even higher value would be expected if all the CO was bridge bonded.

3.3.1.3. Linear Sweep Voltammetry

Linear sweep voltammetry (LSV) is an electrochemical technique that is similar to CV. The current on a working electrode generated by the reduction or oxidation of active species can be monitored during the linear scanning of potential between a working electrode and a reference electrode. The difference between CV and LSV is that the LSV potential scan is one directional, from an initial point to an end point, whereas the CV scan is two-directional, from an initial point to an end point and back to form a potential cycle. LSV is often used to measure the rate of hydrogen crossover, which is the undesirable transport of hydrogen across the membrane in a PEM fuel cell. Hydrogen crossover can have negative effects on PEM fuel cells [12], such as reduced fuel efficiency, catalyst layer and membrane degradation [13–16], and decreased PEM fuel cell OCV

[17]. Thus, the measurement of hydrogen crossover plays an important role in PEM fuel cell diagnosis and the development of new membrane materials.

One example of hydrogen crossover measurement using LSV was presented by Song et al. [18]. Ultrahigh-purity hydrogen was supplied to the anode of a PEM fuel cell (the MEA, with an active area of 25 cm^2, was made from a Nafion$^®$-112 membrane), and ultrahigh-purity nitrogen was supplied simultaneously to the cathode of the fuel cell. The flow rates for both H$_2$ and N$_2$ were controlled at 200 ml min^{-1} with their respective mass flow controllers. The cathode acted as the working electrode, and the anode served as both the reference and counterelectrodes. The potential was swept from 0.01 to 0.50 V vs. RHE (reversible hydrogen electrode) at a scan rate of 4 mV s^{-1} with a potentiostat (Solartron SI 1287), and the current generated at the cathode was monitored. Since no other substance was introduced into the cathode except N$_2$, all the current at the cathode was generated by the electrochemical oxidation of hydrogen that had crossed over from the anode to the cathode. Due to the high overpotential, up to 0.5 V vs. RHE, all the crossed over hydrogen could be oxidized and the current in the potential range of 0.35–0.50 V could be treated as a "limiting current." The LSV tests were performed with four sets of different conditions in terms of cell temperature (T_{cell}), anodic RH (RH_{anode}), and cathodic RH ($RH_{cathode}$). Each condition was abbreviated as $T_{cell}/RH_{anode}/RH_{cathode}$; for example, 80 °C/100%/75% represented 80 °C cell temperature, with 100% and 75% anodic and cathodic RH, respectively. Figure 3.7 shows the LSV results for

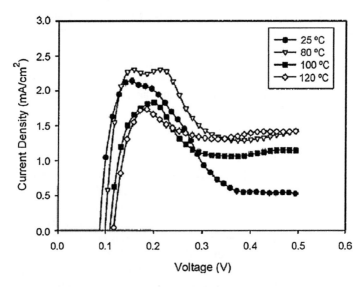

FIGURE 3.7 LSV of Nafion$^®$-112 membrane-based MEA under four conditions [18]. The conditions are denoted using $T_{cell}/RH_{anode}/RH_{cathode}$ and were 25 °C/100%/100%, 80 °C/100%/75%, 100 °C/70%/70%, and 120 °C/35%/35%. Scan rate: 4 mV s^{-1}; potential range: 0.01–0.50 V; gas flow rates: 200 ml min^{-1} pure H$_2$ on the anode and 200 ml min^{-1} N$_2$ on the cathode.

the four conditions. In each case, there is a hydrogen desorption peak on the Pt surface and a plateau on the LSV curve in the potential range of 0.05–0.5 V vs. RHE. The current density caused by hydrogen crossover was determined using the plateau current density at the high potential end, where the current density was limited by the hydrogen transported across the membrane. The contribution of the double layer could be neglected at the high potential end. The current densities caused by hydrogen crossover were 0.6, 1.3, 1.1, and 1.3 mA cm^{-2} under the conditions of 25 °C/100%/100%, 80 °C/100%/75%, 100 °C/70%/70%, and 120 °C/35%/35%, respectively. A similar method was used by Ramani et al. [19] to measure hydrogen crossover through MEAs made from different membranes and to check for electronic shorts. With this same method, Kocha et al. [20] studied the effects of operating conditions, such as temperature, gas pressure, and gas RH, on hydrogen crossover.

This method can also be used to measure the crossover of oxygen across the membrane. But the difficulty here is the lack of a reference electrode. Thus, a special cell design is required to incorporate a reference electrode.

3.3.2. Electrochemical Impedance Spectroscopy

Electrochemical impedance spectroscopy (EIS), also known as AC impedance spectroscopy, is a very powerful technique for characterizing the behaviors of electrode–electrolyte interfaces. Initially, EIS was used to determine double-layer capacity; subsequently, it has been used for more complicated processes, such as metal corrosion [21–24] and electrodeposition [25–27], and to characterize the electrical properties of materials and interfaces. With the developments in PEM fuel cells during recent years, EIS has been widely used for PEM fuel cell diagnosis and the electrochemical characterization of PEM fuel cell materials and components [17,28–35].

EIS measurement includes ex situ and in situ measurement. Ex situ measurement is used mainly to characterize fuel cell materials and components [36–40], such as the catalyst, membrane, and bipolar plates. As such, it is a helpful tool in the screening, designing, and development of PEM fuel cell materials and components. In situ measurement is often used to diagnose a single PEM fuel cell [17,28,31,34,35] or a cell stack [30–32] under actual operating conditions. The information obtained from this diagnosis is helpful for PEM fuel cell designing and the optimization of operating conditions.

The theory and principles of EIS are beyond the scope of this chapter, but they can be found in some textbooks and other references [41]. In this chapter, we focus mainly on the diagnostic applications of EIS in PEM fuel cells. Using typical examples, the following sections will discuss these applications.

3.3.2.1. EIS Measurement and Instrument

As a powerful diagnostic tool, one of the important advantages of EIS is the possibility of using very small AC amplitude signals to determine electrical characteristics without significantly disturbing the properties of the measured system. During EIS measurement, a small AC amplitude signal, usually a voltage or current signal in the range of 1–5% of the DC value, is applied to the DC voltage or current over a frequency range of 0.001–36,000,000 Hz. Then the EIS instrument records the impedance response from the system, from which plots of the signals can be obtained. The most widely used plots are the Nyquist plot and the Bode plot. The Nyquist plot describes the relationship between the imaginary resistance/impedance vs. the real resistance/impedance, as shown in Fig. 3.8 [42]. The Bode plot describes the relationship between the resistance and the phase angle as a function of frequency, as shown in Fig. 3.9 [33].

Depending on the control modes, EIS measurement can be classified into voltage control mode (potentiostatic mode) and current control mode (galvanostatic mode). In the voltage control mode, an AC voltage signal (usually 5–50 mV) is applied to disturb the electrochemical system, and the current response is measured to obtain the system impedance. In this method, a frequency response analyzer (FRA, e.g. Solartron 1260A) and an electrochemical potentiostat (e.g. Solartron SI 1287) are often used. Similarly, in the current control mode, an AC current signal (usually 5–50 mA) is applied to disturb the electrochemical system, and then, the voltage response is measured to obtain the system impedance. Usually, EIS can be measured in both voltage control and current control modes without any significant differences in the results. However, the current and/or voltage are often limited by the potentiostat. For example, the maximum current with the Solartron 1287 potentiostat

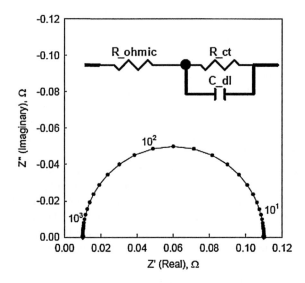

FIGURE 3.8 Nyquist plot for the indicated simple RC circuit, where $R_{ohmic} = 0.01\ \Omega$, $R_{ct} = 0.1\ \Omega$, and $C_{dl} = 0.02$ F. For clarification, three frequencies (10^3, 10^2, 10^1 Hz) are labeled in the plot [42].

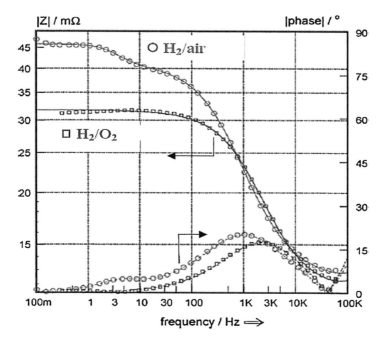

FIGURE 3.9 Comparison of EIS measurements (Bode plots) of a Nafion® 1135-membrane-based MEA made with carbon-supported (60 wt.% Pt) catalyst and 20 wt.% Nafion®, operated with H_2/O_2 ($\square$) and with H_2/air ($\bigcirc$) at 500 mA cm^{-2} [33]. MEA active area: 9 cm^2; cell temperature: 80 °C; pressure: 2.0 atm; H_2 and air flow rates: 100 ml min^{-1} and 300 ml min^{-1}, respectively. (For color version of this figure, the reader is referred to the online version of this book.)

is 2 A, or a 25 A current limitation with a power booster. Thus, an electronic loadbank is required to measure the EIS of a PEM fuel cell stack or an MEA with a large active area under high current. Tang et al. [43] developed a method of EIS measurement under high current using an FRA and an electronic loadbank (TDI RBL 488 series) that can be externally controlled with a signal. The connections between the FRA, loadbank, and fuel cell are depicted in Fig. 3.10 [43].

As shown in Fig. 3.10, the port "Gen Output" on the front panel of the FRA connects with the "REM" and "S−" ports on the rear panel of the loadbank. "V1 HI" and "V1 LO" connect with the "CS" and "S−" ports, respectively. The "V2 HI" port on the FRA and "E+" port on the loadbank connect with the fuel cell cathode, and the "V2 LO" port on the FRA and "E−" port on the loadbank connect with the fuel cell anode. During EIS measurement, the signal generated by the FRA is applied to the loadbank via the connections among the "Gen Output," "REM," and "S−" ports to control the fuel cell. This signal includes AC and DC, with the AC signal superimposed on the DC signal. The AC signals are then fed back to the FRA, and impedance signals are recorded via a program (e.g. ZPlot®).

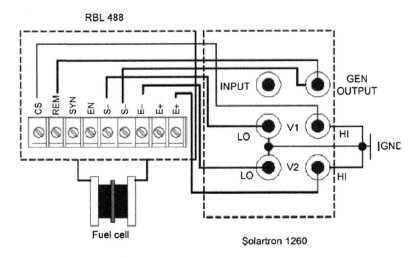

FIGURE 3.10 Connections between FRA (Solartron 1260), programmed electronic loadbank (TDI RBL 488 series), and fuel cell during EIS measurements [43].

EIS data can be analyzed by modeling or fitting the impedance spectrum with an equivalent circuit to extract the physically meaningful properties of the studied system. However, the design of the equivalent circuit is very important, and sometimes, the complexity of the PEM fuel cell system makes this process difficult. Depending on the shape of the EIS spectrum, the equivalent circuit model is usually composed of resistors (R), conductors (L), and capacitors (C), which are connected in series or in parallel, as shown in Fig. 3.11; in this equivalent circuit, R_m, R_t, and R_{mt} represent the membrane resistance, charge transfer resistance, and mass transfer resistance, respectively, and CPE_1 and CPE_2 represent the R_t and R_{mt} associated capacitances, respectively.

After the equivalent circuit is designed, it can be used to fit the EIS spectra, with a program called ZView®. The quality of the fitting can be judged by how well the fitted curve overlaps with the original spectrum.

3.3.2.2. Membrane Conductivity Measurement

As one of the key components in PEM fuel cells, the membrane is a hot topic for research and development. It is well known that membrane conductivity

FIGURE 3.11 An equivalent circuit for EIS spectrum modeling of PEM fuel cells [43].

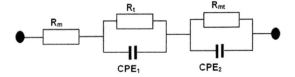

plays an important role in determining PEM fuel cell performance; hence, measurement of membrane conductivity is a necessary step in membrane research and development. A convenient method is to use ex situ AC impedance with two or four probes. This measurement method is given detailed coverage in Chapter 5.

3.3.2.3. EIS Measurement at OCV

It is well known that fuel cell performance is mainly determined by the cathodic ORR, due to its sluggish electrode kinetics as compared to the HOR at the anode. Thus, a study of this kinetics to determine the ORR exchange current density at OCV is necessary, and can be carried out via measuring the impedance at fuel cell OCV.

Zhang et al. [17] developed a method to measure the impedance at fuel cell OCV and then calculated the exchange current density of the ORR. Their study used a Nafion®-112-membrane-based MEA with an active area of 4.4 cm². The fuel cell was operated at different temperatures, 3.0 atm, with fully humidified H_2 and air as the fuel and oxidant, respectively. The flow rates of H_2 and air were $0.1 L min^{-1}$ and $1.0 L min^{-1}$, respectively. Their EIS measurements at OCV involved a Solartron FRA 1260 and a Solartron 1287 potentiostat. During the measurement process, both the working probe and the reference 2 probe of the potentiostat were connected to the cathode, and the counter and reference 1 probes were connected to the anode. A 5 mV AC voltage was used to perturb the fuel cell in the frequency range of 0.1–7000 Hz. The AC impedance spectra were recorded using ZPlot® software. Figure 3.12 presents a typical Nyquist plot obtained at OCV, which shows a semicircle. The resistance demonstrated

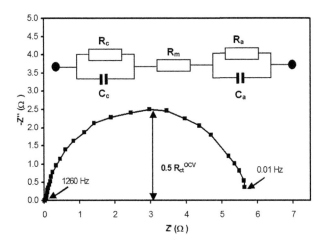

FIGURE 3.12   A Nyquist plot of a PEM fuel cell operated at OCV, 80 °C, 3.0 atm, and 100% RH. MEA: Nafion®-112-membrane based with an active area of 4.4 cm². High-purity hydrogen and air were used for the fuel and oxidant, respectively [17].

by the semicircle represents the charge transfer resistance for the electrode reactions at PEM fuel cell OCV, designated as R_{ct}^{OCV}. This resistance actually includes two portions: the charge transfer resistance for (1) the anodic reaction and (2) the cathodic reaction at OCV. The intercept of the semicircle at the real resistance (Z') axis represents the fuel cell's ohmic resistance, which is dominated by membrane resistance. The equivalent circuit shown in Fig. 3.12 was used to fit the Nyquist plot. R_m represents the membrane resistance; R_c and R_a are the charge transfer resistances for the cathodic and anodic electrode reactions, respectively; and C_c and C_a are the capacitances accompanying R_c and R_a, respectively. As the kinetics of the HOR is much faster than that of the ORR, R_a is much smaller than R_c ($R_a \ll R_c$). So, R_a can be neglected, and R_c can be treated as R_{ct}^{OCV} or $R_{ct-O_2}^{OCV}$ (charge transfer resistance for the ORR at OCV). According to the Butler–Volmer equation, at fuel cell OCV the cathodic overpotential η_c can be described using Eqn (3.6) [17].

$$\eta_c = \frac{RT}{n_{\alpha O}Fi_{O_2}^o}I_c \tag{3.6}$$

where η_c is the cathode overpotential at fuel cell OCV, I_c is the cathodic current density, R is the universal gas constant ($8.314 \text{ J mol}^{-1}\text{ K}^{-1}$), T is the thermodynamic temperature in Kelvins, $n_{\alpha O}$ is the electron transfer number in the rate-determining step of the ORR, F is Faraday's constant ($96,487 \text{ C mol}^{-1}$), and $i_{O_2}^o$ is the apparent exchange current density of the ORR, with a unit of A cm^{-2}.

As the AC impedance measurement at OCV was performed using a very small AC disturbing voltage (5 mV), and based on the above discussion, the obtained $R_{ct-O_2}^{OCV}$ can be expressed as in Eqn (3.7) by differentiating Eqn (3.6):

$$R_{ct-O_2}^{OCV} = \frac{\partial\eta_c}{\partial I_c}\frac{RT}{n_{\alpha O}Fi_{O_2}^o} \tag{3.7}$$

Since $n_{\alpha O}$ is known to be 2, the $i_{O_2}^o$ can be calculated using Eqn (3.7) if $R_{ct-O_2}^{OCV}$ can be obtained through EIS measurement at OCV. Thus, the apparent exchange current densities for the ORR at different temperatures were obtained by this method and are listed in Table 3.1, where it can be seen that the value increased with increasing temperature.

TABLE 3.1 Measured Apparent Exchange Current Densities of the ORR at 3.0 atm, 100% RH, and Different Temperatures [17]. MEA: Nafion-112®-Membrane Based With an Active Area of 4.4 cm^2

Temperature (°C)	23	40	60	80	100	120
$i_{O_2}^o$ (A cm^{-2})	1.22×10^{-4}	2.43×10^{-4}	3.92×10^{-4}	4.60×10^{-4}	3.43×10^{-4}	2.24×10^{-4}

Similarly, the exchange current density of the HOR can also be calculated using EIS measurement at OCV [28]. In this case, during EIS measurement, both the anode and the cathode are flushed with hydrogen. The semicircle on the Nyquist plot at OCV represents twice the charge transfer resistance for the HOR. The apparent exchange current density of the HOR ($i_{H_2}^0$, A cm^{-2}) can be calculated with Eqn (3.8):

$$R_{ct-H_2}^{OCV} = \frac{RT}{n_{\alpha H} F i_{H_2}^0} \qquad (3.8)$$

where $R_{ct-H_2}^{OCV}$ is the charge transfer resistance for the HOR at OCV, and $n_{\alpha H}$ is the electron transfer number of the rate-determining step of the HOR, with a value of 1.

3.3.2.4. EIS Measurement under Load

As shown in Fig. 1.4 of Chapter 1, under a load, PEM fuel cell performance is determined by four voltage losses: the voltage loss caused by mixed potential and hydrogen crossover, which is related to the Pt catalyst status and the membrane properties; the activation loss, which is related to the electrode kinetics; the ohmic loss, which is determined by ohmic resistance; and the voltage loss caused by mass transfer, which is affected by the characteristics of the gas diffusion layer and catalyst layer. The voltage loss caused by mixed potential and hydrogen crossover will be discussed in detail in Chapter 7. The activation loss, ohmic loss, and mass transfer loss can be calculated from the charge transfer resistance, ohmic resistance, and mass transfer resistance, which can be determined by EIS measurement and simulation.

Zhang et al. [28] studied the temperature effect on fuel cell performance with H_3PO_4-doped polybenzimidazole (PBI)-membrane-based MEAs, using EIS as a diagnostic tool. During EIS measurement, a Solartron 1260 FRA and a TDI RBL 488 series programmable loadbank were used to control the PEM fuel cell loads. EIS measurement was performed in the temperature range of 120–200 °C with H_2 and air as fuel and oxidant, respectively. The flow rates of H_2 and air were controlled by their respective mass flow controllers, with stoichiometries of 1.5 and 2.0 for H_2 and O_2, respectively. The impedance spectra at different current densities and temperatures were recorded in the frequency range of 0.01–100,000 Hz using ZPlot® software. Figure 3.13 shows the Nyquist plot of an H_3PO_4-doped PBI-membrane-based PEM fuel cell operated at 140 °C and ambient pressure with a current density of 0.2 A cm^{-2} [28]. Two semicircles appear on the EIS spectrum, one in the high-frequency range and the other in the low-frequency range. The resistance represented by the semicircle in the high-frequency range is the charge transfer resistance (R_t) for electrode reactions, which is dominated by the ORR in the low current density range. The resistance represented by the semicircle in the low-frequency range is the mass transfer resistance (R_{mt}) of the reactant gases. The intercept of the EIS spectrum at the real resistance axis for 3162 Hz represents

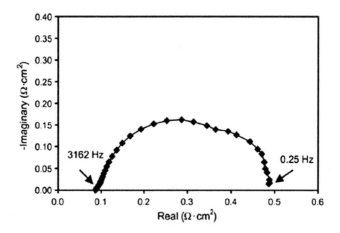

FIGURE 3.13 Nyquist plot of an H_3PO_4-doped PBI-membrane-based PEM fuel cell operated at 140 °C and ambient pressure with a current density of 0.2 A cm^{-2}. MEA active area: 2.6 cm^2; total Pt loading in the MEA: 1.7 mg cm^{-2}; stoichiometry of H_2 and O_2 (in the air stream): 1.5 and 2.0, respectively [28].

the ohmic resistance, which is dominated by the membrane resistance (R_m), and the intercept at 0.25 Hz represents the total resistance for the fuel cell electrode reactions. To simulate the charge transfer resistance and mass transfer resistance values, the equivalent circuit shown in Figure 3.13 was used, along with ZView® software. By this method, the R_m, R_t, and R_{mt} values of the operated PBI-membrane-based PEM fuel cells were determined at different current densities in the temperature range of 120–200 °C. It was found that the membrane resistance decreased with increasing temperature, indicating that the membrane conductivity increased with increasing temperature. It was also found that the charge transfer resistance decreased with increasing temperature, suggesting that the electrode kinetics was enhanced at higher temperatures. However, the mass transfer resistance increased with increasing temperature. The authors attributed this to a tradeoff effect between the gas diffusion change and the gas solubility change in the catalyst layers, both of which are temperature dependent. As the temperature increased, gas diffusion increased, which decreased the gas transfer resistance. Meanwhile, the gas solubility in the proton ionomers of the catalyst layer decreased with rising temperature, which boosted the gas transfer resistance. The increasing effect on mass transfer resistance can exceed the decreasing effect, resulting in an overall increase in mass transfer resistance with temperature.

As discussed above, EIS can be successfully used to diagnose operating PEM fuel cells under loads. Using this method, the authors also studied the effects of temperature [28,43–45], gas stoichiometry [28], RH [34,35], and operating backpressure [46] on fuel cell performance. In situ EIS has also been widely used as a diagnostic method to study PEM fuel cell contamination, such as in studies of toluene contamination [47–49], metal ion contamination [50,51], CO poisoning on Pt catalysts [52,53], and NH_3 contamination [54,55].

3.3.3. Current and Temperature Mapping

The uniform distributions of both temperature and current over the entire MEA of a PEM fuel cell play a critical role in determining performance, durability, and reliability. These uniformities can be measured by mapping the current and temperature distributions of an operating PEM fuel cell across the entire MEA active area. However, current generation is always accompanied by heat production, so it is better to measure the temperature distribution and current distribution simultaneously across the whole MEA active area.

3.3.3.1. Current Mapping

Many approaches have been developed to measure current distribution, such as using a partial MEA, subcells, segmented cells/plates, printed circuit boards (PCBs), and other methods. The partial MEA approach uses several small MEAs across the whole active area, as shown in Fig. 3.14 [56]. Each of them is located at a different section of the entire active area; in this way, the total cell performance is the sum of all the cells' individual performance values. This method is very simple to implement. However, its disadvantage is obvious: all the MEAs must be identical in their electrical, transport, and kinetic properties.

In the subcell approach, several subcells are placed at specific locations in a whole MEA. All the subcells are isolated from each other and from the main cell. Figure 3.15 shows the circuit diagram of a main cell–subcell arrangement [56]. Each subcell and the main cell are controlled by their respective load-banks, allowing independent adjustment of their respective loads during operation. This method can provide a good understanding of the current distribution along the active flow field. However, MEA assembly and flow field manufacture for the subcell approach are complicated and difficult to achieve.

Segmented cell methods combined with other techniques (such as EIS, CV, and others) have been widely used for in situ diagnosis of an operating PEM fuel cell to obtain direct information about not only current distribution but also other phenomena occurring inside the cell. The segmented approach can be carried out with different operating conditions [57], electrode structures [58], and electrode configurations [59]. The basic idea is to divide a cell or cell component into parallel, conductive segments that are electrically insulated from each other. The segmented approach can be classified into several techniques according to their different designs, such as segmented electrodes [56,60,61], segmented current collectors [57,59,62,63], and segmented flow field plates [58,64–68].

Noponen et al. [66] measured the current distribution in a free-breathing PEM fuel cell using a designed cell structure with segmented cathode flow field

Oxidant in

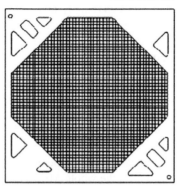

Oxidant out
Whole catalyzed cathode

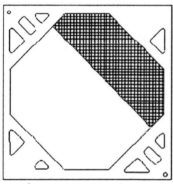

1/3 catalyzed cathode

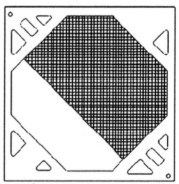

2/3 catalyzed cathode

FIGURE 3.14 Sketch of MEA design showing active cathode areas in the partial MEA method [56].

plates. Figure 3.16 shows the structure of the designed cathode-side flow field plate, and Fig. 3.17 shows the cross-sections of the flow field plate. With this method, the current distribution in the PEM fuel cell was successfully measured at different temperatures. The weakness of this approach is that it is difficult to achieve good contact between the current collector pins and the gas diffusion layer [66,69].

Hakenjos et al. [65] designed a cell in which the anode flow field was segmented and the anode gas diffusion layer was partially segmented. All the segments were connected with their respective current lines and voltage sensors. Infrared (IR) thermography was used to record the temperature of each segment so as to measure the temperature distribution through the active area. With this designed cell, it was possible to measure current, temperature distribution, and flow field water flooding simultaneously.

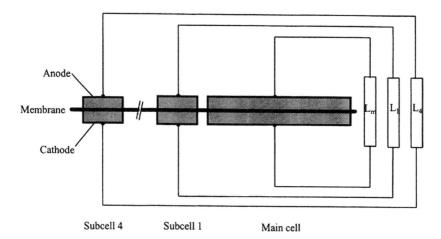

FIGURE 3.15   Circuit diagram of the main cell–subcell arrangement. L_m, L_1, and L_4 denote the loadbanks for the main cell and the subcells, respectively. Additional sensing wires (two for each subcell) are omitted for clarity [56].

The PCB technique was used to make a segmented electrode as early as in 1977 [70]. Brown et al. [71] adopted this method to make a segmented electrode to examine local mass transport in a commercial laboratory electrolyzer. Cleghorn et al. [72] pioneered the method of using the PCB technique to make a segmented cell for measuring the current density in PEM fuel cells. Their designed fuel cell consisted of a segmented anode and a PCB current collector/ flow field plate with serpentine flow channels, as shown in Fig. 3.18. Figure 3.19 shows the PCB segmented electrode with the current and voltage sensing lines fed from the eighteen segments to the edges of the PCB. With this PCB segmented electrode, the current distribution within the active area can be conveniently measured; plus, the PCB segmented cell is simple to assemble and

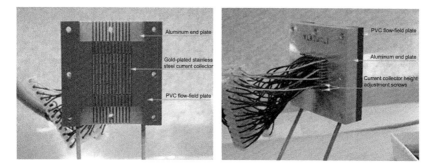

FIGURE 3.16   The left-hand figure illustrates a flow field plate and current collectors, and the right-hand figure shows the corresponding end plate and adjustment screws [66].

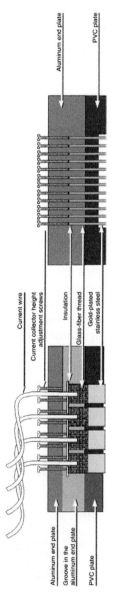

FIGURE 3.17   Cross-sections from a segmented cathode flow field plate. Left is the vertical cut and right is the horizontal cut. [66].

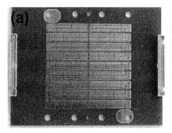 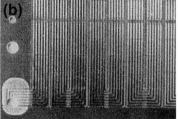

FIGURE 3.18 Photograph of a PCB segmented electrode with a seven-channel serpentine flow field current collector used on the anode side of the cell (a) and a close-up of the PCB (b), showing the flow channels [72].

easy to enlarge, as it does not require the MEA to be segmented into different electronically isolated sections.

Similar to PCB technology, a current distribution measurement gasket was designed by Sun et al. [63] and used to study the influence of operating conditions on the current distribution in a PEM fuel cell. They created a special current distribution gasket with current measuring strips, which consisted of a gasket with the pattern of the flow field used in the fuel cell. Epoxy resin and glass cloth served as the gasket substrates, its top surface was plated with copper and gold, and the measuring strips extended out of the cell, for individual current measurement. The gasket was inserted between the flow field plate and the gas diffusion layer. With this gasket, they successfully investigated the effects of operating temperature, pressure, gas flow rate, and RH on the current distribution.

3.3.3.2. Temperature Mapping

Heat is always generated during fuel cell operation, due to the intrinsic thermodynamics of the electrode reactions, hydrogen crossover, and the fuel cell's

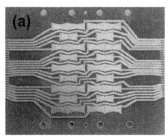 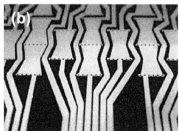

FIGURE 3.19 Photograph of the reverse side (a) and close-up (b) of a PCB segmented electrode, showing the current and voltage sense lines fed from the eighteen segments to the edges of the board [72].

internal resistance. Indeed, temperature plays an important role in determining the performance, durability, and reliability of a PEM fuel cell because it can affect water management, catalyst activity, and mass transfer. If the temperature is too high, dehydration of the membrane and other components can occur, but if the temperature is too low, the reaction kinetics will slow down and water condensation in the catalyst layer or flow channels will become severe. Both these conditions will result in performance degradation and in some cases even fuel cell failure. Therefore, the heat generated in a PEM fuel cell is required to maintain the operating temperature at a safe level. Similar to current distribution, the temperature is usually not distributed homogeneously over the whole active area of the MEA, due to MEA nonuniformity, mass transfer, cell structure, and other factors. Measurement of temperature distribution in a PEM fuel cell is very important for optimizing fuel cell design and MEA fabrication so as to avoid hot spots, which will lead to degradation of the membrane, catalyst, and MEA structure.

In recent years, IR imaging has been successfully used to measure the temperature distribution in an operating PEM fuel cell. Hakenjos et al. [65] used IR thermography to record the temperature distribution over the active area in a segmented cell. Wang et al. [73] designed an experimental fuel cell with a barium fluoride window that was transparent to IR light and was applied to close the anode gas channel, as depicted in Fig. 3.20. Via IR imaging technology, the temperature distribution over the MEA surface beneath the entire flow channel was measured. Similar studies were also conducted by Shimoi et al. [74] and Kondo et al. [75]. In a separate study, Kondo et al. [75] examined the influence of flow field design on temperature distribution using a calcium fluoride crystal as the IR-transparent window.

Wilkinson et al. [76] developed a technique to measure temperature distribution in an operating PEM fuel cell using microthermocouples embedded on a flow field plate. The locations of the microthermocouples were based on the layout of the cathode flow field, as shown in Fig. 3.21. Their results indicated that

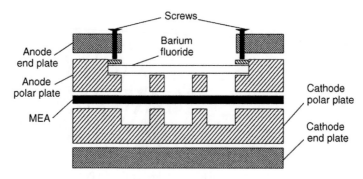

FIGURE 3.20 Structure of an experimental PEM fuel cell with a barium fluoride window for temperature measurement via IR imaging technology [73].

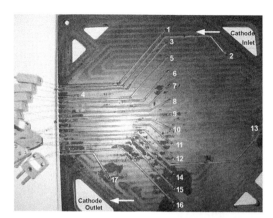

FIGURE 3.21 Locations of thermocouples embedded on an anode plate. The locations of these thermocouples were based on the cathode flow field (e.g. thermocouple no. 2 is near the inlet of the cathode flow field) [76]. (For color version of this figure, the reader is referred to the online version of this book.)

the local temperature at specific experimental conditions correlated well with the local current density determined by the partial MEA approach [56]. This suggested that current distribution over the active area of the MEA in a PEM fuel cell can be determined via the temperature distribution measurement.

3.3.4. Transparent Cell, X-ray Imaging, and Neutron Imaging for Water Flow

Water management plays a critical role in achieving good fuel cell performance. Too much water inside the fuel cell may cause "water flooding," the phenomenon of liquid water or water droplets blocking reactant gas transport, resulting in a sharp decrease in fuel cell performance. However, too little water will lead to dehydration of the PEM, which will also result in low fuel cell performance. Therefore, it is important to characterize the water transport inside a PEM fuel cell to optimize cell operating conditions, MEA fabrication, and flow field design.

3.3.4.1. Transparent Cell

Transparent fuel cells [77–84] are widely used to characterize the water removal process on the flow field of a PEM fuel cell. As they allow optical access to the flow field, one can observe the formation of water droplets and the water removal process in the flow channels. A transparent cell usually includes a transparent plastic end plate, a copper plate to serve as the current collector, and a flow field plate on the anode side, cathode side, or both, as shown in Fig. 3.22. For the test setup, a high-speed camera is required to record the status of liquid water in the flow channels. Thus, water flooding at different stages or under various conditions can be recorded, as shown in Fig. 3.23 [81]. In this way, the visualization of water transport and removal in a transparent cell can be used to optimize the operating conditions, the structural designs of the flow field and gas diffusion layer, and the screening of materials for the gas diffusion layer.

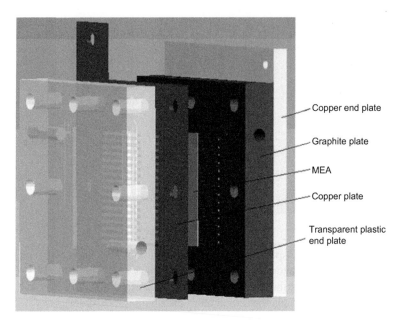

Copper end plate

Graphite plate

MEA

Copper plate

Transparent plastic
end plate

FIGURE 3.22 Depiction of a typical transparent PEM fuel cell [77]. (For color version of this figure, the reader is referred to the online version of this book.)

3.3.4.2. X-ray Imaging

X-ray imaging is a transmission-based technique in which X-rays from a source penetrate the target object and are detected on its opposite side. This technique (which includes X-ray tomography and X-ray radiography) has been widely used in clinical applications to diagnose physical disease. Recently, it has also been used to measure water distribution in a PEM fuel cell because it has strong potential for visualization with high temporal and spatial resolutions. Lee at al.

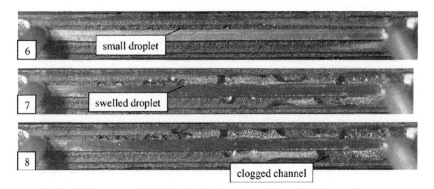

FIGURE 3.23 Photographs of water flooding status for a flow channel at different stages: (6) 5 min, (7) 25 min, and (8) 30 min of operation with a hydrophobic diffusion layer [81].

[85] and Mukaide et al. [86] successfully investigated the water distribution in a PEM fuel cell using X-ray imaging. Sinha et al. [87] quantified the liquid water saturation in PEM fuel cell diffusion media using X-ray micro-tomography. Fluckiger et al. [88] studied liquid water in the gas diffusion layer of a PEM fuel cell using X-ray tomographic microscopy. Kruger et al. [89] visualized water distribution in the gas diffusion layer and flow channels of a PEM fuel cell subsequent to operation, using synchrotron X-ray tomography. They analyzed the water distribution section by section, as shown in Fig. 3.24 [91], and demonstrated that the gas diffusion layer areas under the rib were filled with water, whereas those under the channel were almost free of water. Synchrotron X-ray tomography is a powerful tool to investigate the water distribution in an operating PEM fuel cell on a microscopic scale.

3.3.4.3. Neutron Imaging

Neutron imaging is one of the promising methods developed in recent years to conduct in situ diagnosis of PEM fuel cells. This method is based on the sensitivity of neutrons to hydrogen and is used to monitor hydrogen-containing compounds, such as water. Since Bellow et al. [90], in 1999, successfully demonstrated the use of neutron imaging to study the water distribution in a PEM fuel cell membrane, this technique has been used for in situ monitoring of (1) the production, transport, and removal of water throughout an operating PEM fuel cell [91–94] and (2) water in-plane and through-plane distributions [92], as well as for measuring membrane hydration [95]. Figure 3.25 shows a schematic of the neutron imaging technique for membrane hydration measurement. In this process, a heavy water reactor is used as the neutron

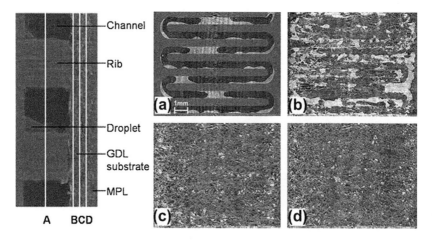

FIGURE 3.24   Sketch with slide positions of a cathodic gas diffusion layer. Section A is located in the flow field structure, whereas sections B–D represent slices in the gas diffusion layer substrate (left). Water agglomerations were visualized in each section [89].

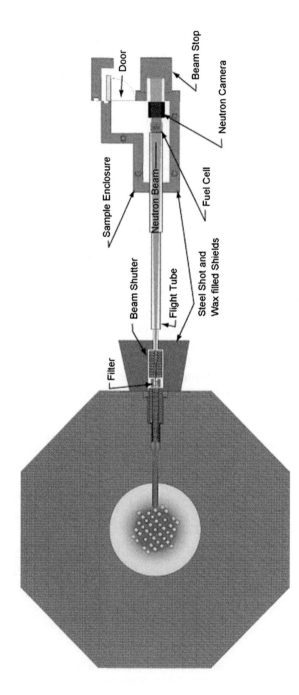

FIGURE 3.25 The BT-6 (beam tube 6) imaging facility at the National Institute of Standards and Technology Center for Neutron Research uses a heavy water reactor as the neutron source [95]. (For color version of this figure, the reader is referred to the online version of this book.)

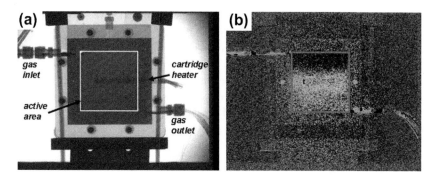

FIGURE 3.26   Neutron images of a 50-cm^2 fuel cell showing (a) the cell construction, gas ports, and active area (outlined in white) and (b) a colorized image showing the active area outlined in red. Red, orange, and green correspond to maximum water, blue and black to minimum water [96]. (For interpretation of the references to color in this figure legend, the reader is referred to the web version of this book.)

source. The neutron beam is processed through collimators, filters, an aperture, and an evacuated flight tube. The neutrons pass through the target and bombard a scintillating screen. Each collision with the screen triggers a nuclear reaction, emitting visible light. Images of the illuminated scintillating screen are captured with a Varian PaxScan 2520 amorphous silicon detector.

Hickner et al. [96] studied the liquid water distribution in an operating PEM fuel cell via a real-time neutron imaging technique. Figure 3.26 shows neutron images of their experimental cell and the MEA active area. The process of water generation and accumulation was recorded using real-time images, as shown in Fig. 3.27. Initially, the color is dark and blue, representing minimal

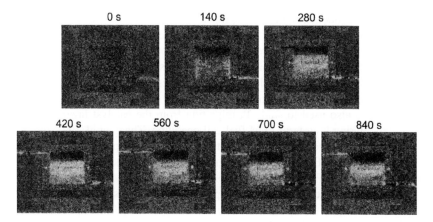

FIGURE 3.27   Images of increasing water content in a PEM fuel cell during a current step from 0 to 1500 mA cm^{-2} occurring at time 30 s. Cell temperature: 80 °C; 100% RH gas feeds; 1.4-bar gauge outlet pressure; 1140 std cm^3 min^{-1} H$_2$; 2700 std cm^3 min^{-1} air [96]. (For color version of this figure, the reader is referred to the online version of this book.)

water. This is because the cell was at open circuit status. At 140 s, liquid water appears. With the cell running continuously, the liquid water accumulates and then reaches a nearly steady state for some time.

The neutron imaging technique has a high temporal and spatial resolution and has been verified as a powerful technique for obtaining qualitative and quantitative information on water content and distribution inside a PEM fuel cell. However, its application in PEM fuel cells is limited by the need for a facility that has a neutron source.

3.3.5. Scanning Electron Microscopy

Scanning electron microscopy (SEM) creates an image of a sample by scanning it with a high-energy beam of electrons. It has been widely used during materials development to characterize samples in terms of morphology, composition, and similar qualities. In the development of PEM fuel cells, it has been used to characterize key materials and components to provide the necessary information for fuel cell designs. It has also been widely used in the ex situ diagnosis of fuel cells, usually for characterizing the key components after fuel cell durability testing to understand fuel cell degradation mechanisms, which are important in designing MEAs and optimizing operating conditions to improve the durability of PEM fuel cells. Prasanna et al. [97] studied the effect of MEA fabrication methods on PEM fuel cell durability by characterizing the cross-sectional surface of the MEAs before and after durability tests. As shown in Fig. 3.28, the membrane thicknesses of the MEAs decreased after the durability tests; for example, the thickness decreased from 42.4 to 29 μm for the conventional MEA after 1000 h of operation. This decrease in membrane thickness is believed to be caused by hydrogen peroxide produced during the fuel cell electrochemical reaction. This reduction in membrane thickness after durability testing was also found by Seo et al. [98] using SEM images of the cross-sectional surfaces of MEAs. Similar phenomena were also observed in other studies [99] using SEM to characterize the cross-sectional surfaces of MEAs after operation. Using SEM images of MEA cross-sectional surfaces, decreases in the thicknesses of both anode and cathode were also observed [98].

SEM was also used to study Pt migration from the catalyst layer into the membrane during long-term PEM fuel cell operation. Zhang et al. [99] studied the effect of hydrogen and oxygen partial pressure on Pt precipitation within the membrane of a PEM fuel cell after long-term operation under OCV. As shown in Fig. 3.29, a bright Pt line or band can be observed on the SEM image of the MEA cross-sectional surface after 2000 h of operation at OCV. A similar Pt band was also observed by others after fuel cell operation at OCV [100,101]. Moreover, SEM also revealed a Pt band during operation of PEM fuel cells under load [97,98,102]. This phenomenon of Pt migration is considered an important mechanism of Pt catalyst degradation, namely, Pt surface area loss and Pt dissolution.

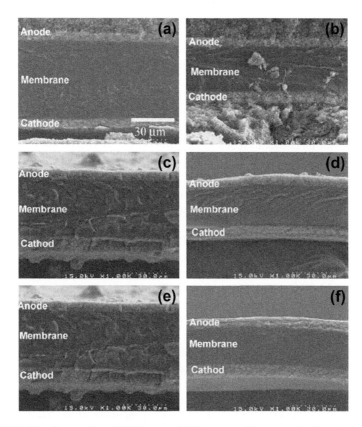

FIGURE 3.28   Cross-sectional SEM images of MEAs prepared by conventional, catalyst-coated membrane (CCM), and CCM hot-pressed methods, taken before and after long-term operation. (a) Conventional—fresh, (b) Conventional—1000 h, (c) CCM—fresh, (d) CCM—1000 h, (e) CCM hot pressed—fresh, and (f) CCM hot pressed—500 h [97].

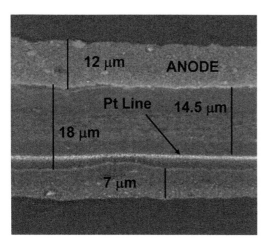

FIGURE 3.29   Cross-sectional SEM photomicrograph of an MEA after operation for 2000 h at OCV conditions with H_2/air. Operating conditions: 80 °C, 100% RH, 150 kPa (absolute) [99].

3.3.6. Transmission Electron Microscopy

Although similar to SEM, TEM has a much higher resolution. It is often used to characterize materials to gain microstructural information. In PEM fuel cells, it is widely used to characterize electrocatalysts and provides useful information such as the mean size of active metal particles and the particle size distribution over a whole size range. In PEM fuel cell diagnosis, TEM is often used to characterize the electrocatalyst after fuel cell operation, and the results are used to compare the electrocatalyst before and after operation to gain information on Pt particle changes and carbon corrosion. In this way, both catalyst and support degradation can be observed. A simple application is the measurement of Pt particle size. Usually, the Pt particles sinter or aggregate together, resulting in larger diameters after fuel cell operation, especially after running at higher temperatures.

As discussed above, after long-term PEM fuel cell operation, Pt dissolution from the catalyst causes Pt to migrate into the membrane and form a band. TEM has also been used to study the Pt band in membranes [98,101,103]. For example, by using TEM, Kim et al. [103] studied the dissolution and migration of Pt after long-term (>1000 h) operation of a PEM fuel cell under various conditions and monitored the cross-sectional surface of the MEA. As shown in Fig. 3.30, after operation at 80 mA cm^{-2} under constant-current conditions, Pt dissolution and migration could be observed at the interface between the cathode and the membrane, in the membrane, as well as at the interface between the membrane and the anode. After investigating the Pt band position and the particle sizes in the band under various conditions, the authors concluded that Pt migrating from the cathode to the membrane phase underwent repeated oxidation/dissolution by crossover oxygen, and reduction by crossover hydrogen [103].

3.3.7. X-ray Diffraction

XRD is based on the elastic scattering of X-rays from the electron clouds of the individual atoms in the system. It is often used to characterize crystal structure, crystallite size, phase, and preferred orientation in polycrystalline or powdered solid samples. It can also be used to identify unknown substances by comparing their characteristic peaks with the standard database for XRD data. In PEM fuel cells, XRD is widely used to characterize carbon-supported catalysts, detect their structural properties, and determine the particle size of metal catalysts. These applications can be found throughout the literature. Figure 3.31 shows a typical peak on an XRD pattern. The most important parameters of an XRD peak are peak position (2θ angle), peak width (W), and peak intensity (I), but the half maximum peak intensity ($I_{max}/2$) and the full width at half maximum peak intensity (FWHM), as shown in Fig. 3.31, are more useful. The relative independent peak on the XRD pattern is usually selected and fit with Gaussian, Lorentzian, or Gaussian–Lorentzian to obtain the 2θ angle, $I_{max}/2$, and FWHM

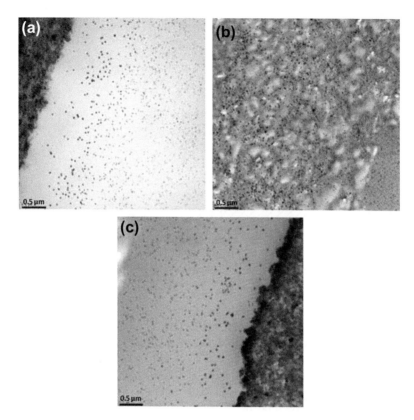

FIGURE 3.30 TEM images of a MEA cross-section after operation at $80 \, mA \, cm^{-2}$ under constant-current conditions (with air for the first 320 h and oxygen for the additional 170 h): (a) between the cathode interface and the membrane; (b) in the polytetrafluoroethylene (PTFE) layer; (c) Between the anode interface and the membrane [103].

values. Then, the particle size can be calculated according to the following Scherrer equation:

$$D = \frac{0.9\lambda}{B\cos(\theta)} \tag{3.9}$$

where D is the average particle diameter in nm; B is the FWHM in radius; and λ is the wavelength, with a value of 0.154 nm for Cu K_α X-ray.

In PEM fuel cell diagnosis, XRD is often used to characterize the catalyst or catalyst layer after fuel cell operation, to determine the Pt particle size and compare it with the size before fuel cell operation [98].

3.3.8. X-ray Photoelectron Spectroscopy

X-ray photoelectron spectroscopy (XPS) is a quantitative spectroscopic technique that can be used to analyze the surface chemistry of a material, or measure

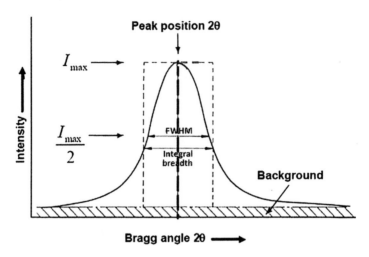

FIGURE 3.31   A typical peak in an XRD pattern. (For color version of this figure, the reader is referred to the online version of this book.)

the elemental composition, empirical formula, chemical state, and electronic state of the elements existing in the material. In PEM fuel cells, it has been widely used in electrocatalyst studies—for example, to determine the electronic structures of catalysts. For more details on the applications of XPS in fuel cell catalyst studies, please refer to the review in [104]. Tan et al. [105] used XPS to study the surface chemistry of elastomeric gasket materials before and after exposure to a simulated fuel cell environment over time. In this study, a commercial silicon G gasket was first exposed to a solution consisting of 1 M H_2SO_4, 10 ppm HF, and reagent-grade water with a resistance of 18 MΩ, for an accelerated durability test (ADT). After the gasket was exposed to the ADT solution for a period of time, it was analyzed using XPS. The results of their study demonstrated that the C/Si ratio in the silicon G gasket was decreased and the O/Si ratio was increased slightly after four weeks of exposure in the ADT solution at 80 °C. They attributed this to the attachment of a methyl group onto the silicon atom, which breaks the chain in the backbone. They also found changes in another commercial gasket, silicon S. The changes in the gasket components indicated the chemical degradation of materials in the simulated fuel cell environment.

3.4. CHAPTER SUMMARY

Testing and diagnosis are very important in evaluating PEM fuel cell performance and understanding degradation and failure mechanisms. They can provide guidelines for optimizing the design and operating conditions of PEM fuel cells to improve their performance, durability, and reliability, as well as to develop degradation mitigation strategies. In this chapter, various methods for PEM fuel cell testing and diagnosis were discussed, and many widely used diagnostic techniques were introduced, especially those pertaining to their applications in PEM fuel cell testing and diagnosis. However, challenges remain for the further development of fuel cell diagnosis, especially for in situ diagnosis, to gain a better understanding of operating PEM fuel cells.

REFERENCES

[1] http://www.answers.com/topic/cyclic-voltammetry.
[2] Pozio A, De Francesco M, Cemmi A, Cardellini F, Giorgi L. J Power Sources 2002;105:13–9.
[3] Watanabe M, Makita K, Usami H, Motoo S. J Electroanal Chem 1986;197:195–208.
[4] Fournier J, Faubert G, Tilquin JY, Cote R, Guay D, Dodelet JP. J Electrochem Soc 1997;144:145–54.
[5] Ciureanu M, Wang H. J Electrochem Soc 1999;146:4031–40.
[6] Vidakovic T, Christov M, Sundmacher K. Electrochim Acta 2007;52:5606–13.
[7] Perez J, Gonzalez ER, Ticianelli EA. Electrochim Acta 1998;44:1329–39.
[8] Eikerling MF, Malek K, Wang Q. In: Zhang J, editor. PEM fuel cell catalyst and catalyst layers: fundamentals and applications. London: Springer; 2008.
[9] Lindstrom RW, Korstdottir K, Lindbergh G. ECS Trans 2009;25:1211–20.
[10] Shinozaki K, Hatanaka T, Morimoto Y. ECS Trans 2007;11:497–507.
[11] Weaver MJ, Chang SC, Leung LWH, Jiang X, Rubel M, Szklarczyk M, et al. J Electroanal Chem 1992;327:247–60.
[12] Cheng X, Zhang J, Tang Y, Song C, Shen J, Song D, et al. J Power Sources 2007;167:25–31.
[13] Inaba M, Kinumoto T, Kiriake M, Umebayashi R, Tasaka A, Ogumi Z. Electrochim Acta 2006;51:5746–53.
[14] Collier A, Wang H, Zi Yuan X, Zhang J, Wilkinson DP. Int J Hydrogen Energy 2006;31:1838–54.
[15] Yu J, Matsuura T, Yoshikawa Y, Islam MN, Hori M. Electrochem Solid State Lett 2005; 8:A156–8.
[16] Teranishi K, Kawata K, Tsushima S, Hirai S. Electrochem Solid State Lett 2006;9:A475–7.
[17] Zhang J, Tang Y, Song C, Zhang J, Wang H. J Power Sources 2006;163:532–7.
[18] Song Y, Fenton JM, Kunz HR, Bonville LJ, Williams MV. J Electrochem Soc 2005; 152:A539–44.
[19] Ramani V, Kunz HR, Fenton JM. J Memb Sci 2004;232:31–44.
[20] Kocha SS, Yang JD, Yi JS. AIChE J 2006;52:1916–25.
[21] Cruz RPV, Nishikata A, Tsuru T. Corros Sci 1996;38:1397–406.
[22] El-Mahdy GA, Nishikata A, Tsuru T. Corros Sci 2000;42:1509–21.
[23] González JA, Otero E, Bautista A, Almeida E, Morcillo M. Prog Org Coat 1998;33:61–7.
[24] Ismail M, Ohtsu M. Constr Build Mater 2006;20:458–69.
[25] Krzewska S. Electrochim Acta 1997;42:3531–40.
[26] Maurin G, Solorza O, Takenouti H. J Electroanal Chem. 1986;202:323–8.
[27] Pauwels L, Simons W, Hubin A, Schoukens J, Pintelon R. Electrochim Acta 2002;47:2135–41.
[28] Zhang J, Tang Y, Song C, Zhang J. J Power Sources 2007;172:163–71.
[29] Alcaraz A, Holdik H, Ruffing T, Ramírez P, Mafé S. J Membr Sci 1998;150:43–56.
[30] Yan X, Hou M, Sun L, Liang D, Shen Q, Xu H, et al. Int J Hydrogen Energy 2007;32:4358–64.
[31] Yuan X, Wang H, Colin Sun J, Zhang J. Int J Hydrogen Energy 2007;32:4365–80.
[32] Zhu WH, Payne RU, Tatarchuk BJ. J Power Sources 2007;168:211–7.
[33] Wagner N, Kaz T, Friedrich KA. Electrochim Acta 2008;53:7475–82.
[34] Zhang J, Tang Y, Song C, Xia Z, Li H, Wang H, et al. Electrochim Acta 2008;53:5315–21.

[35] Zhang J, Tang Y, Song C, Cheng X, Zhang J, Wang H. Electrochim Acta 2007;52:5095–101.

[36] Soboleva T, Xie Z, Shi Z, Tsang E, Navessin T, Holdcroft S. J Electroanal Chem 2008; 622:145–52.

[37] Sistat P, Kozmai A, Pismenskaya N, Larchet C, Pourcelly G, Nikonenko V. Electrochim Acta 2008;53:6380–90.

[38] Kelly MJ, Egger B, Fafilek G, Besenhard JO, Kronberger H, Nauer GE. Solid State Ionics 2005;176:2111–4.

[39] Gavach C, Pamboutzoglou G, Nedyalkov M, Pourcelly G. J Memb Sci 1989;45:37–53.

[40] Chilcott TC, Coster HGL, George EP. J Memb Sci 1995;100:77–86.

[41] Yuan X, Song C, Wang H, Zhang J. In: Yuan X, Song C, Wang H, Zhang J, editors. Electrochmical impedance spectroscopy in PEM fuel cells. London: Springer; 2010.

[42] http://www.scribner.com/files/tech-papers/Scribner%20Associates%20-%20Electrochemical%20Impedance%20Spectroscopy%20for%20Fuel%20Cell%20Research.pdf.

[43] Tang Y, Zhang J, Song C, Liu H, Zhang J, Wang H, et al. J Electrochem Soc 2006;153: A2036–43.

[44] Song C, Tang Y, Zhang JL, Zhang J, Wang H, Shen J, et al. Electrochim Acta 2007;52:2552–61.

[45] Zhang J, Zhang L, Bezerra CWB, Li H, Xia Z, Zhang J, et al. Electrochim Acta 2009;54: 1737–43.

[46] Zhang J, Li H, Zhang J. ECS Trans 2009;19:65–76.

[47] Li H, Zhang J, Fatih K, Wang Z, Tang Y, Shi Z, et al. J Power Sources 2008;185:272–9.

[48] Li H, Zhang J, Shi Z, Song D, Fatih K, Wu S, et al. J Electrochem Soc 2009;156:B252–7.

[49] Li H, Zhang J, Fatih K, Shi Z, Wu S, Song D, et al. ECS Trans 2008;16:1059–67.

[50] Li H, Tsay K, Wang H, Shen J, Wu S, Zhang J, et al. J Power Sources 2010;195:8089–93.

[51] Li H, Gazzarri J, Tsay K, Wu S, Wang H, Zhang J, et al. Electrochim Acta 2010;55:5823–30.

[52] Wagner N, Gülzow E. J Power Sources 2004;127:341–7.

[53] Ciureanu M, Wang H. J New Mater Electrochem Syst 2000;3:107–19.

[54] Zhang X, Serincan MF, Pasaogullari U. J ECS Trans 2009;25:1565–74.

[55] Yuan XZ, Li H, Yu Y, Jiang M, Qian W, Zhang S, et al. Int J Hydrogen Energy 2012; 37:12464–73.

[56] Stumper J, Campbell SA, Wilkinson DP, Johnson MC, Davis M. Electrochim Acta 1998; 43:3773–83.

[57] Hottinen T, Noponen M, Mennola T, Himanen O, Mikkola M, Lund P. J Appl Electrochem 2003;33:265–71.

[58] Wu J, Yi B, Hou M, Hou Z, Zhang H. Electrochem Solid State Lett 2004;7:A151–4.

[59] Natarajan D, Van Nguyen T. J Power Sources 2004;135:95–109.

[60] Noponen M, Ihonen J, Lundblad A, Lindbergh G. J Appl Electrochem 2004;34:255–62.

[61] Rajalakshmi N, Raja M, Dhathathreyan KS. J Power Sources 2002;112:331–6.

[62] Hwnag JJ, Chang WR, Peng RG, Chen PY, Su A. Int J Hydrogen Energy 2008;33:5718–27.

[63] Sun H, Zhang G, Guo L-J, Liu H. J Power Sources 2006;158:326–32.

[64] Geiger AB, Eckl R, Wokaun A, Scherer GG. J Electrochem Soc 2004;151:A394–8.

[65] Hakenjos A, Muenter H, Wittstadt U, Hebling C. J Power Sources 2004;131:213–6.

[66] Noponen M, Mennola T, Mikkola M, Hottinen T, Lund P. J Power Sources 2002;106:304–12.

[67] Mench MM, Wang CY, Ishikawa M. J Electrochem Soc 2003;150:A1052–9.

[68] Bender G, Wilson MS, Zawodzinski TA. J Power Sources 2003;123:163–71.

[69] Wieser C, Helmbold A, Gülzow E. J Appl Electrochem. 2000;30:803–7.

[70] Storck A, Coeuret F. Electrochim Acta 1977;22:1155–60.

[71] Brown CJ, Pletcher D, Walsh FC, Hammond JK, Robinson D. J Appl Electrochem 1992;22:613–9.

[72] Cleghorn SJC, Derouin CR, Wilson MS, Gottesfeld S. J Appl Electrochem 1998;28:663–72.
[73] Wang M, Guo H, Ma C. J Power Sources 2006;157:181–7.
[74] Shimoi R, Masuda M, Fushinobu K, Kozawa Y, Okazaki K. J Energy Res Technol 2004;126:258–61.
[75] Kondo Y, Daiguji H, Hihara E. Proceedings of the international conference on power engineering: ICOPE 2003;(2):463–8, The Japan Society of Mechanical Engineers, Japan, 2003.
[76] Wilkinson M, Blanco M, Gu E, Martin JJ, Wilkinson DP, Zhang JJ, et al. Electrochem Solid State Lett 2006;9:A507–11.
[77] Zhan Z, Wang C, Fu W, Pan M. Int J Hydrogen Energy 2012;34:1094–105.
[78] Ma HP, Zhang HM, Hu J, Cai YH, Yi BL. J Power Sources 2006;162:469–73.
[79] Jiao K, Park J, Li X. Appl Energy 2010;87:2770–7.
[80] Spernjak D, Prasad AK, Advani SG. J Power Sources 2007;170:334–44.
[81] Tüber K, Pócza D, Hebling C. J Power Sources 2003;124:403–14.
[82] Weng F-B, Su A, Hsu C-Y, Lee C-Y. J Power Sources 2006;157:674–80.
[83] Weng F-B, Su A, Hsu C-Y. Int J Hydrogen Energy 2007;32:666–76.
[84] Kim H-S, Min K. J Mech Sci Technol 2008;22:2274–85.
[85] Lee SJ, Lim N-Y, Kim S, Park G-G, Kim C-S. J Power Sources 2008;185:867–70.
[86] Mukaide T, Mogi S, Yamamoto J, Morita A, Koji S, Takada K, et al. J Synchrotron Radiat 2008;15:329–34.
[87] Sinha PK, Halleck P, Wang C-Y. Electrochem Solid State Lett 2006;9:A344–8.
[88] Flückiger R, Marone F, Stampanoni M, Wokaun A, Büchi FN. Electrochim Acta 2011;56:2254–62.
[89] Krüger P, Markötter H, Haußmann J, Klages M, Arlt T, Banhart J, et al. J Power Sources 2011;196:5250–5.
[90] Bellows RJ, Lin MY, Arif M, Thompson AK, Jacobson D. J Electrochem Soc 1999;146:1099–103.
[91] Satija R, Jacobson DL, Arif M, Werner SA. J Power Sources 2004;129:238–45.
[92] Hussey DS, Jacobson DL, Arif M, Owejan JP, Gagliardo JJ, Trabold TA. J Power Sources 2007;172:225–8.
[93] Hickner MA, Siegel NP, Chen KS, Hussey DS, Jacobson DL, Arif M. J Electrochem Soc 2008;155:B294–302.
[94] Hickner MA, Siegel NP, Chen KS, Hussey DS, Jacobson DL, Arif M. J Electrochem Soc 2008;155:B427–34.
[95] Ludlow DJ, Calebrese CM, Yu SH, Dannehy CS, Jacobson DL, Hussey DS, et al. J Power Sources 2006;162:271–8.
[96] Hickner MA, Siegel NP, Chen KS, McBrayer DN, Hussey DS, Jacobson DL, et al. J Electrochem Soc 2006;153:A902–8.
[97] Prasanna M, Cho EA, Lim TH, Oh IH. Electrochim Acta 2008;53:5434–41.
[98] Seo D, Lee J, Park S, Rhee J, Choi SW, Shul Y-G. Int J Hydrogen Energy 2011;36:1828–36.
[99] Zhang J, Litteer BA, Gu W, Liu H, Gasteiger HA. J Electrochem Soc 2007;154:B1006–11.
[100] Yoon W, Huang X. J Electrochem Soc 2010;157:B599–606.
[101] Péron J, Nedellec Y, Jones DJ, Rozière J. J Power Sources 2008;185:1209–17.
[102] Bi W, Fuller TF. J Electrochem Soc 2008;155:B215–21.
[103] Kim L, Chung CG, Sung YW, Chung JS. J Power Sources 2008;183:524–32.
[104] Corcoran CJ, Tavassol H, Rigsby MA, Bagus PS, Wieckowski A. J Power Sources 2010;195:7856–79.
[105] Tan J, Chao YJ, Yang M, Williams CT, Van Zee JW. J. Mater Eng Perform 2008;17:785–92.

Chapter 4

The Effects of Temperature on PEM Fuel Cell Kinetics and Performance

4.1. INTRODUCTION

Temperature is one the most important operating conditions of proton exchange membrane (PEM) fuel cells, and it can significantly influence cell performance. Generally, an increase in the temperature can improve performance. In light of this fact, high-temperature PEM (HT-PEM) fuel cells [1–5] operated above 95 °C (usually 95–200 °C) have recently been developed. This performance improvement at higher temperatures is mainly due to increased membrane proton

PEM Fuel Cell Testing and Diagnosis. http://dx.doi.org/10.1016/B978-0-444-53688-4.00004-8

conductivity [6,7], enhanced electrode kinetics for the oxygen reduction reaction (ORR) and the hydrogen oxidation reaction (HOR) [1,2,8], and improved mass transfer of the reactants. In addition, increasing the temperature can also increase the tolerance of electrocatalysts to contaminants [4,9–12]. However, higher operating temperatures can lead to membrane dehydration [3], increased hydrogen crossover rate [13], and the degradation of components such as electrocatalysts [14,15], gasket materials, and bipolar plates, resulting in a shortened fuel cell lifetime. Chapter 10 presents a detailed discussion of HT-PEM fuel cells. This chapter will focus on conventional (i.e. low-temperature) PEM fuel cells that use perfluorosulfonic acid (PFSA) membranes (e.g. Nafion® membranes) and are usually operated below 95 °C (typically from room temperature to 80 °C).

In general, fuel cell performance can be affected by several operating conditions, such as temperature, pressure, and relative humidity (RH). We will discuss the effects of RH and pressure in Chapters 8 and 9, respectively. In this chapter, only the temperature effects on the performance of PEM fuel cells will be discussed in detail.

4.2. ANODE H_2 OXIDATION ON PT CATALYSTS

Carbon-supported Pt nanoparticles (e.g. Pt/C) are currently the most efficient electrocatalysts for the HOR in PEM fuel cells. If pure hydrogen is used as the fuel, the overpotential of the HOR at the Pt anode is small due to the fast electrode kinetics of the reaction. Normally, the overpotential is <20 mV, even at a current density of 1.0 A cm^{-2}. The mechanism and kinetic behavior of the HOR on a Pt electrode or Pt/Nafion® interface has been studied extensively in both acidic and alkaline media [9,16–20].

The anode electrochemical reaction of a H_2/O_2 PEM fuel cell can be expressed using Reaction (1.I) in Chapter 1. The generally accepted mechanism of the HOR on a Pt catalyst includes three steps [9,17,21–23]: (1) the adsorption of H_2 on the Pt surface, (2) the dissociated chemical adsorption of the adsorbed H_2, which is considered the rate-determining step of the HOR, and (3) the fast electrochemical oxidation of adsorbed hydrogen atoms, producing protons. These three steps are expressed in Reactions (1.IV), (1.V), and (1.VI), respectively, whereas the electrode kinetics of the HOR has been addressed in Section 1.3.1 of Chapter 1.

4.3. CATHODE O_2 REDUCTION ON PT CATALYST

The ORR in PEM fuel cells can be expressed using Reaction (1.II) and is a multielectron transfer process. The generally accepted mechanism for the ORR in acidic media is presented in Fig. 4.1 [24–28]. As shown, there are two pathways: one is a direct 4-electron reaction that produces H_2O, and the other is a 2-electron reaction that produces H_2O_2, which is undesirable in PEM fuel cells. Which pathway the ORR will take strongly depends on the

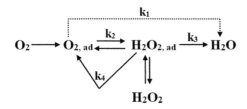

FIGURE 4.1   Observed mechanism of the ORR on a Pt catalyst [24–28].

type and the properties of the electrocatalyst. Currently, Pt and Pt/C are recognized as the most effective electrocatalysts for the ORR in PEM fuel cells, although other catalyst materials have also been developed and investigated; these include Pt-M alloy catalysts (where M is a transition metal, e.g. Fe, Co, Ni, Pd) [29–32]; other precious metal catalysts, such as Au, Ir, and Rh, [33]; transition metal macrocyclic complexes, such as transition metal phthalocyanine and porphyrin [34–41]; as well as transition metal chalcogenides, such as $Mo_4Ru_2Se_8$ [42,43]. On a Pt catalyst surface, the ORR proceeds primarily via the direct 4-electron transfer pathway, with H_2O as the main product.

The kinetics of the ORR has been addressed in Chapter 1. The exchange current density and Tafel slope are two important factors that describe the electrode kinetics. According to Eqns (1.56)–(1.59) and Figure 1.3, the Tafel slope, $\dfrac{2.303RT}{(1 - \alpha_{2O})n_{\alpha2O}F}$, can be obtained, and if $E^r_{O_2/H_2O}$ in Eqn (1.59) is known, the exchange current density $(i^o_{O_2/H_2O})$ for the ORR can be calculated. In the literature, two sets of Tafel slopes can be found [24,44–47]. In the low current density range (corresponding to high cathode potential), the value of the slope is approximately 60 mV decade^{-1}, whereas in the high current density range (corresponding to low cathode potential), the value of the slope is approximately 120 mV decade^{-1}. This disparity results from the differences in the Pt surface. When the cathode potential is high (e.g. >0.8 V vs. standard hydrogen electrode (SHE)), the Pt surface will be partially covered by PtO due to the following reaction:

$$Pt + \frac{1}{2} O_2 \leftrightarrow PtO \quad E^o = 0.88\ V\ vs.\ RHE \qquad (4.I)$$

It has also been reported [48–50] that PtO coverage can be approximately 0.3 at cathode potentials >0.8 V. Only when the cathode potential is <0.8 V is the Pt surface close to being pure. Several studies [24,51,52] have demonstrated that the kinetics of the ORR on a pure Pt surface is different from that on a PtO surface.

At the cathode of a PEM fuel cell, a pure Pt surface is not easy to achieve because oxygen is present, leading to a mixed surface of Pt and PtO. Thus, the fuel cell thermodynamic open circuit voltage (OCV) at 25 °C is always

<1.23 V. The value most often obtained is about 1.06 V, which is a mixture of the thermodynamic OCV of O_2/H_2O and of Pt/PtO due to the coexistence of the ORR and Pt oxidation [53,54]. As discussed in Chapter 7, hydrogen crossover from the anode to the cathode can also reduce the fuel cell OCV.

4.4. POLARIZATION CURVE ANALYSIS USING EIS

In Chapter 1, Figure 1.4 shows a typical polarization curve of a PEM fuel cell. The voltage loss of a cell is determined by its OCV, electrode kinetics, ohmic resistance (dominated by the membrane resistance), and mass transfer property. In experiments, the OCV can be measured directly. If the ohmic resistance (R_m), kinetic resistance (R_t, also known as charge transfer resistance), and mass transfer resistance (R_{mt}) are known, the fuel cell performance is easily simulated. As described in Chapter 3, electrochemical impedance spectroscopy (EIS) has been introduced as a powerful diagnostic technique to obtain these resistances. By using the equivalent circuit shown in Figure 3.3, R_m, R_t, and R_{mt} can be simulated based on EIS data.

It is well known that the electrode kinetics of the HOR is much faster than that of the ORR, and thus, the anodic overpotential is much smaller than the cathodic one. By assuming that the anodic polarization is negligible, one can describe the steady-state polarization curve by using the semiempirical Eqn (4.1) [3,55–57]:

$$E = E_o - b\ln(I) - IR_m - m_{mt}\exp(n_{mt}I) \tag{4.1}$$

where E is the fuel cell voltage; E_o is the fuel cell OCV-related constant, which is determined by many factors, such as fuel cell OCV, exchange current density, hydrogen crossover, and Pt surface composition (i.e. the ratio of Pt to PtO); b is the Tafel slope; I is the current density; R_m is the membrane resistance; m_{mt} is the mass transfer coefficient; and n_{mt} is the simulation parameter for curve fitting. In Eqn (4.1), $b\ln(I)$ is the contribution of the fuel cell reaction kinetics (dominated by the ORR), IR_m is the contribution of ohmic resistance (dominated by membrane resistance), and $m_{mt}\exp(n_{mt}I)$ is the mass transfer contribution to the fuel cell polarization (dominated by oxygen transfer in the electrode). The internal alternating current (AC) resistance of the fuel cell (R_{cell}) can be obtained by differentiating Eqn (4.1) [3]:

$$R_{cell} = -\frac{\partial E}{\partial I} = \frac{b}{I} + R_m + m_{mt}n_{mt}e^{n_{mt}I} \tag{4.2}$$

If b/I in Eqn (4.2) can be defined as the charge transfer resistance (R_t) and $m_{mt}n_{mt}e^{n_{mt}I}$ as the mass transfer resistance (R_{mt}), the fuel cell internal AC impedance can be expressed as Eqn (4.3) [3]:

$$R_{cell} = R_m + R_t + R_{mt} \tag{4.3}$$

In Eqn (4.3), R_m is a function of water content and temperature, and both R_{mt} and R_t are functions of current density and temperature. Therefore, to understand the effect of temperature on the performance of a PEM fuel cell, it is necessary to study its effect on these resistances and on their associated exchange current density, membrane water uptake, and mass transfer property.

Once the values of R_m, R_t, and R_{mt} in Eqn (4.3) are simulated based on the AC impedance spectra and the equivalent circuit in Figure 3.11, the decreases in cell voltage (ΔV_i) caused by the individual resistances can be calculated at a constant current density, and their contributions ($\Delta V_i \%$) to the overall decline in cell voltage (ΔV_{cell}) can be calculated according to Eqn (4.4):

$$\Delta V_i\% = \frac{\Delta V_i}{\Delta V_{cell}} \times 100 \qquad (4.4)$$

This can also be expressed as follows:

$$\Delta V_i\% = \frac{R_i}{R_{cell}} \times 100 \qquad (4.5)$$

As shown in Fig. 4.2, at both 80 °C and 120 °C, in the low current density range (<0.5 A cm^{-2}), the cell polarization is mainly due to the charge transfer process, whereas in the high current density range (>1.5 A cm^{-2}), the cell polarization is dominated by the mass transfer process [3].

4.5. TEMPERATURE EFFECTS ON PEM FUEL CELL KINETICS

The temperature can significantly affect the electrode kinetics in PEM fuel cells. This section will discuss the effects of temperature on (i) fuel cell thermodynamics and OCV, (ii) the kinetics of both the HOR and the ORR, (iii) the proton conductivity and hydration of the membrane, and (iv) mass transfer.

4.5.1. Temperature Effect on PEM Fuel Cell Thermodynamics and OCV

In H_2/air PEM fuel cells, the overall electrochemical reaction is expressed as Reaction (4.II):

$$H_2 + \frac{1}{2}O_2 \rightarrow H_2O \qquad (4.II)$$

For Reaction (4.II), the fuel cell's theoretical OCV or thermodynamic OCV can be expressed as V_{theory}^{OCV} or V_{cell}^{OCV}, as used in Eqn (1.3) in Chapter 1.

$$V_{theory}^{OCV} = V_{cell}^{OCV} = E_{O_2/H_2O}^r - E_{H_2/H^+}^r$$

$$= E_{O_2/H_2O}^o - E_{H_2/H^+}^o + 2.303\frac{RT}{2F}\log\left(\frac{a_{H_2}a_{O_2}^{\frac{1}{2}}}{a_{H_2O}}\right) \qquad (4.6)$$

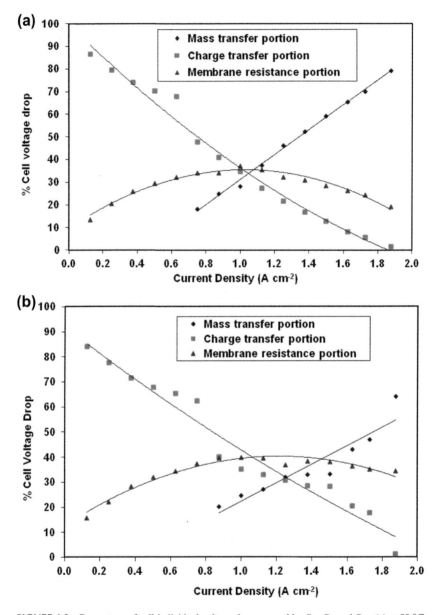

FIGURE 4.2 Percentage of cell individual voltage drops caused by R_m, R_t, and R_{mt}. (a) at 80 °C and (b) at 120 °C. Nafion®-112 membrane-based membrane electrode assembly(MEA) with an active area of 4.4 cm^2; RH for both H$_2$ and air: 100%; 3.0 atm. [3].

where $E^r_{H_2/H^+}$ is the reversible anode potential (V) at temperature T; a_{H_2} and a_{H^+} are the respective activities of H_2 and H^+; $E^o_{H_2/H^+}$ is the electrode potential of the H_2/H^+ redox couple under standard conditions (1.0 atm, 25 °C), which is defined as zero voltage; n_H is the electron transfer number (with a value of 2 for the H_2/H^+ redox couple); R is the universal gas constant (8.314 J K^{-1} mol^{-1}); F is Faraday's constant (96,487 C mol^{-1}); $E^r_{O_2/H_2O}$ is the reversible cathode potential (V) at temperature T; a_{O_2} and a_{H^+} are the respective activities of O_2 and H^+; $E^o_{O_2/H_2O}$ is the electrode potential of the O_2/H_2O redox couple under standard conditions (1.0 atm, 25 °C), which is 1.229 V (vs. the SHE); n_O is the electron transfer number (with a value of 4 for the O_2/H_2O redox couple); and a_{H_2O} is the activity of H_2O.

For an approximate evaluation of theoretical electrode potentials and fuel cell voltage, a_{O_2}, a_{H_2}, and a_{H_2O} can be replaced by their partial pressures, P_{O_2}, P_{H_2}, and P_{H_2O}, in the anode and cathode feed streams, respectively. Thus, Eqn (4.6) can be expressed as Eqn (4.7):

$$V^{OCV}_{cell} = E^o_{O_2/H_2O} - E^o_{H_2/H^+} + 2.303 \frac{RT}{2F} \log \left(\frac{P_{H_2} P^{\frac{1}{2}}_{O_2}}{P_{H_2O}} \right) \qquad (4.7)$$

In Eqn (4.7), $E^o_{H_2/H^+}$ equals zero at standard conditions, whereas $E^o_{O_2/H_2O}$ has been reported to be a function of temperature and can be expressed as Eqn (4.8) [58–60]:

$$E^o_{O_2/H_2O} = 1.229 - 0.000486(T - 298.15) \qquad (4.8)$$

Therefore, the thermodynamic fuel cell OCV can be expressed as follows:

$$V^{OCV}_{cell} = 1.229 - 0.000486(T - 298.15) + 2.303 \frac{RT}{2F} \log \left(\frac{P_{H_2} P^{\frac{1}{2}}_{O_2}}{P_{H_2O}} \right) \qquad (4.9)$$

Our recent study indicated that the fuel cell OCV decreased with increasing temperature, as shown in Fig. 4.3 [58]. It can be seen that both the theoretical and the measured OCV decreased when the temperature increased from 23 to 120 °C. This was mainly because of the effect of temperature on the fuel cell thermodynamics and hydrogen crossover, which will be addressed in detail in Chapters 6 and 7, respectively.

4.5.2. Temperature Effect on the Kinetics of the HOR in PEM Fuel Cells

In Chapter 1, Eqns (1.30) and (1.31) describe the relationship between the overpotential and the exchange current density of the HOR. In Eqn (1.31),

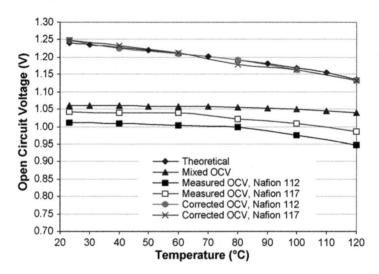

FIGURE 4.3 Fuel cell OCV at different temperatures. Nafion®-112-membrane-based MEA with an active area of 4.4 cm², at 100% RH and 3.0 atm backpressure [58]. (For color version of this figure, the reader is referred to the online version of this book.)

$\dfrac{2.303RT}{\alpha_H n_{\alpha H} F}$ is the Tafel slope, which is a function of temperature; $n_{\alpha H}$ is the electron transfer number for the HOR, whose value is widely reported to be 2.0; and α_H is the charge transfer coefficient, whose value is recognized as 0.5. In this equation, $i^o_{H_2/H^+}$ is the apparent exchange current density for the HOR and is a function of temperature. The relationship between $i^o_{H_2/H^+}$ and T can be expressed using Eqn (1.34), where $I^o_{H_2/H^+}$ is the intrinsic exchange current density, which is the exchange current density at infinite temperature. The relationship between $i^o_{H_2/H^+}$ and $I^o_{H_2/H^+}$ can also be expressed using the following equation [2,8]:

$$i^o_{H_2/H^+} = (EPSA)_a \times I^o_{H_2/H^+} \left(\frac{P_{H_2}}{P^o_{H_2}}\right)^{0.5} \qquad (4.10)$$

where $(EPSA)_a$ is the electrochemical surface area of Pt at the anode, and P_{H_2} and $P^o_{H_2}$ are, respectively, the operating hydrogen partial pressure in the feed stream at the anode and the hydrogen pressure at standard conditions.

Recently, we studied the temperature effect on the exchange current density of the HOR between 23 and 120 °C using a Nafion®-112 membrane-based PEM fuel cell, and between 120 and 200 °C using a phosphoric-acid-doped polybenzimidazole (PBI) membrane-based PEM fuel cell. As shown in Tables 4.1 and 4.2, $I^o_{H_2/H^+}$ increases as the temperature is increased from 23 to 200 °C.

TABLE 4.1 Measured and Simulated Exchange Current Densities for the HOR and ORR in a Nafion®-112-Membrane-Based PEM Fuel Cell at 3.0 atm (Absolute), 100% RH, and Different Temperatures, in a Low Current Density Range ($i^0_{O_2/H_2O}$ is the Measured Apparent Exchange Current Density for the ORR, $i^0_{H_2/H^+}$ is the Simulated Apparent Exchange Current Density for the HOR, and $I^0_{O_2/H_2O}$ and $I^0_{H_2/H^+}$ are the Intrinsic Exchange Current Densities for the ORR and HOR, Respectively)[8]

Temperature (°C)	23	40	60	80	100	120
$i^0_{O_2/H_2O}$ (A cm⁻²)	1.22×10^{-4}	2.43×10^{-4}	3.92×10^{-4}	4.60×10^{-4}	3.43×10^{-4}	2.24×10^{-4}
$I^0_{O_2/H_2O}$ (A cm⁻²)	5.02×10^{-7}	1.85×10^{-6}	3.44×10^{-6}	6.25×10^{-6}	6.88×10^{-6}	1.05×10^{-5}
$i^0_{H_2/H^+}$ (A cm⁻²)	0.134	0.198	0.344	0.607	0.604	0.497
$I^0_{H_2/H^+}$ (A cm⁻²)	1.73×10^{-3}	4.06×10^{-3}	7.21×10^{-3}	1.23×10^{-2}	3.25×10^{-2}	6.13×10^{-2}

TABLE 4.2 Apparent and Intrinsic Exchange Current Densities at Different Temperatures in the Low Current Density Range, Measured at AMBIENT Pressure and Zero RH ($i^0_{O_2/H_2O}$ is the Measured Apparent Exchange Current Density for the ORR, $i^0_{H_2/H^+}$ is the Simulated Apparent Exchange Current Density for the HOR, and $I^0_{O_2/H_2O}$ and $I^0_{H_2/H^+}$ are the Intrinsic Exchange Current Densities for the ORR and HOR, Respectively)[2]

Temperature (°C)	120	140	160	180	200
$i^0_{O_2/H_2O}$ (A cm^{-2})	2.30×10^{-3}	2.05×10^{-3}	2.64×10^{-3}	3.60×10^{-3}	5.43×10^{-3}
$I^0_{O_2/H_2O}$ (A cm^{-2})	8.97×10^{-5}	1.05×10^{-4}	1.51×10^{-4}	3.18×10^{-4}	7.88×10^{-4}
$i^0_{H_2/H^+}$ (A cm^{-2})	0.72	1.24	1.88	2.50	2.71
$I^0_{H_2/H^+}$ (A cm^{-2})	1.02×10^{-2}	2.18×10^{-2}	3.51×10^{-2}	6.82×10^{-2}	1.16×10^{-1}

4.5.3. Temperature Effect on the Kinetics of the ORR in PEM Fuel Cells

As discussed in Chapter 1, the Tafel slope of the ORR is $\dfrac{2.303RT}{(1-\alpha_{2O})n_{\alpha2O}F}$, which is a function of temperature. Here, α_{2O} is the electron transfer coefficient in Reaction (1.XV) and is also dependent on temperature. On a Pt electrode, the electron transfer coefficient for the ORR increases linearly with temperature between 20 and 250 °C, in accordance with Eqn (4.11) [8,44]:

$$\alpha_O = \alpha_O^o T \qquad (4.11)$$

where α_O is the electron transfer coefficient of the ORR on a Pt electrode, α_O^o equals 0.001678, and T is the temperature in Kelvin.

The relationship between the intrinsic exchange current density (I_{O_2/H_2O}^o) and the apparent exchange current density (i_{O_2/H_2O}^o) can be described using the following equation [2,8]:

$$i_{O_2/H_2O}^o = (EPSA)_c \times I_{O_2/H_2O}^o \left(\frac{P_{O_2}}{P_{O_2}^o}\right)^{0.001678T} \qquad (4.12)$$

where $(EPSA)_c$ is the electrochemical surface area of Pt at the cathode, and P_{O_2} and $P_{O_2}^o$ are the operating oxygen partial pressure in the feed stream at the cathode and the oxygen pressure at standard conditions, respectively.

As discussed previously, the Tafel slope at low current density (high potential range) is approximately 60 mV decade^{-1}, while the Tafel slope at a high current density (low potential range) is approximately 120 mV decade^{-1}; thus, the exchange current densities should be different in these two current ranges. It is also expected that the temperature dependencies of these two exchange current densities will differ. For example, Parthasarathy et al. [44] studied the temperature dependence of the ORR at a Pt/Nafion®-117 interface using a microelectrode method and found that the exchange current density increased with increasing temperature from 1.69×10^{-10} A cm^{-2} at 30 °C to 1.87×10^{-8} A cm^{-2} at 80 °C in the low current density range, whereas in the high current density range, it increased from 2.84×10^{-7} A cm^{-2} at 30 °C to 1.39×10^{-6} A cm^{-2} at 80 °C. A recent study [8] of a Nafion®-112-membrane-based PEM fuel cell showed that the intrinsic exchange current densities of the ORR in the high and low potential ranges at 80 °C were 6.25×10^{-6} A cm^{-2} and 3.87×10^{-4} A cm^{-2}, respectively. The temperature effects on the exchange current density of the ORR between 23 and 120 °C using a Nafion®-112-membrane-based PEM fuel cell and in the temperature range of 120–200 °C using a PBI-membrane-based PEM fuel cell have also been reported [2,8]. As shown in Tables 4.1 and 4.2, the I_{O_2/H_2O}^o increases with temperature from 23 to 200 °C.

The relationship between i_{O_2/H_2O}^o and I_{O_2/H_2O}^o can be expressed in an Arrhenius form, as shown in Eqn (1.62). By plotting $\log(i_{O_2/H_2O}^o)$ vs. $1/T$, the reaction activation energy $(E^o_{O_2})$ of the ORR can be calculated based on the slope of this plot. Using a Nafion®-112-membrane-based PEM fuel cell, $E^o_{O_2}$ values of 28.3 kJ mol^{-1} and 57.3 kJ mol^{-1} were obtained in the low current density range and the high current density range, respectively [8].

4.5.4. Temperature Effect on Membrane Conductivity and Hydrogen Crossover

It has been widely reported [61–64] that temperature can significantly affect the proton conductivity (σ) of a membrane. For a PFSA membrane (e.g. Nafion® membrane), the proton conductivity strongly depends on the water content of the membrane. Therefore, when studying the effect of temperature on membrane conductivity, the RH or water content of the membrane must be considered. At a low RH, an increase in the temperature will cause membrane dehydration, resulting in decreased proton conductivity, whereas with a well-hydrated membrane, the proton conductivity will increase with increasing temperatures. For example, the conductivity of Nafion® 117 at 100% RH increases from 0.1 to 0.2 S cm^{-1} when the temperature is raised from 30 to 85 °C [65]. Generally, under well-hydrated conditions, the temperature dependence of conductivity can be expressed in an Arrhenius form [2]:

$$log(\sigma) = log(\sigma_o) - \frac{2.303E^m_a}{R}\left(\frac{1}{T}\right) \qquad (4.13)$$

where σ, σ_o, E^m_a, R, and T are the membrane conductivity (S cm^{-1}), the pre-exponential factor (S K^{-1} cm^{-1}), the proton conducting activation energy (kJ mol^{-1}), the ideal gas constant (J mol^{-1} K^{-1}), and the temperature (K), respectively. Figure 4.4 shows the Arrhenius plots for the ionic conductivities of several X-form (X = H, Li, Na) membranes, along with Nafion® 117 H-form in the fully hydrated state. It can be seen that in a fully hydrated state, the proton conductivity increases with increasing temperatures, regardless of whether the membrane is H-, Li-, or Na-form.

However, for a PFSA membrane operated at high temperatures, the situation can be more complex due to the tradeoff between increased conductivity and membrane dehydration when the temperature is increased. As shown in Fig. 4.5, the resistance of Nafion®-112 membrane decreases slightly when the temperature is increased from 80 to 100 °C, which indicates an increase in proton conductivity, but the resistance increases when the temperature is raised to 120 °C, suggesting a decrease in proton conductivity. This is because the negative effect of dehydration is larger than the positive effect of temperature on membrane conductivity.

Chapter 5 provides a more detailed discussion of the temperature effect on membrane proton conductivity.

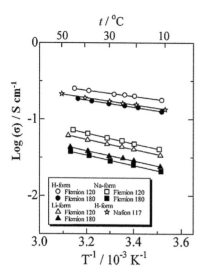

FIGURE 4.4 Arrhenius plots for ionic conductivities (σ) of Flemion X-form (X = H, Li, Na) membranes, along with Nafion® 117 H-form in the fully hydrated state [66].

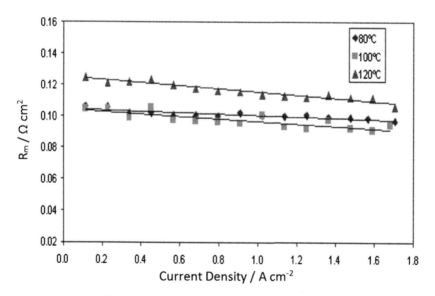

FIGURE 4.5 Nafion® through-plane resistance as a function of current density at three fuel cell operating temperatures: 100% RH for both the anode and the cathode; 3.0 atm backpressure with hydrogen and air feeding; Nafion®-112-based baseline MEA with an active area of 4.4 cm² [3]. (For color version of this figure, the reader is referred to the online version of this book.)

Besides affecting the proton conductivity, the temperature can also influence the hydrogen crossover of PEM fuel cell membranes. During hydrogen crossover, hydrogen diffuses across the membrane from the anode to the cathode, leading to a lower fuel cell efficiency and degradation of the membrane. Obviously, hydrogen crossover is a diffusion-controlled process. It has been reported that hydrogen crossover can be affected by membrane structure [67] and by fuel cell operating conditions such as temperature, backpressure, and the RH of the reactant gases [13,68,69]. Hydrogen crossover will be dealt with in detail in Chapter 6.

Temperature affects hydrogen crossover mainly by affecting the hydrogen permeability coefficient, which is temperature dependent and can be expressed in an Arrhenius form [13]:

$$\ln\Psi_{H_2}^{PEM} = \ln\Psi_{H_2}^{o} + \left(-\frac{E_{H_2}^{PEM}}{R} \right)\frac{1}{T} \tag{4.14}$$

where $\Psi_{H_2}^{PEM}$ is the hydrogen permeability coefficient, $\Psi_{H_2}^{o}$ is the maximum permeability coefficient (e.g. at infinite temperature), $E_{H_2}^{PEM}$ is the activation energy for hydrogen crossover, R is the gas constant, and T is the temperature in Kelvin. As shown in Fig. 4.6, the hydrogen permeability coefficient increases with increasing temperature.

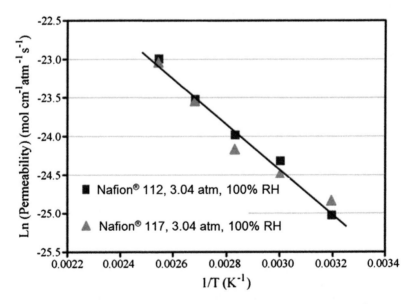

FIGURE 4.6 Arrhenius plots of the H_2 permeability coefficient for both Nafion®-112- and Nafion®-117-membrane-based MEAs at a 3.04 atm backpressure and a 100% RH in the temperature range of 40–120 °C. Both MEAs have an active area of 4.4 cm². Anode H_2 stream flow rate: 0.1 L min⁻¹; cathode N_2 stream flow rate: 0.5 L min⁻¹; cathode potential: 0.5 V vs. anode hydrogen electrode [13]. (For color version of this figure, the reader is referred to the online version of this book.)

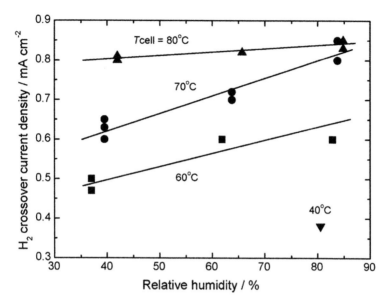

FIGURE 4.7   Effect of cell temperature and humidification on H_2 crossover current density at atmospheric pressure. H_2 and Ar gases were humidified at the same temperature in each case. Nafion®-112 membrane, $H_2/Ar = 300/300$ mL min^{-1} [69].

Teranish et al. [69] showed that the H_2 crossover current density increases with temperature and RH, as shown in Fig. 4.7; this indicates that the H_2 crossover rate increases with temperature and RH.

4.5.5. Temperature Effect on Mass Transfer in PEM Fuel Cells

The temperature can also affect the gas mass transfer (or mass transport) in PEM fuel cells. However, when studying the temperature effect on gas mass transport, the RH, current density, backpressure, and flow rate (or stoichiometry) of the reactant gases should be considered. At low current densities (<0.5 A cm^{-2}), the fuel cell polarization is under the control of activation and kinetics, and temperature has little effect on gas mass transport. At high current densities (>1.5 A cm^{-2}), the fuel cell polarization is under the control of mass transport, and the effect of temperature can be significant, depending on other operating conditions. For example, at a low RH, an increase in the temperature may cause membrane dehydration. At a high RH, a temperature increase will help to remove the water from inside the fuel cell electrode and flow channels and will prevent "water flooding" and reduce the mass transfer resistance. A high gas flow rate, or stoichiometry, will also help to eliminate the "water flooding" phenomenon. Thus, to study the effect of temperature on mass transport, a high current density is often used, and

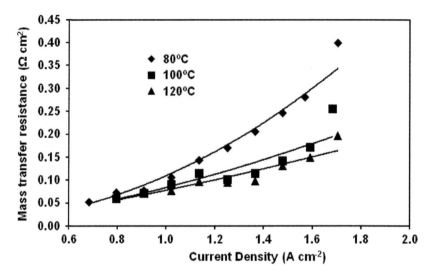

FIGURE 4.8 Mass transfer resistance as a function of current density at 80 °C, 100 °C, and 120 °C. Nafion®-112-membrane-based MEA with an active area of 4.4 cm², 3.0 atm backpressure, and 100% RH; H_2 flow rate: 0.3 L min⁻¹, air flow rate: 1.0 L min⁻¹ [3]. (For color version of this figure, the reader is referred to the online version of this book.)

certain RHs and gas stoichiometries are applied for fuel cell operation. Figure 4.8 shows some results [3] obtained in a Nafion®-112-membrane-based PEM fuel cell. The fuel cell was operated at a 3.0 atm backpressure and 100% RH, and the flow rates of H_2 and air were controlled at 0.3 L min⁻¹ and 1.0 L min⁻¹, respectively. The R_{mt} values at different current densities and temperatures were simulated from the AC impedance spectra. As shown in Fig 4.8, the R_{mt} decreased when the temperature was increased from 80 to 100 °C and then to 120 °C, and the trend was valid in the current density range of 0.8–1.7 A cm⁻², which indicates the benefit of high-temperature operation.

4.6. THE EFFECT OF TEMPERATURE ON THE OVERALL PERFORMANCE OF A PEM FUEL CELL

As discussed above, the temperature can influence the OCV of a PEM fuel cell and thermodynamics, electrode kinetics, membrane conductivity, hydrogen crossover, and mass transfer process, and this influence will be reflected in the overall cell performance. However, the dependence of performance on temperature can be complicated by the fact that other conditions, such as RH, backpressure, gas stochiometry, flow field design, and electrode structure, also affect performance.

The literature contains several studies on how the operating temperature affects PEM fuel cell performance [2,3,70–75]. For example, Yan et al. [71] studied the effect of temperature on performance at different gas humidification temperatures, by using Gore PRIMEA® 57 MEA. Figure 4.9 presents the fuel cell performance at different temperatures with a humidification temperature of 50 °C. As shown, the fuel cell performance improves when the operating temperature is increased from 30 to 50 °C. This can be explained by enhanced electrode kinetics and increased membrane conductivity. However, the fuel cell performance decreases if the temperature is further increased from 50 to 75 °C. This may occur because the additional increase can cause the water inside the membrane and the electrode to evaporate quickly. Any operating temperature higher than the humidification temperature (50 °C in their experiment) will lead to a fast water evaporation rate, causing membrane dehydration and lower membrane conductivity. This negative effect is larger than the positive effect of kinetics enhancement, which results in decreased fuel cell performance. They also studied the temperature effect on fuel cell performance at humidification temperatures of 70 °C, 60 °C, and 30 °C, and achieved better performance at 30 °C than at 60 °C and 70 °C, with a fixed humidification temperature.

Recently, the temperature effect on the performance of an HT-PEM fuel cell with a PBI-membrane-based MEA (purchased from PEMEAS Fuel Cell Technologies) in the temperature range of 120–200 °C was also reported [2]. As the proton conductivity of a PBI membrane does not rely on its water content, the fuel cell was operated without extra humidification (0% RH) at ambient backpressure.

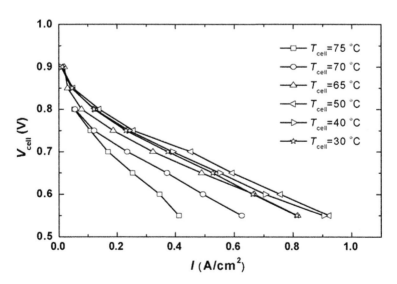

FIGURE 4.9   Effect of cell temperature on cell performance at a humidification temperature of 50 °C [71].

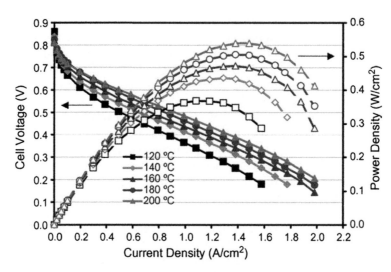

FIGURE 4.10 Polarization curves obtained at ambient backpressure, 0% RH, and different temperatures (as marked). The stoichiometries of H_2 and air are 1.5 and 2.0, respectively [2]. (For color version of this figure, the reader is referred to the online version of this book.)

As shown in Fig. 4.10, the fuel cell performance increased with increasing temperature. At a current density of 1.0 A cm^{-2}, the cell voltages were 0.366 V (120 °C), 0.415 V (140 °C), 0.44 V (160 °C), 0.465 V (180 °C), and 0.485 V (200 °C). The maximum power density of the fuel cell increased linearly with temperature, which indicates the benefit of high-temperature operation.

4.7. CHAPTER SUMMARY

In this chapter, the effects of operating temperature on PEM fuel cell kinetics and performance were addressed in detail. In general, an increase in the temperature will enhance the kinetics of both the HOR and the ORR. Fuel cell thermodynamics and OCV can be significantly affected by the operating temperature; raising the temperature will lower the fuel cell OCV. The temperature also influences the proton conductivity and hydrogen crossover of membranes. Because the water and proton diffusion coefficients increase with increasing temperature, the proton conductivity will increase for a well-hydrated PEM. Thus, the temperature affects the fuel cell kinetics, fuel cell OCV, membrane conductivity, hydrogen crossover, and mass transfer process. These effects are reflected in the overall fuel cell performance, but the relationship of fuel cell performance to temperature can be complicated by the fact that the flow field design, the fuel cell structure, and operating conditions such as RH, gas flow rate, and backpressure can also influence fuel cell performance.

By using AC impedance spectroscopy, the membrane resistance (R_m), charge transfer resistance (R_t), and gas mass transfer resistance (R_{mt}) of an operating PEM fuel cell, and the associated temperature dependencies of these resistances, can be separated and simulated by means of a suitable equivalent circuit.

REFERENCES

[1] Zhang J, Xie Z, Zhang J, Tang Y, Song C, Navessin T, et al. J Power Sourc 2006;160:872–91.

[2] Zhang J, Tang Y, Song C, Zhang J. J Power Sourc 2007;172:163–71.

[3] Tang Y, Zhang J, Song C, Liu H, Zhang J, Wang H, et al. J Electrochem Soc 2006;153: A2036–43.

[4] Yang C, Costamagna P, Srinivasan S, Benziger J, Bocarsly AB. J Power Sourc 2001;103:1–9.

[5] Li Q, He R, Jensen JO, Bjerrum NJ. Chem Mater 2003;15:4896–915.

[6] Zhang J, Tang Y, Song C, Xia Z, Li H, Wang H, et al. Electrochim Acta 2008;53:5315–21.

[7] Zawodzinski JTA, Derouin C, Radzinski S, Sherman RJ, Smith VT, Springer TE, et al. J Electrochem Soc 1993;140:1041–7.

[8] Song C, Tang Y, Zhang JL, Zhang J, Wang H, Shen J, et al. Electrochim Acta 2007; 52:2552–61.

[9] Vogel W, Lundquist L, Ross P, Stonehart P. Electrochim Acta 1975;20:79–93.

[10] Dhar HP, Christner LG, Kush AK. J. Electrochem Soc 1987;134:3021–6.

[11] Li Q, He R, Gao J-A, Jensen JO, Bjerrum NJ. J Electrochem Soc 2003;150:A1599–605.

[12] Murthy M, Esayian M, Lee W-k, Van Zee JW. J Electrochem Soc 2003;150:A29–34.

[13] Cheng X, Zhang J, Tang Y, Song C, Shen J, Song D, et al. J Power Sourc 2007;167:25–31.

[14] Bi W, Fuller TF. J Electrochem Soc 2008;155:B215–21.

[15] Lobato J, Cañizares P, Rodrigo MA, Linares JJ. Electrochim Acta 2007;52:3910–20.

[16] Jiang J, Kucernak A. J Electroanal Chem 2004;567:123–37.

[17] Li H, Lee K, Zhang J. In: Zhang J, editor. PEM fuel cell electrocatalysts and catalyst layers: fundamentals and applications. Springer; 2008. p. 135–64.

[18] Mello RMQ, Ticianelli EA. Electrochim Acta 1997;42:1031–9.

[19] Breiter MW. In: Vielstich W, Lamm A, H.A.G., editors. Handbook of fuel cells: fundamentals, technology, applications. New York: John Wiley & Sons; 2003.

[20] Conway BE. In: Wieckowski A, editor. Interfacial electrochemistry: theory: experiment, and applications. New York: CRC Press; 1999 [Chapter 9].

[21] Stonehart P, Kohlmayr G. Electrochim Acta 1972;17:369–82.

[22] Watanabe M, Igarashi H, Yosioka K. Electrochim Acta 1995;40:329–34.

[23] Springer TE, Raistrick ID. J Electrochem Soc. 1989;136:1594–603.

[24] Damjanovic A, Brusic V. Electrochim Acta 1967;12:615–28.

[25] Yeager E. Electrochim Acta 1984;29:1527–37.

[26] Mustain WE, Prakash J. J Power Sourc 2007;170:28–37.

[27] Markovic NM, Ross PN. Surf Sci Rep 2002;45:117–229.

[28] Song C, Zhang J. In: Zhang J, editor. PEM fuel cell electrocatalysts and catalyst layers: fundamentals and applications. Springer; 2008. p. 89–134.

[29] Drillet JF, Ee A, Friedemann J, Kötz R, Schnyder B, Schmidt VM. Electrochim Acta 2002;47:1983–8.

[30] García-Contreras MA, Fernández-Valverde SM, Vargas-García JR, Cortés-Jácome MA, Toledo-Antonio JA, Ángeles-Chavez C. Int J Hydrogen Energy 2008;33:6672–80.

[31] Jeon MK, Zhang Y, McGinn PJ. Electrochim Acta 2010;55:5318–25.

[32] Zeng J, Liao S, Lee JY, Liang Z. Int J Hydrogen Energy 2010;35:942–8.

[33] Nørskov JK, Rossmeisl J, Logadottir A, Lindqvist L, Kitchin JR, Bligaard T, et al. J Phys Chem B 2004;108:17886–92.

[34] Ahmed J, Yuan Y, Zhou L, Kim S. J Power Sourc 2012;208:170–5.

[35] Morozan A, Campidelli S, Filoramo A, Jousselme B, Palacin S. Carbon 2011;49:4839–47.

[36] Müller K, Richter M, Friedrich D, Paloumpa I, Kramm UI, Schmeißer D. Solid State Ionics 2012;216:78–82.

[37] Charreteur F, Jaouen F, Dodelet J-P. Electrochim Acta 2009;54:6622–30.

[38] Goubert-Renaudin SNS, Zhu X, Wieckowski A. Electrochem Commun 2010;12:1457–61.

[39] Ji Y, Li Z, Wang S, Xu G, Yu X. IntJ Hydrogen Energy 2010;35:8117–21.

[40] Liu H, Zhang L, Zhang J, Ghosh D, Jung J, Downing BW, et al. J Power Sourc 2006;161: 743–52.

[41] Zhang L, Lee K, Bezerra CWB, Zhang J, Zhang J. Electrochim Acta 2009;54:6631–6.

[42] Chiao S-P, Tsai D-S, Wilkinson DP, Chen Y-M, Huang Y-S. Int J Hydrogen Energy 2010;35: 6508–17.

[43] Feng Y, Gago A, Timperman L, Alonso-Vante N. Electrochim Acta 2011;56:1009–22.

[44] Parthasarathy A, Srinivasan S, Appleby AJ, Martin CR. J Electrochem Soc. 1992; 139:2530–7.

[45] Parthasarathy A, Martin CR, Srinivasan S. J Electrochem Soc 1991;138:916–21.

[46] Appleby AJ, Baker BS. J Electrochem Soc 1978;125:404–6.

[47] Damjanovic A, Genshaw MA. Electrochim Acta 1970;15:1281–3.

[48] Nagel K, Dietz H. Electrochim Acta 1961;4:1–11.

[49] Dietz H, Göhr H. Electrochim Acta 1963;8:343–59.

[50] Thacker R, Hoare JP. J Electroanal Chem Interfacial Electrochem 1971;30:1–14.

[51] Damjanovic A, Bockris JOM. Electrochim Acta 1966;11:376–7.

[52] Sawyer DT, Day RJ. Electrochim Acta 1963;8:589–94.

[53] Hoare JP. J Electrochem Soc 1962;109:858–65.

[54] Wroblowa H, Rao MLB, Damjanovic A, Bockris JOM. J Electroanal Chem Interfacial Electrochem 1967;15:139–50.

[55] Amphlett JC, Baumert RM, Mann RF, Peppley BA, Roberge PR, Harris TJ. J Electrochem Soc 1995;142:1–8.

[56] Kim J, Lee S-M, Srinivasan S, Chamberlin CE. J Electrochem Soc 1995;142:2670–4.

[57] Laurencelle F, Chahine R, Hamelin J, Agbossou K, Fournier M, Bose TK, et al. Fuel Cells 2001;1:66–71.

[58] Zhang J, Tang Y, Song C, Zhang J, Wang H. J Power Sourc 2006;163:532–7.

[59] Amphlett JC, Baumert RM, Mann RF, Peppley BA, Roberge PR, Harris TJ. J Electrochem Soc 1995;142:9–15.

[60] Xu H, Song Y, Kunz HR, Fenton JM. J Electrochem Soc 2005;152:A1828–36.

[61] Sumner JJ, Creager SE, Ma JJ, DesMarteau DD. J Electrochem Soc 1998;145:107–10.

[62] Sone Y, Ekdunge P, Simonsson D. J Electrochem Soc 1996;143:1254–9.

[63] Ma YL, Wainright JS, Litt MH, Savinell RF. J Electrochem Soc 2004;151:A8–A16.

[64] Cappadonia M, Erning JW, Stimming U. J Electroanal Chem 1994;376:189–93.

[65] Ma C, Zhang L, Mukerjee S, Ofer D, Nair B. J Memb Sci 2003;219:123–36.

[66] Saito M, Hayamizu K, Okada T. J Phys Chem B 2005;109:3112–9.

[67] Yeager HL, Steck A. J Electrochem Soc 1981;128:1880–4.

[68] Inaba M, Kinumoto T, Kiriake M, Umebayashi R, Tasaka A, Ogumi Z. Electrochim Acta 2006;51:5746–53.

[69] Teranishi K, Kawata K, Tsushima S, Hirai S. Electrochem Solid State Lett 2006;9:A475–7.

[70] Yuan W, Tang Y, Pan M, Li Z, Tang B. Renew Energy 2010;35:656–66.

[71] Yan W-M, Wang X-D, Mei S-S, Peng X-F, Guo Y-F, Su A. J Power Sourc 2008;185:1040–8.

[72] Williamson Z, Kim D, Chun D-K, Lee T, Squibb C. Appl Therm Eng 2011;31:3761–7.

[73] Oono Y, Fukuda T, Sounai A, Hori M. J Power Sourc 2010;195:1007–14.

[74] Niu JJ, Wang JN. Acta Mater 2010;58:408–14.

[75] Yan W-M, Chen C-Y, Mei S-C, Soong C-Y, Chen F. J Power Sourc 2006;162:1157–64.

Chapter 5

Membrane/Ionomer Proton Conductivity Measurements

5.1. INTRODUCTION

The proton exchange membrane (PEM) is a key component of PEM fuel cells. It separates the anodic and cathodic compartments and at the same time acts as a proton conductor by transporting protons generated at the anode to the cathode. The protons in the membrane are the main charge carriers. Hence, the conductivity induced by this proton transport is called proton conductivity. Because protons can get transported in two directions, both across and through the membrane, there are two types of conductivity: in-plane and through-plane conductivities. The two types are theoretically different unless the membrane is isotopic in these two dimensions. In reality, the PEM is not an absolute electronic isolator. The electronic conductivity, normally much smaller than the

PEM Fuel Cell Testing and Diagnosis. http://dx.doi.org/10.1016/B978-0-444-53688-4.00005-X
143

proton conductivity, can also contribute to the overall measured conductivity and is difficult to distinguish from the proton-mediated contribution.

Normally, proton conductivity can be measured by detecting the voltage drop across a membrane that is induced by the proton current flow when two H_2/H^+ metal electrodes, such as Pt black, are separately attached onto each side of the membrane. On these Pt black electrodes, the reversible electrochemical reactions are H_2/H^+ redox processes, which act as proton supplier and acceptor. The reversibility of these redox reactions or the activities of the electrodes, which can change the polarization potentials of the reactions, will affect the accuracy of voltage detection and conductivity measurements. The operating temperature and the membrane water content significantly influence the proton conductivity by affecting hydrogen ionization as well as proton concentration and mobility in the membrane. Therefore, an accurate measurement of the PEM proton conductivity still presents a nontrivial experimental challenge. Many techniques have been suggested and used to obtain proton conductivity; these include the current interruption method [1–5], the two-point probe and four-point probe methods [6–8], electrochemical impedance spectroscopy (EIS) [5,9], electrochemical atomic force microscopy (EC-AFM) [10,11], and the solid-state pulsed field gradient nuclear magnetic resonance (NMR) technique [12]. In addition, a new attempt has been made to screen and measure proton conductivity based on proton transport visualization [13].

This chapter discusses proton conduction mechanisms and then presents an overview of the various methods for measuring proton conduction, including typical measurement examples, based on a comparative analysis of published studies.

5.2. PROTON CONDUCTION MECHANISMS

A review of the literature on the development of proton-conductive materials reveals a variety of proton carriers, including water (e.g. in hydrated acidic polymers), oxo-acid anions (e.g. in $CsHSO_4$) or oxo-acids (e.g. phosphoric acid in adducts of basic polymers), heterocycles (e.g. intercalated into acidic polymers or immobilized via flexible spacers), and oxide ions (forming a hydroxide on the oxygen site of an oxide lattice). The essential characteristic of all these species is their involvement in hydrogen bonding, which enables long-range proton transport within materials [14]. Proton conduction mechanisms that occur in water when it is present in a homogeneous phase or as a component of hydrated PEMs will be described briefly in the following sections.

5.2.1. Proton Conduction in Water

In bulk water, protons are transported mainly via two different mechanisms: the "vehicle mechanism" and the "Grotthuss (hopping) mechanism," as illustrated in Fig. 5.1.

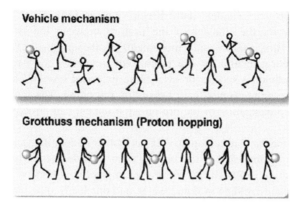

FIGURE 5.1 Schematic of the two types of proton conduction mechanisms [15]. (For color version of this figure, the reader is referred to the online version of this book.)

In the vehicle mechanism, proton movement occurs with the aid of moving carriers (e.g. H_2O, classically in the form of hydronium ions, H_3O^+), and both water and protons diffuse at a similar rate. The overall proton conductivity is mainly determined by the diffusion coefficient of the vehicles.

In the Grotthuss-type mechanism, protons migrate via a highly concerted process that involves the breaking and formation of hydrogen bonds and proton translocation (hopping) within the "Zundel" and "Eigen" ions. Because proton transport involves the "diffusion" of the hydrogen-bond breaking and formation processes, this mechanism is also termed "structural diffusion" [14]. The two short-life forms of the hydrated protons are explained as a complex that has two water molecules centered at a proton, $H_5O_2^+$, and three water molecules in coordination with a center hydronium ion, $H_9O_4^+$ [14]. Figure 5.2 illustrates

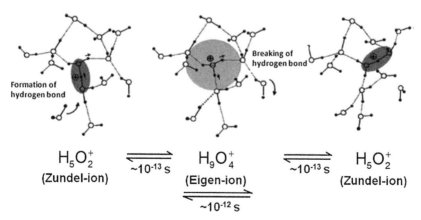

FIGURE 5.2 Schematic of the structure of "Zundel" and "Eigen" ions, and the hydrogen bond formation and breaking processes that occur in the two short-life complexes during proton transport [14]. (For color version of this figure, the reader is referred to the online version of this book.)

proton transport by "Zundel" and "Eigen" ions in bulk water, along with the timescales for the transfer of the protons between the two short-life structures. In this mechanism, movement of the proton solvent is not needed, but reorganization of the proton environment—consisting of reorientation of individual species or even more extended ensembles—is necessary to form an uninterrupted path for proton migration. The rates of proton transfer and reorganization of the proton environment can directly affect this mechanism.

These two mechanisms compete with each other under different conditions: Increased temperature and acid concentration and decreased pressure will attenuate structural diffusion [14,16]. For example, the H-bonded networks in ice are more extensive than in liquid water, and this leads to better conduction in ice [17].

5.2.2. Proton Conduction in the Proton Exchange Membrane

As mentioned in Chapter 1, the commonly used PEMs in fuel cells, such as perfluorosulfonic acid (PFSA) and sulfonated hydrocarbon membranes, have two functional groups in the polymer structure: the hydrophobic polymer backbone and the hydrophilic sulfonic acid groups. These properties lead to the formation of two kinds of domains in the hydrated PEM. One is the polymer matrix, which gives the membrane morphological stability, whereas the other is the continuous water channel network [18], which facilitates proton conduction. These two domains are separated by sulfonic acid groups, as shown in Fig. 5.3 [19]. The proton conduction mechanisms in the hydrated PEM can be understood as comprising the dissociation of the protons from the acidic sites, the subsequent transfer of the protons to the aqueous medium, the screening by water of the hydrated protons from the conjugate base (e.g. the sulfonate anion), and finally the diffusion of the protons in the confined water channels within the polymer matrix [20].

Proton dissociation is a result of the excess positive charge being stabilized in the hydrogen-bonding network, and the excess electron density (due to the breaking of the $-SO_3-H$ bond) being sufficiently delocalized by the neighboring group of $-CF_3$ or $-C_6H_4CH_3$. The combination of these two effects results in a minimum energy configuration. Proton dissociation is virtually affected by the structure and local chemistry of the acidic (and hydrophilic) sites in the PEM, such as the side-chain fragments and the hydration degrees. Electronic structure calculations indicate that on inclusion of the electrostatic interactions in water, the free-energy barrier is substantially reduced for rotation of the acid groups in water media [21]. On the basis of an optimization of the side-chain fragment CF_3OCF_3 with a single water molecule, no hydrogen bond forms with the water molecule because of the strong electron withdrawing effect of the two $-CF_3$ groups, and the reduction in the electron density on the ether oxygen [22,23]. Paddison et al. [24,25]

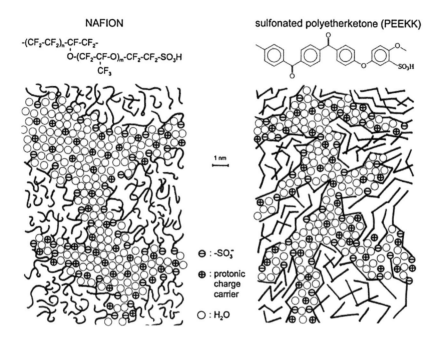

FIGURE 5.3 Schematic of the two zones in hydrated Nafion® and sulfonated poly(ether ether ketone ketone) (PEEKK). This scheme illustrates the distinctions in the hydrophilic/hydrophobic separation, the connectivity of the water and ion domains, and the separation of the $-SO_3^-$ groups [19]. (For color version of this figure, the reader is referred to the online version of this book.)

also conducted a series of explicit water electronic structure calculations with both triflic (as an analog for Nafion®) and para-toluene sulfonic (for PEEKK) acids. No dissociation of the proton can be observed with either CF_3SO_3H or $CH_3C_6H_4SO_3H$ until three water molecules are added; the clusters formed with four and five water molecules [24,26] are found to be similar to those observed with three water molecules in that the hydronium ion forms a contact ion pair with the sulfonate anion. However, the hydronium ion tends to be farther away from the anion as the number of water molecules in the cluster increases from three to five. When the sixth water molecule is added, the excess proton (as a hydronium ion) is completely separated from the anion. Here, the hydronium ion forms a true Eigen ion because it is hydrogen bonded to three water molecules. The differences in the strength of the conjugate base (sulfonate anion) also affect the hydrogen bonding in the water molecules and the consequent transport of protons, especially when the water content of a membrane is minimal. The effects of the distribution of sulfonic acid groups and the neighboring side chains have been examined as well [20,27]. If a fragment of the PEM consists of two side chains, a cluster consisting of six water molecules is required to dissociate both protons, where one of the dissociated protons has formed a hydronium ion in which hydrogen is bonded to both sulfonate anions and the other proton has formed a Zundel ion [20].

Regarding proton transport and diffusion in the water channel network of a PEM, many physical and theoretical models have been developed in an effort to elucidate water and ion transport in ionomer membranes [14,28–38]. With high degrees of hydration, that is, >13 water molecules per sulfonic acid group [14], the majority of excess protons in the PEM are located in the central part of the hydrated hydrophilic nanochannels. The water is bulklike, with local proton transport properties similar to those described for water in Section 5.2.1. Both Grotthuss and vehicular mechanisms are believed to be the predominant modes of proton conduction in the water channel [33,39]. It has been suggested that the contribution of the Grotthuss mechanism to conductivity occurs mainly in the center of a water-swollen pore, and consequently, proton mobility is higher in this region. The self-diffusion coefficient of a proton and the pore dimensions will affect proton transport and diffusion. For low degrees of hydration, the distribution of fixed sites has a substantial impact on proton diffusion in the PEM. But proton transport is actually more complex. Because of the side-chain structure of Nafion® membrane, there is also a third transition region, between the aqueous domain and the hydrophobic polymer backbone [14]. Proton

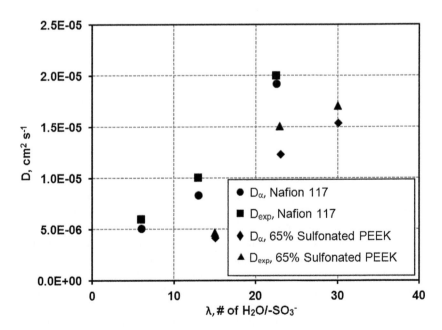

FIGURE 5.4 Computed and experimentally determined proton self-diffusion coefficients in Nafion® 117% and 65% sulfonated PEEKK membranes as a function of water content [20].

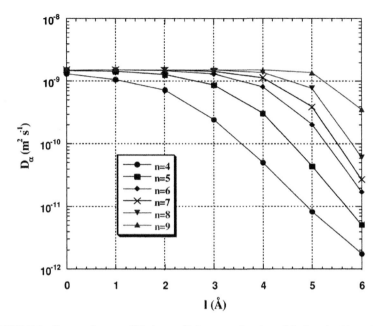

FIGURE 5.5 Computed proton diffusion coefficients as a function of the length of intrusion of the side chain (l) and as a function of the number (n) of axially positioned arrays of fixed sites for an arbitrary membrane pore with a fixed length, diameter, total number of anionic groups, and water content. The pore with the most uniform distribution of anionic groups (i.e. where $n = 9$) shows the smallest, in fact very little, decrease in the proton diffusion coefficient as the length of protrusion of the anionic groups is increased [20].

mobility at intermediate and low degrees of hydration is essentially vehicular in nature. Nonequilibrium statistical mechanics-based calculations of the proton self-diffusion coefficients in Nafion® and PEEKK membranes over a range of hydration conditions have addressed this conductivity contribution, as shown in Figs 5.4 and 5.5 [20]. The diffusion of water (vehicle) and hydrated protons (H_3O^+) is retarded by confinement in an environment in which the water and protons are perturbed by the presence of a substantial density of sulfonate groups.

5.3. METHODS FOR MEASURING CONDUCTIVITY

On the same principle as the routine measurement of electronic resistivity, Ohm's law is used to analyze the resistivity of a proton-conductive membrane against the flow of either alternating current (AC) or direct current (DC). The proton conductivity can be calculated according to Eqn (5.1):

$$\sigma = \frac{l}{RA} \tag{5.1}$$

where σ is the membrane conductivity (S cm^{-1}), R is the measured resistance, l is the length between the two voltage probes for in-plane measurement or the thickness of the membrane for through-plane measurement, and A is the cross-sectional area of the membrane. The measurement methodologies can be found in the literature [1–13,40,41]. The in-plane and/or through-plane proton conductivities of a membrane can be obtained by using different approaches, such as the two-probe method, the four-probe method, and the coaxial-probe method, under in situ or ex situ conditions. These methods and their applications will be presented in detail in the following sections.

5.3.1. Current Interruption

The current interruption (or current interrupt) method is generally used for measuring ohmic losses caused mainly by proton or anion transport resistance in batteries, fuel cells, and other electrochemical cells. The principle of this technique is that the ohmic losses vanish much faster than do the electrochemical overpotentials on the electrodes when the current is interrupted [42]. As shown in Figs 5.5 and 5.6, the cell is operated at a constant current (i); it is

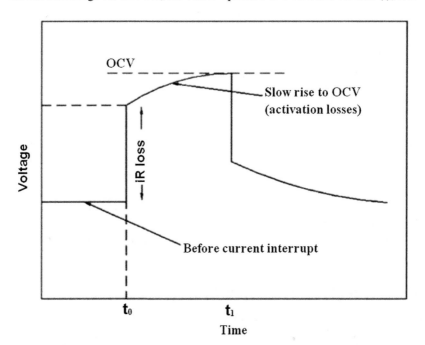

FIGURE 5.6 Ideal voltage transient during a current interruption measurement. The cell is first operated at a fixed current. At t = t$_0$, the current flow (i) is interrupted, and the ohmic losses vanish immediately. After the current interruption, electrochemical overpotentials start to relax, and the voltage increases logarithmically toward the open circuit voltage (OCV). At t = t$_1$, the current is again switched on, and the overpotential reappears [1].

then suddenly interrupted by setting the cell current to zero, and the voltage transient change is immediately recorded. Because the current is suddenly set to zero, the voltage drop (ΔV) induced by the resistance immediately disappears, and this ΔV can be expressed as follows:

$$\Delta V = iR \quad \text{or} \quad R = \frac{\Delta V}{i} \tag{5.2}$$

From Eqn (5.2), the resistance can be obtained.

In an ideal case, the iR loss should be determined simply by measuring the voltage immediately after the current interruption. However, the accuracy of this method depends on the delay between the current interruption and the measurement of the OCV. Therefore, it is very important that the data acquisition for the voltage transient be as rapid as possible to adequately separate the ohmic and activation losses [1]. The equivalent circuit for a current interruption measurement of a PEM fuel cell system (Fig. 5.7) consists of a couple of resistors and a capacitor: The first resistor represents the cell resistance, including electronic resistance and ionic resistance; the second is in a loop with a capacitor of the membrane electrode assembly, and this represents the charge transfer resistance determining activation polarization. When (i) the cell has a current i with a cell voltage V_{cell}, (ii) the voltage drop at the first resistor is ΔV,

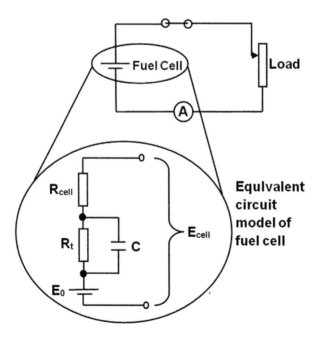

FIGURE 5.7   The equivalent circuit for a fuel cell that is used to perform circuit interruption measurements: R_{cell}, fuel cell resistance including electronic resistance and ionic resistance and R_t, the charge transfer resistance for fuel cell reactions [43].

and (iii) the voltage drop in the loop is ΔV_{act}, which represents activation polarization. Thus, the cell voltage can be expressed as follows: $V_{cell} = E_o - \Delta V - \Delta V_{act}$ (here E_o is the OCV, as presented in Fig. 5.6). When the current suddenly becomes zero, the ΔV portion of the cell voltage immediately disappears, and the cell voltage decreases to $E_o - \Delta V_{act}$. The ΔV_{act} will gradually drop to zero as the capacitor gets discharged through R_t. Finally, the cell voltage reaches the OCV, as shown in Fig. 5.6. In a real system, many features of the idealized curve in Fig. 5.6 are concealed by disturbances that arise from the inductance of the measurement system or from the nonideal properties of the current switches in the measurement devices. The real voltage curve after current interruption is overlapped by a voltage overshoot and subsequent oscillations, and the whole transient can also be changed by noise from electronics in the measurement system. The problems caused by inductance and nonideal switches are typically worse at high currents, when the current loop and the load need to be of a considerable size, and therefore, the inductance is high. This leads to slow current changes and thus less accurate results.

To conduct proton conductivity measurements, Buchi et al. [3] designed a current interruption device that used an auxiliary current pulse method and an instrument for generating fast current pulses (i.e. currents >10 A), and determined the time resolution for the appropriate required voltage acquisition by considering the relaxation processes in the membrane of a PEM fuel cell [3]. They estimated that the dielectric relaxation time, or the time constant for the spontaneous discharge of the double-layer capacitor, τ, is about 1.4×10^{-10} s. They found that the potential of a dielectric relaxation process decreased to <1% of the initial value after 4.6τ (6.4×10^{-10} s) and that the ohmic losses almost vanished about half a nanosecond after the current changes. Because there is presently no theory about the fastest electrochemical relaxation processes in PEM fuel cells, the authors assumed a conservative limit of 10^{-8} s, based on observations of water electrolysis membranes. They concluded that the time window for accurate current interruption measurements on a membrane is between 0.5 and 10 ns. Another typical application of the current interruption method was demonstrated by Mennola et al. [1], who used a PEM fuel cell stack and identified a poorly performing individual cell in the stack.

An example of a voltage transient for an individual cell is shown in Fig. 5.8. At high air flow rates, cells in the middle of the stack showed up to 21% higher ohmic losses than average, which was attributed to membrane dehydration. Mennola et al.'s results showed a good agreement between the ohmic losses in the entire stack and the sum of the ohmic losses in each individual cell. Recently, this technique has also been extended to the study of electrode processes in PEM fuel cells, such as Tafel kinetics, oxygen diffusion, proton migration, and double-layer capacitance [41,42], and the effects of operating conditions (e.g. relative humidity, RH) on cell performance [4].

One of the advantages of current interruption is the single data value, which is easy to interpret. Further, no additional equipment is required because the

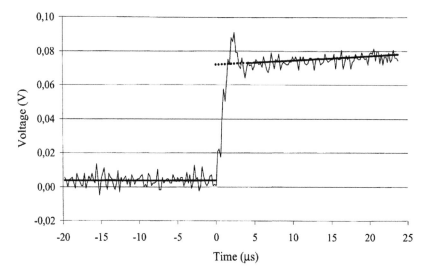

FIGURE 5.8 Voltage transient for an individual cell in a PEM fuel cell stack during current interruption. Extrapolation is indicated by the dotted line. Current density before interruption is 400 mA cm^{-2} [1].

interruption is brought about by the cell load. The primary disadvantage of this method is that it is difficult to determine the exact point of instantaneous voltage gain. The voltage behavior may be monitored in real time with an oscilloscope, or several discreet voltage readings may be taken at different interruption times and then extrapolated back to zero time. This method also imposes a significant perturbation on the cell, even when the interruption is of a short duration (i.e. tens of microseconds); this potentially has some undesirable and irreversible side effects on the cell performance. Further, the data are degraded when long cell cables are used, as a result of the excessive "ringing" caused by cable inductance.

5.3.2. Electrochemical Impedance Spectroscopy

As noted in Chapter 3, EIS, also called AC impedance spectroscopy, is a powerful technique for fuel cell studies and has been widely used for PEM fuel cell testing and diagnosis [9,44–46]. During EIS measurement, an AC signal—usually very small—is applied to perturb the system, and then the EIS data in the applied frequency range are obtained. Using EIS, the membrane conductivity can be measured in situ during fuel cell operation, and this gives more information than in a steady-state experiment. As a powerful diagnostic tool for evaluating fuel cells, the main advantage of EIS is its ability to resolve the individual contributions of the various factors that determine the overall PEM fuel cell performance: ohmic voltage loss, kinetic voltage loss, and voltage loss caused by mass transfer. Such a separation provides useful information for fuel cell design optimization and for selection of the most appropriate operating conditions.

For EIS measurement, a frequency response analyzer (FRA, e.g. Solartron 1260) and a load bank are usually required. In our previous work [47], we used a Solartron 1260 FRA and an RBL 488 series (100-60-400) load bank from TDI Power. The Solartron 1260 FRA can generate DC, AC, and their combined control signals. For an AC sinusoid signal, the analyzed frequency range can be from 10 µHz to 32 MHz. The RBL 488 series is a single-channel load bank and can be controlled with signals via an external program. The magnitude of the voltage signal input to the load bank, which is a DC, AC, or a combination signal, can be in the range of 0–10 V. Figure 5.9 shows the connections between the Solartron 1260 FRA, RBL 488 load bank, and fuel cell for EIS measurements. The FRA port, "GEN OUTPUT," sends a software-command signal to the load bank through ports "REM" and "S–," located on the load bank rear panel, to control the fuel cell load. The cell voltage response goes to FRA "V1" and "V2" for analysis. The impedance information obtained is collected by means of a computer system and is then monitored and analyzed by using ZView and ZPlot software.

As an example, Fig. 5.10 shows the in situ EIS result for a PBI-membrane-based PEM fuel cell operated at 140 °C and ambient pressure at a current density of 0.2 A cm^{-2}, obtained in our previous work [46]. As can be seen, there are two semicircles on the spectrum, one in the high-frequency domain and the other in the low-frequency domain. The first semicircle represents the resistance from the fuel cell reaction kinetics, including the cathodic ORR and anodic HOR processes, although the main contribution is from the ORR. The

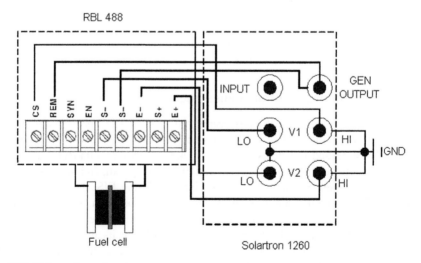

FIGURE 5.9 Connections between a Solartron 1260 FRA, RBL 488 load bank, and fuel cell for EIS measurement [47].

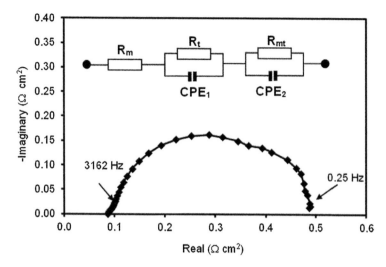

FIGURE 5.10   Nyquist plot for a phosphoric-acid-doped polybenzimidazole- (PBI) membrane-based PEM fuel cell operated at 140 °C and ambient pressure at 0.2 A cm^{-2}. The insert shows the proposed equivalent circuit mode for a PEM fuel cell. $ST_{H2} = 1.5$, $ST_{air} = 2.0$ [46].

second semicircle represents the mass transfer process. The intercept in the high-frequency domain on the real axis represents the ohmic resistance of the fuel cell, which is dominated by membrane resistance. Thus, the membrane resistance under real fuel cell operating conditions can be measured, and the proton conductivity of the membrane can be calculated according to Eqn (5.1). The insert in Fig. 5.10 shows an equivalent circuit used to simulate the impedance data in our work. R_m is the high-frequency resistance (HFR, the intercept on the real axis at the high-frequency end), which represents the membrane resistance. R_t is the charge transfer resistance (or kinetic resistance), dominated by the ORR process; CPE_1 (constant phase element) represents the R_t-associated catalyst layer capacitance properties; R_{mt} is the resistance related to the mass transfer processes of gas (O_2 and H_2) diffusion in the catalyst layers; and CPE_2 represents the R_{mt}-associated capacitance.

If only the proton conductivity in a PEM fuel cell is of interest, it is not necessary to measure the EIS across the whole frequency range; it suffices to measure the high-frequency range. The HFR, R_m, as shown in Fig. 5.10, represents membrane resistance. Typical HFR measurement frequencies range from 500 Hz to 3 kHz. In any case, the same frequency must be used for data comparisons to be valid.

5.3.3. Ex Situ AC Impedance Spectroscopy

In the literature, AC impedance spectroscopy has been widely used for ex situ measurements of the tangential direction conductivity (TDC) [48–50], normal direction conductivity [7,48,51–53], and transverse direction conductivity [54,55] of membranes. The TDC method is known to be relatively insensitive to the contact impedance at the current-carrying electrodes [49] and has been used as a conventional method of evaluating membrane materials for their proton conductivity. The use of AC impedance spectroscopy to measure proton conductivity usually requires an FRA and a specially designed conductivity cell. During measurement, the conductivity cell is usually put in a controlled environmental chamber to keep the membrane at a certain temperature and RH.

Regarding the conductivity cell geometry, two configurations have been developed to measure proton conductivity: the two-point probe method [52–60] and the four-point probe method [7,49,50,56,60–65]. These will be discussed in the following sections.

5.3.3.1. Two-Point Probe Method

In the two-point probe configuration, two electrodes serve as both current and voltage sensing probes, as shown in Fig. 5.11 [60]; in the diagram, "open window A" allows the membrane to be exposed to the environment. For this method, the current and voltage are measured from two identical probes. Because the electrode/membrane interfacial impedance is always included in

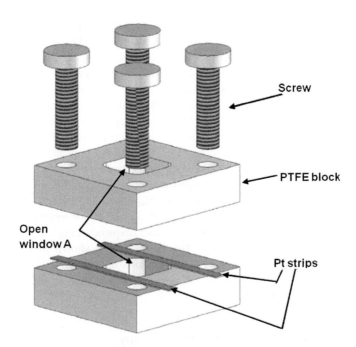

FIGURE 5.11 Schematic diagram of a conductivity cell for proton conductivity measurement with the two-point probe method. The membrane size is about 2.4 cm × 1.0 cm; the distance between the two Pt strips is 0.4 cm [60].

the measurement loop, its contribution to the overall membrane impedance may cause some error. However, our recent study [60] showed that the experimental results obtained with the two-point probe method are reliable compared with those obtained with the four-point probe method. Ma et al. [8] also reported that the membrane resistance can be separated from the interface impedance by equivalent circuit fitting, and stable and reliable experimental results can be achieved using the two-point probe method.

Figure 5.12 shows the impedance spectra of the Nafion® 115 membrane, obtained by the two-point probe method with different distances between voltage sensing probes (Pt strips), at room temperature and under fully hydrated conditions [60]. Each impedance spectrum shows a semicircle in the high-frequency domain and a straight line with an angle of 45° in the lower-frequency domain. The membrane resistance is extracted from the lower-frequency intercept of the semicircle at the Z_{real} axis. It can be seen from Fig. 5.12 that at low frequency, the behavior of the Pt/Nafion® interface is blocked by the 45° line, indicating that interface impedance has very little impact on the results of membrane conductivity.

5.3.3.2. Four-Point Probe Method

To effectively eliminate interfacial impedance from conductivity measurements, a four-point probe method has been widely used [7,49,50,60–62,64]. Compared with the two-point probe configuration, two additional Pt probes

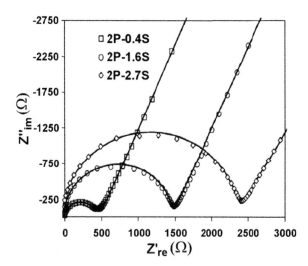

FIGURE 5.12 AC impedance spectra of Nafion®115 membrane obtained by the two-point probe method with different distances between voltage sensing probes (Pt strips), at room temperature and under fully hydrated conditions. 2P = two-probe; S = strip; 0.4, 1.6, and 2.7 represent the distance in centimeters between the two probes [60].

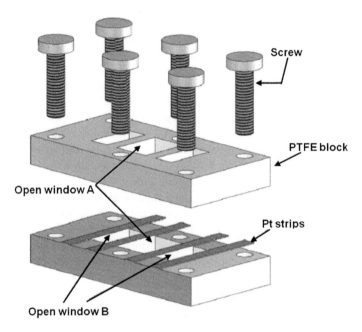

FIGURE 5.13 Schematic of a conductivity cell for proton conductivity measurement with the four-point probe method. The membrane size is about 4.6 cm × 1.0 cm [60].

(Pt wires or Pt strips, designated as inner probes) are laid between the two outer Pt probes that serve as voltage sensors, as shown in Fig. 5.13; the two outer Pt strips serve as AC current injectors. In this configuration, the current is passed between the two outer Pt strips, and the conductance of the membrane is calculated from the AC potential difference between the two inner probes. This method is relatively insensitive to contact impedance at the current-carrying electrodes and is therefore well suited to membrane conductivity measurements.

Figure 5.14 shows the AC impedance spectra of the Nafion® 115 membrane, obtained in the four-point probe method by using Pt strips as both inner and outer probes, at room temperature and under fully hydrated conditions [60]. Each AC impedance spectrum contains one semicircle in the high-frequency domain and an arc in the low-frequency domain. Here, the low-frequency arc shrinks with increasing distance between the two voltage probes, and the whole AC impedance spectrum tends toward an ideal semicircle, which indicates that the distance between the two inner probes affects the measurement results, and that the low-frequency impedance can be reduced by increasing this distance.

In principle, the results measured by the four-point probe method should be more accurate and reliable than those measured by the two-point probe method. However, the latter method is easier. In addition, it has been reported [56] that different measurement frequency ranges should be applied for different

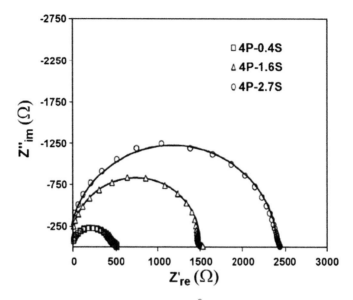

FIGURE 5.14 AC impedance spectra of the Nafion® 115 membrane obtained by the four-point probe method with Pt strips as both inner and outer probes, at room temperature and under fully hydrated conditions. 4P = four-probe; S = strip; 0.4, 1.6, and 2.7 represent the distance in centimeters between the two voltage probes [60].

measurement methods to separate out the low-frequency impedance. Our recent study [60] showed that the distance between the two voltage probes could influence the measurement accuracy, and a relatively larger distance is desirable to obtain reliable results. At a certain distance, similar measurement results were obtainable with both the two-point and the four-point probe methods [60].

5.3.4. Electrochemical Atomic Force Microscopy

More recently, EC-AFM [11,66,67] has been applied to study the morphology and proton conductivity of Nafion® membranes. Using EC-AFM, the spatially resolved proton current driven by the electrochemical reactions occurring on the two sides of the Nafion® membrane can be measured. The experimental setup is shown in Fig. 5.15. A commercial AFM is equipped with an electrochemical cell and a biopotentiostat. The AFM tip is modified with a platinum electrode to act as the cathode catalyst for the fuel cell reaction, whereas the Nafion® membrane coated with Pt catalyst on the side opposite the AFM tip serves as the anode of the electrochemical cell. The applied voltage leads to water oxidation and generates protons on the electrode with the Pt catalyst. The current can then be detected when the AFM tip is in contact with an ion channel that is connected with the ionic network in the membrane. Because water is required for the electrochemical reactions, the experiment must be performed

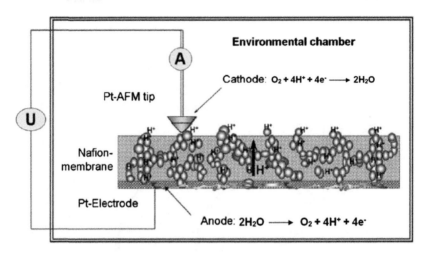

FIGURE 5.15   In situ method for measuring proton conductivity in Nafion® membrane by using EC-AFM. The applied voltage induces water oxidation at the electrode. The protons are only transported through the membrane when the conductive AFM tip makes contact with an ion channel. These protons are used for the ORR at the AFM tip. [11,66,67]. (For color version of this figure, the reader is referred to the online version of this book.)

under hydrated membrane conditions. As shown in Fig. 5.15, the experiment is conducted in an environmental chamber to control the humidity and temperature.

The EC-AFM technique allows one to record images of membrane topography and current simultaneously, as shown in Fig. 5.16. Both the images are clear and stable under a constant RH and voltage. In Fig. 5.16(a), the hydrophilic and hydrophobic domains of the membrane cannot be distinguished. However, in Fig. 5.16(b), the ionically conductive areas in the membrane can be identified by the white spots. Thus, the number of image spots can be profiled against the current. This technique also allows the measurement of the PEM proton conductivity when the water content is varied by changing the RH values. Figure 5.17 shows a histogram of the number of image points with a certain current against current threshold values for the Nafion®-112 membrane at three different relative humidities: 35%, 65%, and 80%. At low RH (35%), there is only a small peak at 0.7 nA, indicating that the ionic channel size has decreased and only a few ionically active areas exist on the surface. At 65% RH, a pronounced maximum peak and a small peak are visible at 1.3 and 0.47 nA, respectively. At high RH (80%), the major peak has moved from 1.3 to 1.7 nA and has broadened. This broadening indicates that the ionically conductive area has increased, which suggests that the membrane channel has swollen and that the surface hydrophilic regions have grown considerably.

Aleksandrova et al. [10] recently employed the EC-AFM technique to measure the proton conductivity of Nafion® with a high spatial resolution, by

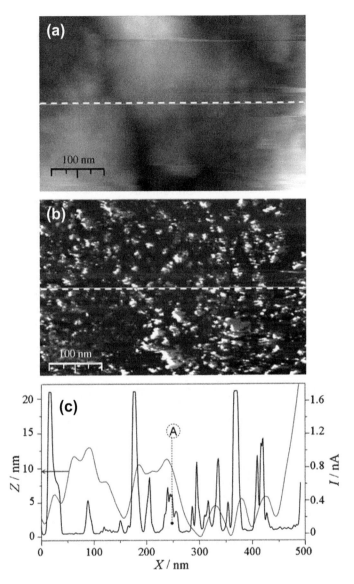

FIGURE 5.16 Simultaneously recorded (a) AFM topography and (b) current images of the Nafion®-112 membrane coated on one side with the 1.0 mg cm^{-2} Pt catalyst (Nafion® side to the tip; 300 × 500 nm^2, scan rate = 2 Hz, RH = 65%); (c) Line profiles of topographic (gray) and current (black) images [11].

using an identical atmosphere on both sides of the membrane instead of different atmospheres. The measured current reflected the distribution of the conductive areas at the membrane surface; in addition, the time-resolved conductivity mapping provided an indirect insight into the connectivity of the

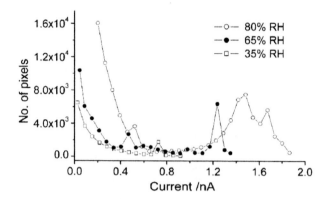

FIGURE 5.17 Histogram of the number of image points with a certain current against current threshold values for Nafion®-112 membrane at three different relative humidities: 35%, 65%, and 80%. The sharp peaks suggest that at a given RH, there may be an optimum pore size for proton conduction [11].

hydrophilic paths in the membrane and their dynamic behavior, including the diffusion processes within the pore network. This further indicates that EC-AFM is a powerful tool for measuring proton conductivity. However, this technique is basically still limited to surface studies.

5.4. TEMPERATURE EFFECT ON PROTON CONDUCTIVITY

Temperature is one of the key operating parameters in PEM fuel cells, and it can greatly influence the membrane proton conductivity, as has been reported in the literature [49,50,61]. Temperature affects the conductivity mainly by changing the membrane's water uptake and proton diffusion coefficient. It is well known that proton conduction in a PEM is associated with water transport. Indeed, the proton conductivity of a membrane is largely dependent on its water content [68,69]. As shown in Fig. 5.18, the conductivity of the Nafion® 117 membrane at 30 °C increases with its water content. A conductivity of 0.06 S cm^{-1} can be obtained at a λ value of 14 [69].

Normally, experimental results show that the water uptake of a membrane is greater at higher temperatures [70–72]. Figure 5.19 shows the water uptake of Nafion® 117 and 125 membranes on immersion in liquid water at different temperatures after they are dried under vacuum at 80 °C (N-form) and 105 °C (S-form). As shown in the figure, the water uptake increases with increasing temperature for both membranes immersed in liquid water, regardless of whether the membrane is the N-form or S-form. The water uptake also increases with temperature in water vapor, as shown in Fig. 5.20. As water uptake increases, so does the membrane's proton conductivity, as illustrated in Fig. 5.18.

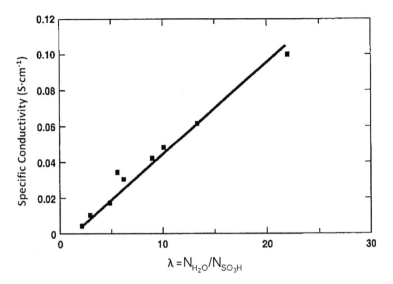

FIGURE 5.18   Nafion® 117 membrane conductivity as a function of water content at 30 °C [69].

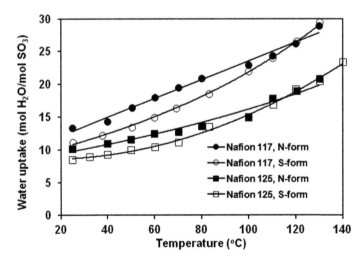

FIGURE 5.19   Water uptake of two Nafion® membranes on immersion in liquid water at different temperatures after drying under vacuum at 80 °C (N-form) and 105 °C (S-form) [71].

Besides water uptake, an increase in the temperature can also increase both the proton diffusion coefficient [73] and the water self-diffusion coefficient of membranes. Because proton transfer across a membrane is associated with water diffusion, an increase in the temperature will further improve the proton conductivity.

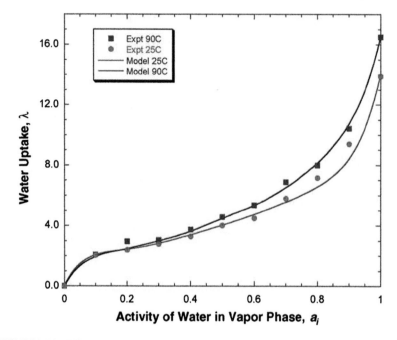

FIGURE 5.20   Effect of temperature on water uptake vs. activity of water vapor for the Nafion®
membrane [72]. (For color version of this figure, the reader is referred to the online version of this
book.)

Zawodzinski et al. [69] studied the proton conductivity of Nafion® 117
membrane in the temperature range of 25–90 °C under constant water content
($\lambda = N_{H_2O}/N_{SO_3H} = 22$), and they found that the temperature dependence of
proton conductivity showed a correlation that could be described using an
Arrhenius plot; Fig. 5.21 clearly indicates that the proton conductivity
increased with temperature. Recently, our group [46] studied the temperature
effect on the performance of a PBI-membrane-based fuel cell in the temperature range of 120–200 °C and proved that the relationship between temperature
and the proton conductivity of a PBI membrane could be expressed using the
following Arrhenius equations, a finding that has been reported by others as
well [61,74–76]:

$$\sigma = \frac{A}{T} \exp\left(-\frac{E_a}{RT}\right) \tag{5.3}$$

$$\ln(\sigma T) = \ln(\sigma_0) - \frac{E_a^m}{R}\left(\frac{1}{T}\right) \tag{5.4}$$

where σ, σ_0, E_a^m, R, and T are, respectively, the membrane conductivity
(S cm^{-1}), the preexponential factor (S K^{-1} cm^{-1}), the proton conducting

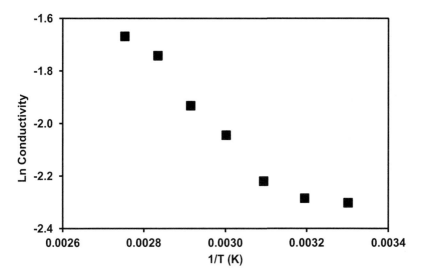

FIGURE 5.21   Arrhenius plot showing temperature dependence of membrane conductivity. Nafion® 117 membrane; temperature range: 25–90 °C; water content in the membrane: $\lambda = N_{H_2O}/N_{SO_3H} = 22$ [69].

activation energy (kJ mol^{-1}), the ideal gas constant, and the temperature (K). Based on the membrane resistances measured by in situ AC impedance at different current densities in the temperature range of 120–200 °C, and using Eqn (5.4), the calculated average values of E_a^m and $\ln(\sigma_0)$ are 19.9 kJ mol^{-1} and 10.8 S K^{-1} cm^{-1} [46], respectively.

5.5. RELATIVE HUMIDITY/WATER CONTENT EFFECT ON PROTON CONDUCTIVITY

It is well known [49,50,69] that the proton conductivity of a PFSA membrane strongly depends on its water content. Similar to temperature, RH influences proton conductivity mainly by affecting the membrane's water uptake and the proton and water self-diffusion coefficients.

In the literature, the water content or water uptake ratio is often expressed as λ [69,70] or N [68,77], the ratio of the number of water molecules in the membrane to the number of sulfonic acid groups (–SO$_3$H) in the membrane. In two studies [68,77], N was evaluated from the peak areas of H$^+$ NMR spectra and plotted as a function of RH, revealing that N increased with RH. The water uptake ratio, N, was expressed using Eqn (5.5) [68]:

$$N = 0.8486 + 0.24954x - 0.01127x^2 - 0.0003x^3 - 3.5878$$
$$\times 10 - 6x^4 + 1.6277 \times 10^{-8}x^5 \tag{5.5}$$

where x represents RH; x must be $>2\%$, otherwise Eqn (5.5) is invalid. This is because at $<2\%$ RH, the membrane is too dry and the proton conduction mechanism is different—for example, it changes from a vehicular mechanism-dominated process above 2% RH to a hopping-mechanism-dominated one when $<2\%$. Other experimental results [49,69,70] also show that water uptake increases with RH. As shown in Fig. 5.22, both the liquid water and water vapor

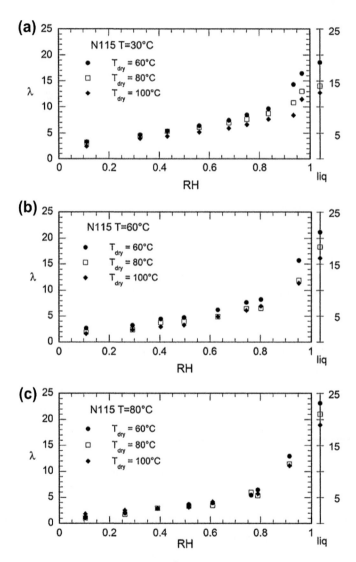

FIGURE 5.22 Water vapor sorption in Nafion® 115 membrane measured at (a) 30 °C, (b) 60 °C, and (c) 80 °C with different drying temperatures (T_{dry}). Liquid water sorption is plotted as a guide to the right of each graph [70].

sorption of the Nafion® 115 membrane increased with increasing RH, from 0.1 to 0.95, regardless of the membrane drying temperature. The same trend was also obtained for membranes immersed at three different temperatures: 30 °C, 60 °C, and 80 °C.

The water diffusion coefficient can be measured by using ^{1}H NMR [69,78] or optical methods [79] in the absence of a concentration gradient and dependence of the water diffusion coefficient on the membrane's water content has been reported. As shown in Fig. 5.23, when the extent of membrane hydration is increased, the membrane's water self-diffusion coefficient increases. Because proton diffusion in the membrane is associated with water transfer, a higher water diffusion coefficient aids proton conduction.

The effects of water content or RH on proton conductivity are described in the literature [45,49,50,61,69]. As shown earlier in Fig. 5.18, the proton conductivity of the Nafion® 117 membrane increases with water content. At low water content values, not all the acid sites in the membrane are dissociated and the interactions among water molecules via hydrogen bonding are low, leading to a low rate of proton transfer that results in relatively low conductivity. Our recent study [45] investigated the RH effect on the performance of a Nafion®-112 membrane-based PEM fuel cell at 120 °C with ambient back-pressure. The AC impedance spectra at different RHs were recorded in situ, and the membrane resistances were simulated; the membrane conductivities at 120 °C with different RHs were then calculated according to Eqn (5.1). As

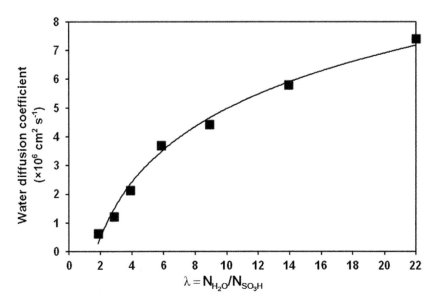

FIGURE 5.23 Self-diffusion coefficient of water in Nafion® as a function of the extent of membrane hydration [69].

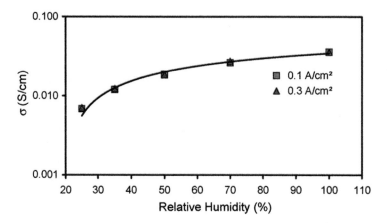

FIGURE 5.24 Membrane conductivity as a function of current density at 120 °C and 1.0 atm backpressure, with different RHs. Nafion®-112-based membrane electrode assembly with an active area of 4.4 cm². The thickness of the Nafion®-112 membrane was taken to be 50 μm [45]. (For color version of this figure, the reader is referred to the online version of this book.)

shown in Fig. 5.24, the proton conductivity increased with RH, the result of increased water content in the membrane at higher RHs.

5.6. CHAPTER SUMMARY

PEMs are key materials in PEM fuel cells, and the proton conductivity of a PEM can largely determine the performance of a fuel cell. Thus, an understanding of proton conduction and a knowledge of how to measure the proton conductivity of PEMs are essential prerequisites for developing novel membrane materials for PEM fuel cells. This chapter first addressed the various proton conduction mechanisms, then discussed in detail the methods for measuring proton conductivity, including current interruption, EIS, two-point probe and four-point probe AC impedance methods, and EC-AFM. The influences of both temperature and RH/water content on the proton conductivity of PEMs were also discussed. Generally, an increase in the temperature and/or RH can improve the proton conductivity of a PEM.

REFERENCES

[1] Mennola T, Mikkola M, Noponen M, Hottinen T, Lund P. J Power Sourc 2002;112:261–72.
[2] Wruck WJ, Machado RM, Chapman TW. J Electrochem Soc 1987;134:539–46.
[3] Buchi FN, Marek A, Scherer GG. J Electrochem Soc 1995;142:1895–901.
[4] Abe T, Shima H, Watanabe K, Ito Y. J Electrochem Soc 2004;151:A101–5.
[5] Cooper KR, Smith M. J Power Sourc 2006;160:1088–95.
[6] Siroma Z, Ioroi T, Fujiwara N, Yasuda K. Electrochem Commun 2002;4:143–5.
[7] Slade S, Campbell SA, Ralph TR, Walsh FC. J Electrochem Soc 2002;149:A1556–64.

[8] Ma S, Kuse A, Siroma Z, Yasuda K. Measuring conductivity of proton conductive membranes in the direction of thickness. In: Espec technology report No20.

[9] Yuan X, Song C, Wang H, Zhang J, editors. Electrochemical impedance spectroscopy in PEM fuel cells fundamentals and applications. New York: Springer; 2010.

[10] Aleksandrova E, Hink S, Hiesgen R, Roduner E. J Phys Condens Matter 2011;23:234109.

[11] Aleksandrova E, Hiesgen R, Eberhard D, Friedrich KA, Kaz T, Roduner E. Chem Phys Chem 2007;8:519–22.

[12] Ye G, Hayden CA, Goward GR. Macromolecules 2007;40:1529–37.

[13] Ivanovskaya A, Fan J, Wudl F, Stucky GD. J Membr Sci 2009;330:326–33.

[14] Kreuer K-D, Paddison SJ, Spohr E, Schuster M. Chem Rev 2004;104:4637–78.

[15] Colomban P. Proton conductors: solids, membranes and gels - materials and devices. Cambridge: Cambridge University Press; 1992.

[16] Dippel T, Kreuer KD. Solid State Ionics 1991;46:3–9.

[17] Alkorta I, Elguero J. Org Biomol Chem 2006;4:3096–101.

[18] Khalatur PG, Talitskikh SK, Khokhlov AR. Macromol Theory Simul 2002;11:566–86.

[19] Kreuer KD. J Memb Sci 2001;185:29–39.

[20] Paddison SJ. Annu Rev Mater Res 2003;33:289–319.

[21] Paddison SJ, Pratt LR, Zawodzinski T, Reagor DW. Fluid Phase Equilib 1998;150–151: 235–43.

[22] Paddison SJ, Zawodzinski Jr TA. Solid State Ionics 1998;113–115:333–40.

[23] Paddison SJ, Pratt LR, Zawodzinski Jr TA. J New Mater Electrochem Syst 1999;2:183–8.

[24] Paddison SJ. J New Mater Electrochem Syst 2001;4:197–207.

[25] Paddison SJ, Pratt LR, Zawodzinski TA. J Phys Chem A 2001;105:6266–8.

[26] Eikerling M, Paddison SJ, Zawodzinski Jr TA. J New Mater Electrochem Syst 2002;5: 15–23.

[27] Habenicht BF, Paddison SJ. J Phys Chem B 2011;115:10826–35.

[28] Hsu WY, Gierke TD. J Memb Sci 1983;13:307–26.

[29] Capeci SW, Pintauro PN, Bennion DN. J Electrochem Soc 1989;136:2876–82.

[30] Verbrugge MW, Hill RF. J Electrochem Soc 1990;137:886–93.

[31] Pintauro PN, Verbrugge MW. J Memb Sci 1989;44:197–212.

[32] Bontha JR, Pintauro PN. Chem Eng Sci 1994;49:3835–51.

[33] Eikerling M, Kornyshev AA, Kuznetsov AM, Ulstrup J, Walbran S. J Phys Chem B 2001;105:3646–62.

[34] Eikerling M, Kornyshev AA. J Electroanal Chem 2001;502:1–14.

[35] Din X-D, Michaelides EE. AIChE J 1998;44:35–47.

[36] Paddison SJ, Paul R, Zawodzinski JTA. J Electrochem Soc 2000;147:617–26.

[37] Paddison SJ, Paul R, Kreuer K-D. Phys Chem Chem Phys 2002;4:1151–7.

[38] Paddison SJ, Paul R. Phys Chem Chem Phys 2002;4:1158–63.

[39] Kreuer KD. Solid State Ionics 1997;94:55–62.

[40] Radev I, Georgiev G, Sinigersky V, Slavcheva E. Int J Hydrogen Energy 2008;33:4849–55.

[41] Jaouen F, Lindbergh G, Wiezell K. J Electrochem Soc 2003;150:A1711–7.

[42] Jaouen F, Lindbergh G. J Electrochem Soc 2003;150:A1699–710.

[43] [Chapter 8] Barbir F. Fuel cell diagnostics. In: Barbir F, editor. PEM fuel cells: theory and practice. Elsevier Academic Press; 2005. p. 253.

[44] Zhang J, Tang Y, Song C, Cheng X, Zhang J, Wang H. Electrochim Acta 2007;52: 5095–101.

[45] Zhang J, Tang Y, Song C, Xia Z, Li H, Wang H, et al. Electrochim Acta 2008;53:5315–21.

[46] Zhang J, Tang Y, Song C, Zhang J. J Power Sourc 2007;172:163–71.

[47] Tang Y, Zhang J, Song C, Liu H, Zhang J, Wang H, et al. J Electrochem Soc 2006;153: A2036–43.

[48] Silva RF, De Francesco M, Pozio A. J Power Sourc 2004;134:18–26.

[49] Sumner JJ, Creager SE, Ma JJ, DesMarteau DD. J Electrochem Soc 1998;145:107–10.

[50] Sone Y, Ekdunge P, Simonsson D. J Electrochem Soc 1996;143:1254–9.

[51] Verbrugge MW, Hill RF. J Electrochem Soc 1990;137:3770–7.

[52] Haufe S, Stimming U. J Memb Sci 2001;185:95–103.

[53] Cappadonia M, Erning JW, Stimming U. J Electroanal Chem 1994;376:189–93.

[54] Nouel KM, Fedkiw PS. Electrochim Acta 1998;43:2381–7.

[55] McLin MG, Wintersgill MC, Fontanella JJ, Chen RS, Jayakody JP, Greenbaum SG. Solid State Ionics 1993;60:137–40.

[56] Cahan BD, Wainright JS. J Electrochem Soc 1993;140:L185–6.

[57] Wang F, Hickner M, Kim YS, Zawodzinski TA, McGrath JE. J Memb Sci 2002;197:231–42.

[58] Kim YS, Wang F, Hickner M, Zawodzinski TA, McGrath JE. J Memb Sci 2003;212:263–82.

[59] Sumner MJ, Harrison WL, Weyers RM, Kim YS, McGrath JE, Riffle JS, et al. J Membr Sci 2004;239:199–211.

[60] Xie Z, Song C, Andreaus B, Navessin T, Shi Z, Zhang J, et al. J Electrochem Soc 2006;153: E173–8.

[61] Ma YL, Wainright JS, Litt MH, Savinell RF. J Electrochem Soc 2004;151:A8–A16.

[62] Doyle M, Choi SK, Proulx G. J Electrochem Soc 2000;147:34–7.

[63] Woo Y, Oh SY, Kang YS, Jung B. J Memb Sci 2003;220:31–45.

[64] Ma C, Zhang L, Mukerjee S, Ofer D, Nair B. J Membr Sci 2003;219:123–36.

[65] Jung DH, Cho SY, Peck DH, Shin DR, Kim JS. J Power Sourc 2002;106:173–7.

[66] Aleksandrova E, Hiesgen R, Andreas Friedrich K, Roduner E. Phys Chem Chem Phys 2007;9:2735–43.

[67] Hiesgen R, Aleksandrova E, Meichsner G, Wehl I, Roduner E, Friedrich KA. Electrochim Acta 2009;55:423–9.

[68] Ochi S, Kamishima O, Mizusaki J, Kawamura J. Solid State Ionics 2009;180:580–4.

[69] Zawodzinski JTA, Derouin C, Radzinski S, Sherman RJ, Smith VT, Springer TE, et al. J Electrochem Soc 1993;140:1041–7.

[70] Maldonado L, Perrin J-C, Dillet J, Lottin O. J Memb Sci 2012;389:43–56.

[71] Hinatsu JT, Mizuhata M, Takenaka H. J Electrochem Soc 1994;141:1493–8.

[72] Jalani NH, Datta R. J Membr Sci 2005;264:167–75.

[73] Choi P, Jalani NH, Datta R. J Electrochem Soc 2005;152:E123–30.

[74] Bouchet R, Miller S, Duclot M, Souquet JL. Solid State Ionics 2001;145:69–78.

[75] Bouchet R, Siebert E. Solid State Ionics 1999;118:287–99.

[76] He R, Li Q, Xiao G, Bjerrum NJ. J Membr Sci 2003;226:169–84.

[77] Kawamura J, Hattori K, Mizusaki J. In: Hashmi SA, Chandra A, Khare N, Chandra A, editors. Electroactive polymer: materials and devices. Allied Publishers; 2007. p. 144.

[78] Zawodzinski TA, Neeman M, Sillerud LO, Gottesfeld S. J Phys Chem 1991;95:6040–4.

[79] Zelsmann HR, Pineri M, Thomas M, Escoubes M. J Appl Polym Sci 1990;41:1673–84.

Hydrogen Crossover

6.1. INTRODUCTION

Proton exchange membrane (PEM) fuel cells usually employ PEMs (e.g. per-fluorosulfonic acid membranes such as the Nafion® membrane) to conduct protons and simultaneously separate the anode compartment from that of the cathode. When using pure hydrogen as the fuel in a H_2/air PEM fuel cell, nearly 100% fuel efficiency and zero emissions can be achieved. However, the PEMs

PEM Fuel Cell Testing and Diagnosis. http://dx.doi.org/10.1016/B978-0-444-53688-4.00006-1
171

are porous and allow a finite amount of gas permeation, in particular H_2. This phenomenon of hydrogen diffusion from the anode, across the membrane, to the cathode is called "hydrogen crossover" and leads to a reduction in fuel efficiency as well as a drop in cathode potential. Hydrogen crossover is an undesirable phenomenon in operating PEM fuel cells.

The performance of an operating membrane electrode assembly (MEA) can be expressed by using Eqn (6.1) [1]:

$$V_{cell} = E^{OCV} - \eta_a - \eta_c - I_{cell}R_{cell} \qquad (6.1)$$

where V_{cell} is the fuel cell voltage at the current density of I_{cell}, E^{OCV} is the fuel cell open circuit voltage (OCV), η_a and η_c are, respectively, the anodic and cathodic overpotentials, and R_{cell} is the internal resistance of the fuel cell, including membrane resistance, material bulk resistances, and all the contact resistances inside the fuel cell components; the internal resistance is dominated by the membrane resistance.

It has been estimated that the voltage loss caused by this membrane resistance accounts for about 30% of the total voltage loss [2]. From Eqn (6.1), it can be seen that fuel cell performance can be increased by reducing the membrane resistance. The simple and effective way to reduce the membrane resistance in a PEM fuel cell is to use a thinner membrane. For example, in the early stages of PEM fuel cell development, Nafion®-117 membranes (thickness: ~175 µm) were used, followed by Nafion®-115 membranes (thickness: ~125 µm), Nafion®-1135 membranes (thickness: ~87 µm), and Nafion®-112 membranes (thickness: ~50 µm). Most recently, Nafion®-211 and Nafion®-111 membranes (thickness: ~25 µm) have been developed and used in PEM fuel cells. Although fuel cell performance can be improved under the same operating conditions with a thinner membrane, a higher hydrogen crossover rate can be induced, which results in fuel efficiency reduction. In addition, hydrogen crossover can also cause a series of other problems, such as fuel cell OCV reduction, membrane degradation due to attack by hydrogen peroxide radicals, formation of pinholes in the membrane, and even catalyst degradation. Besides membrane thickness, other factors such as operating conditions (e.g. temperature, gas relative humidity, RH) can also affect the hydrogen crossover rate.

The following sections of this chapter will discuss in detail the effects of hydrogen crossover on fuel cell performance and the measurement of hydrogen crossover.

6.2. HYDROGEN CROSSOVER THEORY (MODEL)

In a PEM fuel cell, hydrogen molecules can diffuse from the anode and then cross over the membrane to the cathode, which is a diffusion-controlled process. As discussed in Chapter 2, a typical MEA usually consists of an anode gas diffusion layer ($GDL_{(a)}$), anode catalyst layer ($CL_{(a)}$), PEM, cathode

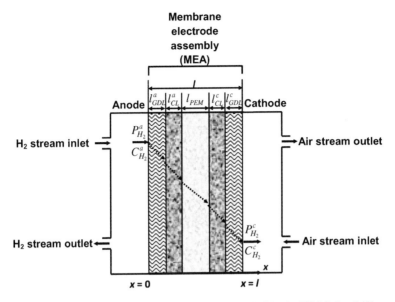

FIGURE 6.1 Schematic of hydrogen crossover in the MEA of a PEM fuel cell [3].

catalyst layer ($CL_{(c)}$), and cathode gas diffusion layer ($GDL_{(c)}$). During the crossover through the MEA, hydrogen diffuses across the PEM and all the four other layers. This process is illustrated in Fig. 6.1. This theory of hydrogen crossover has been proposed in our previous work [3].

According to Fick's first law, the steady-state H_2 crossover rate ($J_{H_2}^{x-over}$) can be treated as the hydrogen diffusion (or crossover) rate through the whole MEA and can be expressed by using Eqn (6.2):

$$J_{H_2}^{x-over} = \left(\frac{D_{H_2}}{l_{MEA}}\right)\left(C_{H_2}^a - C_{H_2}^c\right) \tag{6.2}$$

where $C_{H_2}^a$ and $C_{H_2}^c$ are, respectively, the concentration of H_2 at the interface between the anode bulk H_2 gas phase and the $GDL_{(a)}$ surface, and the concentration at the interface between the cathode air (or nitrogen) gas phase and the $GDL_{(c)}$ surface; D_{H_2} is the overall hydrogen diffusion coefficient across the MEA; and l_{MEA} is the thickness of the whole MEA. Based on the MEA structure, l_{MEA} is the sum of the five layers and can be expressed as in Eqn (6.3):

$$l_{MEA} = l_{GDL}^a + l_{CL}^a + l_{PEM} + l_{CL}^c + l_{GDL}^c \tag{6.3}$$

where l_{GDL}^a, l_{CL}^a, l_{PEM}, l_{CL}^c, and l_{GDL}^c are the thicknesses of the $GDL_{(a)}$, $CL_{(a)}$, PEM, $CL_{(c)}$, and $GDL_{(c)}$, respectively.

Based on the above hydrogen crossover theory as depicted in Fig. 6.1, the overall resistance for H_2 diffusion (l_{MEA}/D_{H_2}) through an MEA is the sum of the resistances for H_2 diffusion through the five layers, and can be expressed as in Eqn (6.4):

$$\frac{l_{MEA}}{D_{H_2}} = \frac{l^a_{GDL}}{D^a_{GDL}} + \frac{l^a_{CL}}{D^a_{CL}} + \frac{l_{PEM}}{D_{PEM}} + \frac{l^c_{CL}}{D^c_{CL}} + \frac{l^c_{GDL}}{D^c_{GDL}} \tag{6.4}$$

where D^a_{GDL}, D^a_{CL}, D_{PEM}, D^c_{CL}, and D^c_{GDL} are the diffusion coefficients of H_2 in the $GDL_{(a)}$, $CL_{(a)}$, PEM, $CL_{(c)}$, and $GDL_{(c)}$, respectively. D_{H_2} can be deduced from the rearrangement of Eqn (6.4) and expressed as in Eqn (6.5):

$$D_{H_2} = \frac{l^a_{GDL} + l^a_{CL} + l_{PEM} + l^c_{CL} + l^c_{GDL}}{\left(\frac{l^a_{GDL}}{D^a_{GDL}}\right) + \left(\frac{l^a_{CL}}{D^a_{CL}}\right) + \left(\frac{l_{PEM}}{D_{PEM}}\right) + \left(\frac{l^c_{CL}}{D^c_{CL}}\right) + \left(\frac{l^c_{GDL}}{D^c_{GDL}}\right)} \tag{6.5}$$

Equation (6.5) clearly indicates that the diffusion from every layer depicted in Fig. 6.1 contributes to the overall H_2 diffusion across the MEA. Because H_2 that diffuses from the anode to the cathode would fully react with O_2 electrochemically or chemically at the PEM/$CL_{(c)}$ interface, the H_2 that diffuses along the paths of l^c_{CL} and l^c_{GDL} should be equal to zero. Thus, Eqn (6.5) can be simplified as Eqn (6.6):

$$D_{H_2} = \frac{l^a_{GDL} + l^a_{CL} + l_{PEM}}{\left(\frac{l^a_{GDL}}{D^a_{GDL}}\right) + \left(\frac{l^a_{CL}}{D^a_{CL}}\right) + \left(\frac{l_{PEM}}{D_{PEM}}\right)} \tag{6.6}$$

For the $GDL_{(a)}$ and $CL_{(a)}$, the diffusion coefficients of D^a_{GDL} and D^a_{CL} can be expressed as in Eqns (6.7) and (6.8), respectively, based on the Bruggeman correlation [4,5]:

$$D^a_{GDL} = D^g_{H_2} \left[\varepsilon^a_{GDL} \left(1 - s^a_{GDL} \right) \right]^\tau \tag{6.7}$$

$$D^a_{CL} = D^g_{H_2} \left[\varepsilon^a_{CL} \left(1 - s^a_{CL} \right) \right]^\tau \tag{6.8}$$

where $D^g_{H_2}$ is the vapor-phase H_2 diffusion coefficient (with a typical value of 2.63×10^{-2} $cm^2\,s^{-1}$ at 80 °C [4,5]); ε^a_{GDL} and ε^a_{CL} are the porosities of the $GDL_{(a)}$ and $CL_{(a)}$, respectively; s^a_{GDL} and s^a_{CL} are the water saturation degrees inside the $GDL_{(a)}$ and $CL_{(a)}$, respectively, and have typical values in the range of 0–1; and τ is the tortuosity, which is often assumed to be 1.5.

Although in the literature, D_{PEM} for Nafion®-membranes has been formulated empirically as a function of temperature (T) [6], the effects of backpressure (P) and RH should also be considered when measuring hydrogen crossover in real situations.

The widely used Nafion®-112 membrane is approximately 50 μm thick, whereas the GDL is about 200 μm thick. The diffusion coefficients of H_2 crossover, D^a_{GDL}, D^a_{CL}, and D_{PEM}, through the $GDL_{(a)}$, $CL_{(a)}$, and PEM in a conventional Nafion®-112 membrane-based MEA at 80 °C are in the order of approximately $10^{-2}\,\mathrm{cm^2\,s^{-1}}$, $10^{-4}\,\mathrm{cm^2\,s^{-1}}$, and $10^{-6}\,\mathrm{cm^2\,s^{-1}}$, respectively. Combining these values into Eqn (6.6), it is obvious that H_2 crossover through an MEA is dominated by hydrogen diffusion in the membrane, which is mainly due to the smallest among the above values, that is, D_{PEM}. Thus, the contributions from hydrogen diffusion in both the $GDL_{(a)}$ and $CL_{(a)}$ are negligible. Therefore, Eqn (6.6) can be simplified further to Eqn (6.9) by ignoring those contributions:

$$\frac{l_{MEA}}{D_{H_2}} = \frac{l_{PEM}}{D_{PEM}} \tag{6.9}$$

Equation (6.9) indicates that the overall diffusion coefficient value of H_2 crossover (or diffusion) through the MEA, obtained experimentally, is equal to that through the PEM. This has been recognized widely in the literature. For example, the dependence of H_2 crossover on the membrane thickness was confirmed by Kocha et al. [7], who compared the H_2 crossover currents measured with both Nafion®-112 and Nafion®-111 membranes in the temperature range of 25–80 °C; their results suggested that the dominant factor limiting the H_2 crossover rate is the diffusion coefficient through the membrane.

In Eqn (6.2), the value of $C_{H_2}^c$ is equal to zero if one assumes that the hydrogen diffused from the anode can be oxidized completely at the PEM/$CL_{(c)}$ interface. Thus, Eqn (6.2) can also be written as Eqn (6.10):

$$J_{H_2}^{x\text{-}over} = \left(\frac{D_{H_2}}{l_{MEA}}\right) C_{H_2}^a \tag{6.10}$$

and according to Eqn (6.9), it can be rewritten as Eqn (6.11), indicating that the H_2 crossover rate through the MEA is equal to that through the PEM.

$$J_{H_2}^{x\text{-}over} = \left(\frac{D_{PEM}}{l_{PEM}}\right) C_{H_2}^a \tag{6.11}$$

An alternative expression for Eqn (6.11) is Eqn (6.12):

$$J_{H_2}^{x\text{-}over} = \left(\frac{K_{H_2}^{PEM} D_{PEM}}{l_{PEM}}\right) P_{H_2}^a \tag{6.12}$$

where $K_{H_2}^{PEM}$ is the H_2 partial-pressure-related solubility coefficient in the PEM, with a unit of $\mathrm{mol\,cm^{-3}\,atm^{-1}}$, and $P_{H_2}^a$ is the H_2 partial pressure at the interface of the $CL_{(a)}$ and PEM, with a unit of atm (this can be approximately treated as the H_2 partial pressure in the anode feed stream). The product of the solubility coefficient ($K_{H_2}^{PEM}$) and diffusion coefficient (D_{PEM}) can be defined as the permeability coefficient of H_2 in the PEM, $\psi_{H_2}^{PEM}$, with a unit of $\mathrm{mol\,cm^{-1}\,atm^{-1}\,s^{-1}}$, expressed as in Eqn (6.13):

$$\psi_{H_2}^{PEM} = K_{H_2}^{PEM} D_{PEM} \tag{6.13}$$

By combining Eqns (6.12) and (6.13), one can express the hydrogen crossover rate across the MEA as in Eqn (6.14):

$$J_{H_2}^{x-over} = \left(\frac{\psi_{H_2}^{PEM}}{l_{PEM}}\right) P_{H_2}^a \qquad (6.14)$$

Similar to D_{H_2} (or D_{PEM}), $\psi_{H_2}^{PEM}$ is also a function of T, P, RH, and the nature of the membrane [8]. For a defined fuel cell system, the value of l_{PEM} in Eqn (6.14) can be considered to be fixed; thus, $\psi_{H_2}^{PEM}$ and $P_{H_2}^a$ may become the more dominant factors that influence the H_2 crossover rate.

By rearranging Eqn (6.14), the permeability coefficient, $\psi_{H_2}^{PEM}$, can then be determined by Eqn (6.15):

$$\psi_{H_2}^{PEM} = \frac{J_{H_2}^{x-over} l_{PEM}}{P_{H_2}^a} \qquad (6.15)$$

In Eqn (6.15), the value of l_{PEM} is known for a defined PEM fuel cell system (e.g. $l_{PEM} = 50\ \mu m$ for a Nafion®-112-membrane-based MEA); the values of $J_{H_2}^{x-over}$ and $P_{H_2}^a$ can be measured experimentally, and then $\psi_{H_2}^{PEM}$ can be calculated at different temperatures, pressures, and RHs.

6.3. IMPACTS OF HYDROGEN CROSSOVER ON FUEL CELL PERFORMANCE AND DURABILITY

6.3.1. Hydrogen Crossover Effect on Fuel Efficiency

Because the membranes used in PEM fuel cells are porous materials, hydrogen that has crossed over from the anode to the cathode can react with oxygen at the PEM/CL$_{(c)}$ interface and produce heat and water. Obviously, this process reduces the fuel efficiency. Moreover, diffused hydrogen that reacts with oxygen can also lead to a reduction in the oxygen concentration at the PEM/CL$_{(c)}$ interface, resulting in a decreased cathode potential.

6.3.2. Hydrogen Crossover Effect on Fuel Cell OCV

Hydrogen that diffuses from the anode to the cathode will be oxidized electrochemically at the PEM/CL$_{(c)}$ interface under the cathodic potential, according to Reaction (6.I):

$$H_2 \rightarrow 2H^+ + 2e^- \qquad (6.I)$$

At the same time, the oxygen reduction reaction proceeds at the cathode at the PEM/CL$_{(c)}$ interface, and this process is often expressed as in Reaction (6.II):

$$\frac{1}{2}O_2 + 2H^+ + 2e^- \rightarrow H_2O \qquad (6.II)$$

Reactions (6.I) and (6.II) form a local cell at the cathode, which depresses the cathode potential, resulting in the reduction of fuel cell OCV. This has been proven by our recent study in the temperature range of 23–120 °C [9]; the effect of hydrogen crossover on fuel cell OCV will be discussed in detail in Chapter 8. In addition, it is also possible for the chemical reaction between the crossed hydrogen and the oxygen at the cathode Pt catalyst surface to produce hydrogen peroxide or water, $1/2\,O_2 + H_2 \rightarrow H_2O$ and $O_2 + H_2 \rightarrow H_2O_2$, which leads to a reduction in O_2 concentration and OCV. However, the electrochemical reactions shown in Reactions (6.I) and (6.II) are the dominant reactions.

6.3.3. Hydrogen Crossover Effect on Membrane Degradation

As discussed above, when hydrogen diffuses from the anode to the cathode, the hydrogen can react chemically with the oxygen at the cathode Pt catalyst to produce hydrogen peroxide radicals [10–12]. These radicals have been confirmed to attack the membrane, which leads to its chemical degradation. Meanwhile, the highly exothermic chemical reaction between the crossed H_2 and the O_2 at the cathode's Pt catalyst can also produce heat and often generates local hot points, which can lead to local pinholes in the membrane [12]; pinhole generation in turn increases the hydrogen crossover rate. Thus, a destructive cycle of membrane degradation and accelerated membrane failure is established.

6.3.4. Hydrogen Crossover Effect on Cathode Degradation

It has been confirmed that the hydrogen crossover can also cause cathode degradation [11,12]. The hydrogen peroxide radicals produced by the reaction between hydrogen and oxygen at the cathode attack the membrane and the ionomers (e.g. Nafion® ionomer) in the $CL_{(c)}$, and this leads to their chemical degradation and loss of proton conductivity. Thus, the three-phase reaction boundary in the $CL_{(c)}$ will be decreased due to the lack of proton conductors. Moreover, the formation of local hot points because of the reaction of H_2 and O_2 at the cathode will accelerate the sintering of Pt catalyst particles due to the local high temperature, decreasing the electrochemical surface area (ESA) of the Pt catalyst. This has been proven by Yu et al. [12].

6.4. TECHNIQUES FOR HYDROGEN CROSSOVER MEASUREMENTS

During the development of PEM fuel cells, many techniques have been devised to measure the gas permeation rate and/or hydrogen crossover rate of membranes. These techniques include ex situ and in situ methods. Among them, the in situ electrochemical technique is a direct and effective method to measure the hydrogen crossover rate of membranes in PEM fuel cells.

6.4.1. Membrane Permeability Measurement

In a H_2/air PEM fuel cell, the gradient of the hydrogen concentration across the membrane drives H_2 to permeate from one side of the membrane to the other. The hydrogen crossover rate $(J_{H_2}^{x-over})$ through a membrane can be expressed as in Eqn (6.2) or Eqn (6.14). The hydrogen permeability coefficient, $\psi_{H_2}^{PEM}$, can be expressed as in Eqn (6.15).

Several techniques can be used to measure the hydrogen crossover rate, including the volumetric method, time-lag technique, gas chromatography method, and the electrochemical monitoring technique. For example, by applying higher pressure at one side of the membrane, Sakai et al. [13] measured the H_2 permeability coefficient in Nafion membranes by using the volumetric method.

Instead of measuring H_2 flow rate for a given sample area via the volumetric method, the time-lag technique measures the time to fill a fixed volume downstream of the membrane [14–16]. The gas chromatography method measures the concentration change downstream of the membrane when the same total pressure but different gas concentrations are applied across the membrane [17]. For example, when measuring the hydrogen crossover rate, hydrogen gas is applied on one side of the membrane and another gas (e.g. N_2) is made to flow into the other side of the membrane, while the pressures on both sides are kept the same. The H_2 crossover rate can be determined by measuring the hydrogen concentration in the N_2 flow.

The electrochemical monitoring technique is another effective way of measuring the gas permeability coefficient of PEMs. In this method, one side of the membrane is attached to a working electrode and is exposed to an electrolyte, and then the current generated by H_2 diffusion through this membrane is measured over time to estimate the H_2 permeability coefficient of the membrane. A fuel cell setup can be used for this technique. In this way, the hydrogen crossover rate can be measured in situ for an operating PEM fuel cell, which will be discussed in the following two sections.

6.4.2. Linear Scan Voltammetry

Linear scan voltammetry (LSV) is an electrochemical method that can be used to measure hydrogen crossover in situ [10,18]. In this method, a fuel cell setup is used. Humidified H_2 is made to flow into the anode (or cathode), which serves as both reference electrode and counterelectrode. An inert gas (e.g. N_2 or Ar) is made to flow into the cathode (or anode), which serves as the working electrode. The potential of the cathode (or anode) is swept using a potentiostat instrument (e.g. Solartron 1287) from the rest potential (normally <200 mV) to 500 mV, with a scan rate of $1–5$ mV s^{-1}. The hydrogen permeating from the anode (or cathode) is oxidized electrochemically on the cathode (or anode) to produce a current density. This current density can reach a plateau, which demarcates a limiting current density, which can be considered to be the maximum current density obtainable by hydrogen crossover. Thus, based on the measured hydrogen crossover current density, the hydrogen crossover rate $(J_{H_2}^{\text{x-over}})$ can be calculated using Eqn (6.16):

$$J_{H_2}^{\text{x-over}} = \frac{I_{H_2}^{\text{x-over}}}{nF} \tag{6.16}$$

where $I_{H_2}^{\text{x-over}}$ is the measured hydrogen crossover current density (A cm^{-2}), n is the electron transfer number of hydrogen oxidation ($n = 2$), and F is Faraday's constant (96,487 C mol^{-1}). Therefore, the unit of $J_{H_2}^{\text{x-over}}$ is mol cm^{-2} s^{-1}.

Figure 6.2 shows the LSV recorded on an H$_2$/Ar fuel cell in the potential range of 120–500 mV at different temperatures from 40 to 80 °C. As shown in this figure, the current density generated by the hydrogen oxidation increased with potential and reached a constant value at potentials >170 mV in each case. In this study, the authors treated limiting current density in the potential range of 300–350 mV, which is the double-layer region of platinum. In this potential range, the effect of hydrogen adsorption/desorption on platinum has an insignificant effect on the measured hydrogen crossover.

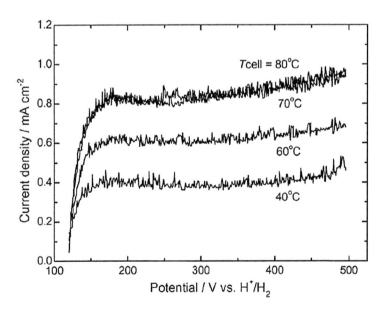

FIGURE 6.2 Linear sweep voltammograms for H$_2$/Ar cell at various cell temperatures. H$_2$ and Ar gases were humidified at $T_{\text{cell}} = -4$ °C in each case. Atmospheric pressure, H$_2$/Ar $= 300/300$ mL min^{-1} [10].

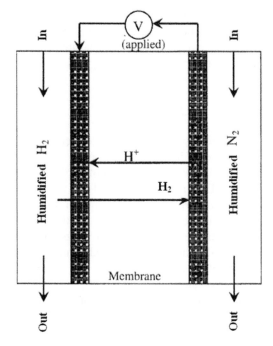

FIGURE 6.3 Schematic of the setup for hydrogen crossover measurement.

6.4.3. Steady-State Electrochemical Method

The limiting current density can also be measured with a potentiostat by applying a fixed potential for a time span [3,7,9]. Figure 6.3 shows a schematic of the fuel cell setup for hydrogen crossover measurement. Before measurement, a fuel cell assembled with a normal MEA is usually conditioned at the operating conditions. After that, a humidified inert gas (e.g. N_2) stream is introduced into the cathode to remove the air. Once the fuel cell OCV is <0.1 V, the humidified H_2 and N_2 are made to flow into the fuel cell anode (or cathode) and cathode (or anode), respectively. A potentiostat (e.g. Solartron 1287) is then connected to the fuel cell for H_2 crossover measurements, with the working electrode probe connected to the cathode (or anode) and the counter/reference electrode probes connected together to the anode (or cathode). A steady-state electrochemical method is used to record the current generated from the oxidation of crossed H_2 from the anode (or cathode) at an applied potential of 0.5 V relative to the potential of the H_2-flushed electrode. At this potential, all H_2 that permeated across the membrane should be completely oxidized. The current is usually recorded over a period of 5–20 min, till it attains a limiting current under a steady state. The limiting current density is indicative of the amount of hydrogen that has crossed over under a steady state.

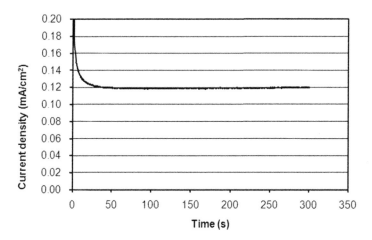

FIGURE 6.4 Hydrogen crossover current density measured on a Nafion®-112-membrane-based MEA at a fixed potential of 0.5 V vs. H⁺/H₂. Measuring conditions: 80 °C; 100% RH; 1.0 atm backpressure.

Figure 6.4 shows the hydrogen crossover current density measured on a Nafion®-112-membrane-based MEA using the steady-state electrochemical method. It can be seen that the limiting current density can reach a steady level after 50 s. With this limiting current density, the hydrogen crossover rate, $J_{H_2}^{x-over}$, can be calculated easily based on Eqn (6.16). Then, the permeability coefficient, $\psi_{H_2}^{PEM}$, can be obtained according to Eqn (6.15) if the hydrogen pressure and membrane thickness are known.

6.5. DEPENDENCE OF HYDROGEN CROSSOVER ON *T*, RH, AND *P*

As stated in Section 6.1, the hydrogen crossover rate is dependent on the fuel cell operating conditions. The following sections will address the influences of these operating conditions on hydrogen crossover, based on some typical examples.

6.5.1. Temperature Dependence of Hydrogen Crossover

As one of the important operating conditions for PEM fuel cells, temperature has a significant effect on hydrogen crossover. Our recent study [3] indicated that the hydrogen crossover rate increases with increasing temperature. As shown in Table 6.1, when the temperature increases from 80 to 120 °C, the hydrogen crossover rate increases from 2.04×10^{-8} mol cm^{-2} s^{-1} at 80 °C with 100% RH and 3.04 atm backpressure, to 2.69×10^{-8} mol cm^{-2} s^{-1} at 100 °C and 3.05×10^{-8} mol cm^{-2} s^{-1} at 120 °C, with the same RH and

TABLE 6.1 Measured H_2 Crossover Rate at Various Temperatures, Backpressures, and Relative Humidities Using a Nafion®-112-Based MEA. The Thickness of the Nafion® 112 was Adopted as 50 μm [3].

| Temperature (°C) | RH (%) | Measured H_2 Crossover Rate (mol cm^{-2} s^{-1}) | | |
| | | Backpressure (atm) | | |
		3.04	2.02	1.00
80	100	2.04E-08	1.30E-08	3.78E-09
	70	1.91E-08	1.20E-08	4.14E-09
	50	1.80E-08	1.16E-08	4.51E-09
	25	1.74E-08	1.14E-08	5.30E-09
100	100	2.69E-08	1.48E-08	4.13E-09
	70	2.82E-08	1.43E-08	5.91E-09
	50	2.57E-08	1.35E-08	7.00E-09
	25	2.23E-08	1.42E-08	8.23E-09
120	100	3.05E-08	1.65E-08	8.13E-09
	70	4.16E-08	2.33E-08	1.19E-08
	50	5.02E-08	2.90E-08	1.61E-08
	25	6.28E-08	4.26E-08	2.17E-08

backpressure. The same trends were observed for other RHs, from 25 to 75%, and with backpressures of 2.04 atm and 1.0 atm. Other studies have found the same trends [7,10,11,18]. For example, as shown in Fig. 6.5, the hydrogen crossover current density increased when the temperature was increased from 40 to 80 °C. The authors attributed this to the increased flexibility of the membrane with increased temperature [10].

6.5.2. RH Dependence of Hydrogen Crossover

An increase in the RH also leads to an increased hydrogen crossover rates, which has been proven by both Inaba et al. [10] and Teranishi et al. [11]. Figure 6.5 clearly indicates this trend. The authors attributed this result to the increase in membrane flexibility with increasing RH. The membrane water content can also increase with increasing RH, and both the solubility coefficient

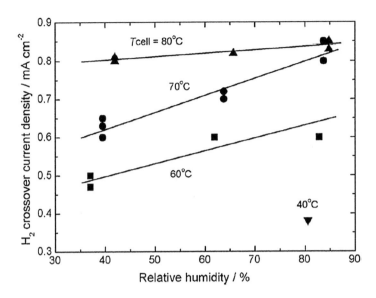

FIGURE 6.5 Effect of cell temperature and humidification on H_2 crossover current density at atmospheric pressure. H_2 and Ar gases were humidified at the same temperature in each case. H_2/Ar $= 300/300$ mL min^{-1} [10].

$(K_{H_2}^{PEM})$ and the diffusivity coefficient $(D_{H_2}^{PEM})$ increase as well. However, at high temperatures ($>80\ °C$), the effect of RH on hydrogen crossover is more complicated. As shown in Table 6.1, the hydrogen crossover actually decreases with increasing RH at cell temperatures of 100 and 120 °C, a reversal of the trend at 80 °C. This is because when the temperature is increased from 80 to 100 °C and to even 120 °C, the increase in RH can cause an increase in water partial pressure (P_{H_2O}). As the inlet total pressure of the cell (P_{total}^{inlet}) is the sum of P_{H_2O} and P_{H_2} $(P_{total}^{inlet} = P_{H_2O} + P_{H_2})$, the increase in P_{H_2O} means that P_{H_2} has to decrease. According to Eqn (6.14), when the hydrogen crossover rate decreases with increasing RH, this is due to the decrease in P_{H_2}. Although both the solubility coefficient $(K_{H_2}^{PEM})$ and the diffusivity coefficient $(D_{H_2}^{PEM})$ increase when the membrane's water content increases, the positive effect of the increase in the permeability coefficient might be smaller than the negative effect of decreased P_{H_2}. Thus, the overall effect is that the hydrogen crossover rate decreases with increasing RH at temperatures of 100–120 °C.

6.5.3. Pressure Dependence of Hydrogen Crossover

Pressure can also have an important effect on hydrogen crossover. Table 6.1 shows that the hydrogen crossover rate increased with increasing backpressure at all the cell temperatures and RHs. As mentioned above, the total inlet pressure is the sum of P_{H_2O} and P_{H_2}, and increasing the backpressure will cause

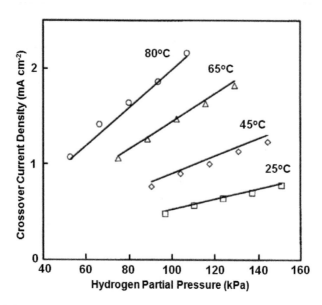

FIGURE 6.6 In situ hydrogen crossover current density versus hydrogen partial pressure in Nafion®-112 membrane [7].

an increase in P_{H_2} at the same cell temperature and RH. According to Eqn (6.14), the hydrogen crossover rate will increase with increasing P_{H_2}. Kocha et al. [7] proved that the hydrogen crossover current density can be increased by increasing the partial pressure of hydrogen. As shown in Fig. 6.6, at all the temperatures studied from 25 to 80 °C, the changes in hydrogen crossover current density follow this trend.

6.6. SUMMARY

Hydrogen crossover is an undesirable permeation of fuel in H_2/air PEM fuel cells that is inevitable due to the porosity of PEMs. In PEM fuel cells, hydrogen crossover can not only decrease both the fuel efficiency and the cathode potential drop but it can also induce fuel cell failure by accelerating membrane degradation, even to the point of membrane failure. When hydrogen crosses over from the anode to the cathode, it reacts with oxygen to form hydrogen peroxide radicals, which then attack the membrane and create pinholes, resulting in membrane failure. These radicals can also attack the ionomer inside the CL, resulting in severe catalyst layer degradation. Hydrogen crossover can also accelerate cathode degradation by the formation of hotspots that lead to decreased Pt ESA.

In this chapter, the theory of hydrogen crossover and the techniques for determining the hydrogen crossover rate have been presented. Among these different techniques, the in situ electrochemical method is the most simple and effective. The experimental and theoretical models indicate that the steady-state hydrogen crossover rate is a function of operating temperature, RH, and backpressure.

REFERENCES

[1] Song C, Tang Y, Zhang JL, Zhang J, Wang H, Shen J, et al. Electrochim Acta 2007;52:2552–61.
[2] Tang Y, Zhang J, Song C, Liu H, Zhang J, Wang H, et al. J Electrochem Soc 2006;153: A2036–43.
[3] Cheng X, Zhang J, Tang Y, Song C, Shen J, Song D, et al. J Power Sources 2007;167:25–31.
[4] Yao KZ, Karan K, McAuley KB, Oosthuizen P, Peppley B, Xie T. Fuel Cells 2004;4:3.
[5] West AC, Fuller TF. J Appl Electrochem 1996;26:557.
[6] Yeo RS, McBreen J. J Electrochem Soc 1979;126:1682–7.
[7] Kocha SS, Deliang Yang J, Yi JS. AIChE J 2006;52:1916–25.
[8] Weber AZ, Newman J. J Electrochem Soc 2004;151:A311–25.
[9] Zhang J, Tang Y, Song C, Zhang J, Wang H. J Power Sources 2006;163:532–7.
[10] Inaba M, Kinumoto T, Kiriake M, Umebayashi R, Tasaka A, Ogumi Z. Electrochim Acta 2006;51:5746–53.
[11] Teranishi K, Kawata K, Tsushima S, Hirai S. Electrochem Solid State Lett 2006;9:A475–7.
[12] Yu J, Matsuura T, Yoshikawa Y, Islam MN, Hori M. Electrochem Solid State Lett 2005;8: A156–8.
[13] Sakai T, Takenaka H, Wakabayashi N, Kawami Y, Torikai E. J Electrochem Soc 1985;132: 1328–32.
[14] Sakai T, Takenaka H, Torikai E. J Electrochem Soc 1986;133:88–92.
[15] Sakai T, Takenaka H, Torikai E. J Memb Sci 1987;31:227–34.
[16] Chiou JS, Paul DR. Ind Eng Chem Res 1988;27:2161–4.
[17] Broka K, Ekdunge P. J Appl Electrochem 1997;27:117–23.
[18] Song Y, Fenton JM, Kunz HR, Bonville LJ, Williams MV. J Electrochem Soc 2005;152: A539–44.

Fuel Cell Open Circuit Voltage

7.1. OPEN CIRCUIT VOLTAGE THEORY

A proton exchange membrane (PEM) fuel cell directly converts the chemical energy stored in its fuel (e.g. hydrogen, or low-carbon fuels such as methanol) into electrical energy through anodic and cathodic electrochemical reactions. For H_2/air or H_2/O_2 PEM fuel cells, the electrochemical reactions are usually expressed as Reactions (1.I) to (1.III) in Chapter 1. The thermodynamic electrode potentials for half-electrochemical Reactions (1.I) and (1.II) can be expressed as in Eqns (1.1) and (1.2). The overall reaction, expressed in Reaction (1.III) $(1/2 O_2 + H_2 \leftrightarrow H_2O)$, is a combination of Reactions (1.I) and (1.II), so the theoretical fuel cell voltage can be obtained by combining Eqns (1.1) and (1.2). In fact, the theoretical cell voltage is the fuel cell open circuit voltage (OCV), V_{theory}^{OCV}, which can be expressed as in Eqn (1.3). As V_{theory}^{OCV} is the theoretical thermodynamic voltage at OCV, Eqn (1.3) can also be written as follows:

$$E_{theory}^{OCV} = E_{O_2/H_2O}^o - E_{H_2/H^+}^o + 2.303 \frac{RT}{2F} \log \left(\frac{a_{H_2} a_{O_2}^{\frac{1}{2}}}{a_{H_2O}} \right) \qquad (7.1)$$

PEM Fuel Cell Testing and Diagnosis. http://dx.doi.org/10.1016/B978-0-444-53688-4.00007-3

where E_{theory}^{OCV} is the theoretical OCV for an H_2/O_2 or H_2/air PEM fuel cell; E_{H_2/H^+}^o and E_{O_2/H_2O}^o are the standard anodic and cathodic potentials, respectively; E_{H_2/H^+}^o equals zero at any temperature; E_{O_2/H_2O}^o is a temperature-dependent parameter; a_{H_2}, a_{O_2}, and a_{H_2O} are the activities of H_2, O_2, and H_2O, respectively; and R and T are the universal gas constant and the temperature, respectively. Here, E_{O_2/H_2O}^o can be expressed as in Eqn (7.2) [1,2]:

$$E_{O_2/H_2O}^o = 1.229 - 0.000846 \times (T - 298.15) \tag{7.2}$$

Therefore, Eqn (7.1) can be rewritten as Eqn (7.3):

$$E_{theory}^{OCV} = 1.229 - 0.000846 \times (T - 298.15) + 2.303 \frac{RT}{2F} \log \left(\frac{a_{H_2} a_{O_2}^{\frac{1}{2}}}{a_{H_2O}} \right) \tag{7.3}$$

Thus, at standard conditions, the value of E_{theory}^{OCV} is 1.229 V.

For an approximate calculation of theoretical OCV using Eqn (7.3), a_{O_2}, a_{H_2}, and a_{H_2O} can be replaced by their partial pressures in PEM fuel cells, as P_{O_2}, P_{H_2}, and P_{H_2O}, respectively. Then, Eqn (7.3) can be written as Eqn (7.4):

$$E_{theory}^{OCV} = 1.229 - 0.000846 \times (T - 298.15) + 2.303 \frac{RT}{2F} \log \left(\frac{P_{H_2} P_{O_2}^{\frac{1}{2}}}{P_{H_2O}} \right) \tag{7.4}$$

In an operating PEM fuel cell, the temperature (T) is known, and P_{O_2}, P_{H_2}, and P_{H_2O} inside the fuel cell are measurable. Then, E_{theory}^{OCV} can be calculated according to the above parameters. At temperatures close to the standard temperature, the value of E_{theory}^{OCV} should be close to 1.229 V. However, in practice, the measured OCV value is always lower than that theoretically calculated using Eqn (7.4). This is because several factors can affect the OCV, including temperature, the status of the Pt catalyst surface, and hydrogen crossover, which will be discussed in the following sections.

7.2. MEASURED OCV

During the practical operation of a PEM fuel cell, the OCV of the cell is its voltage at zero current density, meaning that at OCV, the circuit is open without any power output. OCV values can be measured directly with a voltage meter, and are designated as $E_{measured}^{OCV}$. As shown earlier, in Fig. 1.8, the $E_{measured}^{OCV}$ of a practical operated single PEM fuel cell is around 1.0 V, usually 0.95–1.05 V, which is much smaller (~20% less) than the E_{theory}^{OCV} calculated according to Eqn (7.4) (which, as indicated above, is ~1.229 V). The reasons for this large difference between the values of E_{theory}^{OCV} and $E_{measured}^{OCV}$ were not clear for several decades of fuel cell development, although there were many hypotheses. Vilekar and Datta [3] summarized some explanations in the introduction to their article. Hoare [4] and Bockris and Srinivasan [5] suggested that one or more side reactions accompanied the oxygen reduction reaction (ORR) on the Pt surface at extremely low current densities. The possible side reactions, involving O_2, H_2, the carbon support C, impurities CH_x, and Pt at the PEM fuel cell cathode, are listed in Table 7.1. The presence of side reactions can result in a mixed potential between these reactions and the ORR, leading to a significant decrease in OCV that makes it much lower than the E_{theory}^{OCV}. The formation of H_2O_2 through side reaction 7 in Table 7.1 is the common explanation for the lower OCV. However, the concentration of H_2O_2 is far too small to account for the big difference between the measured and theoretical OCVs [5]. Another explanation includes the formation of Pt surface oxides due to one or more Pt corrosion reactions, as expressed in Table 7.1. However, in a PEM fuel cell system, the Pt oxidation reaction cannot go on indefinitely, and hence, it cannot be considered the only reason for the lower measured OCV [3]. Bockris and Srinivasan [5] studied the half-cell open circuit potential of the ORR in a liquid electrolyte and attributed the lower measured OCV to the mixed potential arising from the existence of organic impurities. But such organic impurities would not be present at the same level during long-term operation of a fuel cell, despite operating conditions such as temperature, pressure, relative humidity (RH).

TABLE 7.1 Possible Side Reactions Involving O_2, H_2, Carbon Support C, Impurities CH_x, and Pt at the PEM Fuel Cell Cathode [3,5]

Reaction No., p	Overall Reaction	Standard Electrode Potential, $\Phi_{p,0}^0$ (V)
1	$H_2O_2 + 2H^+ + 2e^- \leftrightarrows 2H_2O$	1.77
2	$PtO_3 + 2H^+ + 2e^- \leftrightarrows PtO_2 + H_2O$	1.48
3	$O_2 + 4H^+ + 4e^- \leftrightarrows 2H_2O$	1.229
4	$PtO_2 + 2H^+ + 2e^- \leftrightarrows Pt(OH)_2$	1.11
5	$Pt(OH)_2 + 2H^+ + 2e^- \leftrightarrows Pt + 2H_2O$	0.98
6	$PtO + 2H^+ + 2e^- \leftrightarrows Pt + H_2O$	0.88
7	$O_2 + 2H^+ + 2e^- \leftrightarrows H_2O_2$	0.68
8	$C + 2H_2O \leftrightarrows CO_2 + 4H^+ + 4e^-$	0.207
9	$CH_x + 2H_2O \leftrightarrows CO_2 + (x+4)H^+ + (x+4)e^-$	?
10	$2H^+ + 2e^- \leftrightarrows H_2$	0.00

Laraminie and Dicks [6] attributed the lower $E^{OCV}_{measured}$ to H_2 crossover and/or to internal electrical short circuiting caused by the small electronic conductivity of the electrolyte membrane. Similarly, Sompalli et al. [7] suggested that the lower $E^{OCV}_{measured}$ is derived from the parasitic current caused by a combination of H_2 crossover and ohmic shorting via the membrane, the former being the main contributor and the latter the minor. But the results of Cleghorn et al. [8] and Vilekar and Datta [3] indicated that ohmic shorting did not noticeably affect the observed OCV drop. Vilekar and Datta [3] concluded that the large difference between the values of E^{OCV}_{theory} and $E^{OCV}_{measured}$ was caused exclusively by hydrogen crossover, based on a theoretical model of mixed potential with parameters. However, if this were true, the OCV difference would decrease significantly if hydrogen crossover could be suppressed by using a thicker PEM. Yet in fact, the difference was still about 200–250 mV, even when a Nafion®-117 membrane (with a thickness of ~175 μm) was used.

Zhang et al. [2] studied the PEM fuel cell OCV in the temperature range of 23–120 °C and explained the large difference between the values of E^{OCV}_{theory} and $E^{OCV}_{measured}$ on the basis of their experimental results and theoretical calculations. They concluded that at certain operating conditions, the decrease in the OCV of a PEM fuel cell compared to its theoretical value is mainly caused by hydrogen crossover and the mixed potential induced by the Pt oxidation and ORR reactions.

7.3. FACTORS AFFECTING OCV

Mixed potential, hydrogen crossover, and temperature are the main factors that affect the OCV of a PEM fuel cell. The followings sections will discuss each of these in detail.

7.3.1. Effect of Mixed Potential on OCV

Mixed potential on the cathode side has been considered one of the most important causes of the large difference between the values of E^{OCV}_{theory} and $E^{OCV}_{measured}$. This mixed cathode potential results from the 4-electron ORR and one or more side reactions that occur on the cathode. These side reactions involve O_2, H_2, the carbon support C, impurities, contaminants, and the Pt on the cathode, and have been addressed in section 7.2 and listed in Table 7.1. Among these side reactions, the Pt oxidation reaction on the cathode plays a dominant role, as discussed in our previous article [2]. The mechanism of a Pt electrode reaction in an O_2-saturated acidic solution can be explained by a local-cell mechanism [2,9,10]. This local cell is composed of two electrode reactions, the O_2/H_2O cathode reaction ($O_2 + 4H^+ + 4e^- \leftrightarrow 2H_2O$, $E^o_c = 1.229V$ (vs. normal hydrogen electrode, NHE)) and the Pt/PtO anode reaction (Pt + $H_2O \leftrightarrow$ PtO + $2H^+ + 2e^-$, $E^o_{Pt/PtO} = 0.88V$ (vs. NHE)). The local electrochemical reaction on the Pt surface can lead to a PtO surface coverage of up to 30%, with 70% of the surface remaining as pure Pt. At the steady-state mixed potential, a complete PtO layer is never achieved, which keeps the Pt oxidation reaction moving forward. The reported and widely accepted mixed potential caused by the above-described local cell is around 1.06 V (vs. NHE) at standard conditions (25 °C, 1.0 atm) [9,10]. The mixed potential (V_{mixed}) is a function of the oxygen partial pressure (P_{O_2}) in an operating PEM fuel cell, and their relationship can be expressed using Eqn (7.5):

$$\frac{dV_{mixed}}{dP_{O_2}} = \frac{2.3RT}{n_{\alpha O}\alpha_O F} \tag{7.5}$$

where α_O and $n_{\alpha O}$ are the electron transfer coefficient and the electron transfer number in the rate-determining step of ORR, respectively. Based on Eqn (7.5), the temperature-dependent values of E^{OCV}_{mixed} can be calculated after measuring the P_{O_2} at different temperatures. Table 7.2 lists the E^{OCV}_{mixed} at temperatures from 23 to 120 °C, with 3.0 atm total air pressure and 100% RH. As shown in Table 7.2, the E^{OCV}_{mixed} decreases from 1.060 V at 23 °C to 1.040 V at 120 °C. To

TABLE 7.2 OCVs at 3.0 atm, 100% RH, and Different Temperatures [2]

Temperature (°C)		23	40	60	80	100	120
Theoretical OCV (E^{OCV}_{theory}), (V)		1.241	1.228	1.210	1.192	1.169	1.136
Measured OCV ($E^{OCV}_{measured}$) (V)	Nafion-117-based membrane electrode assembly (MEA)	1.011	1.009	1.005	1.000	0.975	0.948
	Nafion 112-based MEA	1.042	1.041	1.039	1.021	1.008	0.985
Mixed OCV (E^{OCV}_{mixed}) (V)		1.060	1.060	1.059	1.056	1.051	1.040
OCV drop caused by surface Pt–O_2 reaction ($\Delta E^{OCV}_{Pt-O_2}$) (V)		0.182	0.168	0.152	0.135	0.119	0.096
OCV drop caused by H_2 crossover ($\Delta E^{OCV}_{H_2-x-over}$) (V)	Nafion-117-based MEA	0.025	0.024	0.022	0.024	0.037	0.052
	Nafion-112-based MEA	0.054	0.049	0.053	0.056	0.071	0.088
Corrected OCV ($E^{OCV}_{corrected}$) (V)	Nafion-117-based MEA	1.249	1.234	1.213	1.180	1.164	1.133
	Nafion-112-based MEA	1.247	1.226	1.210	1.191	1.164	1.133

quantify the effect of mixed potential on the OCV of a PEM fuel cell, the values of the voltage drop caused by the surface Pt–O_2 reaction at OCV, $\Delta E_{O_2-Pt}^{OCV}$, were also calculated, according to Eqn (7.6):

$$\Delta E_{O_2-Pt}^{OCV} = E_{theory}^{OCV} - E_{mixed}^{OCV} \tag{7.6}$$

where E_{theory}^{OCV} is the theoretical OCV calculated according to Eqn (7.4). The values of $\Delta E_{O_2-Pt}^{OCV}$ at different temperatures <3.0 atm total operating pressure and 100% RH are listed in Table 7.2 as well. As shown there, the voltage loss caused by the mixed potential at 23 °C is as high as 182 mV, suggesting that mixed potential is one of the main factors that influences fuel cell OCV. Table 7.2 also shows that the voltage loss caused by mixed potential decreases with increasing temperature, from 182 mV at 23 °C to 96 mV at 120 °C.

The calculated mixed OCV (E_{mixed}^{OCV}) and the voltage drop caused by the surface Pt–O_2 reaction at OCV ($\Delta E_{O_2-Pt}^{OCV}$) as a function of temperature are also plotted in Fig. 7.1 for comparison.

7.3.2. H$_2$ Crossover Effect

H_2 crossover is the phenomenon of hydrogen crossing from the anode to the cathode through the PEM in a PEM fuel cell. This is an inevitable but undesirable phenomenon in operating PEM fuel cells. As discussed in Chapter 6, H_2 crossover not only decreases fuel efficiency but it also leads to membrane decay and even fuel cell failure. Hydrogen that crosses over from the anode will be oxidized electrochemically on the cathode; this reaction can be expressed using Reaction (1.I) in Chapter 1. Similar to Pt oxidation on the cathode, this

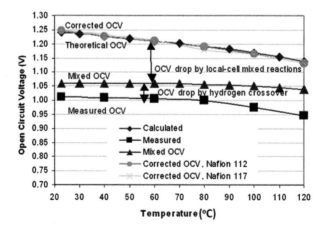

FIGURE 7.1 Fuel cell OCVs as a function of temperature. Operating conditions: H$_2$/air, 3.0 atm backpressure, and 100% RH [2]. (For color version of this figure, the reader is referred to the online version of this book.)

hydrogen oxidation reaction will form a local cell with the ORR on the cathode, which will lead to decreased fuel cell OCV [2,3] by forming a mixed potential, as the Pt oxidation reaction does. The voltage loss at OCV caused by H_2 crossover can be calculated if the H_2 crossover rate or the current density generated by the oxidation of crossover H_2 can be measured/calculated.

Hydrogen crossover current density has been measured by using a steady-state electrochemical method, as addressed in Chapter 3 and in our previous work [2], during which the fuel cell cathode was used as the working electrode, flushed with nitrogen to remove O_2, and then set up at a potential of 0.5 V relative to the H_2-flushed anode, which was used as the reference electrode and counter-electrode. At this potential (vs. the hydrogen reference electrode), all H_2 atoms that cross over from the anode to the cathode will be oxidized, which enables one to measure the hydrogen crossover current density. The measured current densities generated due to the electrochemical oxidation of hydrogen that crossed the membrane are plotted in Fig. 7.2 as a function of temperature. As shown there, for both the Nafion®-112- and the Nafion®-117-membrane-based MEAs, the hydrogen crossover current density increases with increasing temperature.

According to the Butler–Volmer equation, the cathode kinetic current density can be expressed as in Eqn (7.7):

$$I_c = i^o_{O_2/H_2O}\left(e^{\frac{n_{\alpha O-Pt/PtO}\,\alpha^o_{O-Pt/PtO}\,TF\eta_c}{RT}} - e^{-\frac{n_{\alpha O-Pt/PtO}(1-\alpha^o_{O-Pt/PtO})T)F\eta_c}{RT}}\right) \quad (7.7)$$

where I_c is the cathode current density; $n_{\alpha O-Pt/PtO}\,(=2)$ is the electron transfer number for the rate-determining step of the ORR on a Pt/PtO electrode surface; $i^o_{O_2/H_2O}$ is the apparent exchange current density for the cathode O_2

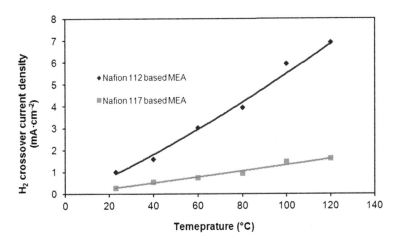

FIGURE 7.2 H_2 crossover current densities as a function of temperature at OCV with different MEAs. Operating conditions: 3.0 atm backpressure, 100% RH [2]. (For color version of this figure, the reader is referred to the online version of this book.)

TABLE 7.3 Apparent Exchange Current Densities for the ORR, $i^o_{O_2/H_2O'}$ at Different Temperatures, 3.0 atm, and 100% RH [2]

Temperature (°C)	23	40	60	80	100	120
$i^o_{O_2/H_2O}$ O_2 reduction (A cm^{-2})	1.2×10^{-4}	2.4×10^{-4}	3.9×10^{-4}	4.6×10^{-4}	3.8×10^{-4}	2.2×10^{-4}

reduction (Table 7.3), which can be calculated using Eqn (1.61) by measuring the AC impedance at OCV; and $\alpha^o_{O-Pt/PtO} = 0.00168 \text{ K}^{-1}$ in the temperature range of 25–250 °C.

The cathodic potential loss caused by H$_2$ crossover, $\Delta E^{OCV}_{H_2-x\text{-over}}$, can be calculated according to Eqn (7.7) by taking the H$_2$ crossover current density to be I_c. In this calculation, it is assumed that the crossover H$_2$ could react with O$_2$ to produce a corresponding cathode current density of the same magnitude, resulting in a cathode potential depression. The calculated values of $\Delta E^{OCV}_{H_2-x\text{-over}}$ are listed in Table 7.2. As shown there, the value of $\Delta E^{OCV}_{H_2-x\text{-over}}$ increased with increasing temperature. For example, the value increased from 54 mV at 23 °C to 88 mV at 120 °C for a Nafion®-112-membrane-based MEA at 3.0 atm pressure and 100% RH. Table 7.2 also shows that the voltage drop caused by hydrogen crossover is always smaller than that caused by mixed potential, across the whole temperature range from 23 to 120 °C; this indicates that mixed potential plays a more dominant role than does hydrogen crossover.

Membrane thickness can affect the fuel cell OCV by influencing the hydrogen crossover rate. A thinner membrane has a higher gas permeability, which leads to a larger hydrogen crossover current density and thereby results in a larger voltage loss at OCV. As shown in Table 7.2, the values of $\Delta E^{OCV}_{H_2-x\text{-over}}$ for the MEA containing Nafion®-112-membrane (thickness ~50 μm) MEA are larger than those of the membrane containing Nafion® 117 (thickness ~175 μm), across the entire temperature range from 23 to 120 °C.

7.3.3 Temperature Effect

Temperature is an important operating parameter and can also affect the fuel cell OCV by influencing the partial pressures of H$_2$ (P_{H_2}), O$_2$ (P_{O_2}), and H$_2$O (P_{H_2O}) in the fuel and oxidant gas streams. Figure 7.3 shows the temperature-dependent partial pressures of P_{H_2}, P_{O_2}, and P_{H_2O} in the feed streams of a fuel cell operated at 100% RH and 3.0-atm backpressure with pure H$_2$ as fuel and air as oxidant. P_{H_2O} increases dramatically with increasing temperature, whereas P_{H_2} and P_{O_2} decrease quickly with increasing temperature, especially once the

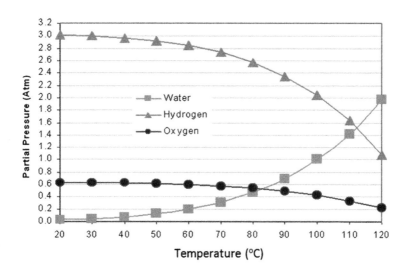

FIGURE 7.3   Partial pressures of O_2, H_2, and H_2O in fuel cell feed streams as a function of operating temperature. Operating conditions: 3.0 atm backpressure; 100% RH [2]. (For color version of this figure, the reader is referred to the online version of this book.)

temperature is $>80\,^\circ$C [2]. According to Eqn (7.4), the theoretical OCV of a PEM fuel cell will decrease when the operating temperature increases due to the increase in both temperature and P_{H_2O}, and to the decrease in both P_{H_2} and P_{O_2}.

Temperature can also affect the OCV by influencing the hydrogen crossover rate. Usually, a higher temperature results in the membrane having greater gas permeability, which can be measured using the crossover current density caused by the electrochemical oxidation of the crossover hydrogen on the cathode. Thus, a larger voltage loss in the OCV will result. This has been proven by the values of $\Delta E^{OCV}_{H_2-x\text{-over}}$ at different temperatures, as listed in Table 7.2.

The table also lists the theoretical OCV (E^{OCV}_{theory}), measured OCV $(E^{OCV}_{measured})$, corrected OCV $(E^{OCV}_{corrected})$, $\Delta E^{OCV}_{H_2-x\text{-over}}$, and $\Delta E^{OCV}_{Pt-O_2}$ values for a Nafion®-112-membrane-based MEA at a backpressure of 3.0 atm, with 100% RH, in the temperature range of 23–120 $^\circ$C. The $E^{OCV}_{corrected}$ can be expressed by Eqn (7.8):

$$E^{OCV}_{corrected} = E^{OCV}_{measured} + \Delta E^{OCV}_{Pt-O_2} + \Delta E^{OCV}_{H_2-x\text{-over}} \tag{7.8}$$

As shown in Table 7.2, the values of $E^{OCV}_{corrected}$ at all the tested temperatures are very close to the E^{OCV}_{theory} values. Figure 7.1 also shows that the values of $E^{OCV}_{corrected}$ are very close to those of E^{OCV}_{theory} across the whole temperature range. Thus, the main factors that influence the OCV of a PEM fuel cell are the sum of the hydrogen crossover, and the mixed potential caused by the formation of a local cell due to the ORR and Pt oxidation electrochemical reactions on the cathode.

7.4. APPLICATIONS OF OCV MEASUREMENT

7.4.1. OCV Measurement as a Diagnostic Method

OCV measurement is an effective method of monitoring MEA failure, especially membrane failure (e.g. via the formation of pinholes) [11]. After a PEM fuel cell is assembled, it is usual to check the OCV of the cell by flowing a little H_2 and air before practical operation. A low OCV value (e.g. <0.8 V) could indicate potential defects in the membrane, gaskets, or even bipolar plates. For an operating PEM fuel cell, a low OCV may suggest a problem in the cell, such as a broken membrane. This means that gasket failure, cracks in bipolar plates, and broken membranes are easy to detect, as they usually lead to an abnormally low OCV. However, pinholes in the membrane cannot be straightforwardly discerned during practical fuel cell operation. Lu et al. [11] developed a simple, effective method to confirm the presence of membrane pinholes by increasing the anodic pressure and measuring the resulting reduction in fuel cell OCV. They used two MEAs, one without and one with membrane pinholes, and followed a five-step testing procedure to measure the variations in the fuel cell OCVs. The testing procedure was as follows (each step lasted about 1000 s): (1) balance the anode and cathode pressure at 0.1 MPa; (2) raise the cathode pressure to 0.2 MPa; (3) restore the cathode pressure to 0.1 MPa; (4) raise the anode pressure to 0.2 MPa; and (5) decrease the anode pressure to restore it to 0.1 MPa. The changes in OCV with time during the procedure were recorded for the fuel cells with and without pinholes in their respective MEAs. As shown in Fig. 7.4(a), the OCV of the fuel cell without membrane pinholes decreases slightly (~20 mV) for both H_2/O_2 and H_2/air operation when the anode pressure is increased to 0.2 MPa from 0.1 MPa, which could be caused by the increase in hydrogen crossover due to the pressure difference between the anode and the cathode. In the case of the fuel cell with pinholes in the MEA (Fig. 7.4(b)), when the anode pressure is increased to 0.2 MPa, the fuel cell OCV decreases significantly to 200 mV for H_2/O_2 operation and 700 mV for H_2/air operation. This is because the pressure difference causes more hydrogen to penetrate the membrane from the anode to the cathode through the pinholes.

CV measurement can also help localize the position of leakage or membrane failure in a fuel cell stack during lifetime testing. Stumber et al. [12] used OCV profiling as a diagnostic tool during a 10-cell stack lifetime test. They combined OCV profile measurement with current mapping and obtained useful information regarding crossover leakage formation and location.

7.4.2. Accelerated Durability Testing at OCV

As an accelerated testing method, the OCV hold test plays an important role in accelerated durability tests [13–18]. The OCV hold test is believed to accelerate the chemical degradation of the membrane, which results in higher gas

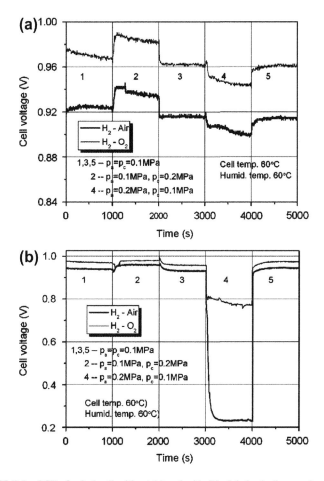

FIGURE 7.4 OCV of a fuel cell without (a) and with (b) pinholes in the membrane [11].

permeability. Recently, it has also been suggested that the operation of a PEM fuel cell at OCV for an extended time could lead to the formation of a Pt band in the membrane, due to the electrochemical deposition of Pt from the catalyst. The membrane will then further degrade within and near the Pt band location.

Ohma et al. [16] studied membrane degradation behavior during an OCV hold test and found that Pt bands were formed in their three membrane samples, as shown in Fig. 7.5. The OCV hold test conditions are listed in Table 7.4. The results indicated that the location of the Pt band was mainly determined by the gas composition at both the anode and the cathode, as well as by the gas permeability of the membrane. The formation of a Pt band enhanced membrane decomposition during the OCV hold test, which increased the fluoride ion emission rates.

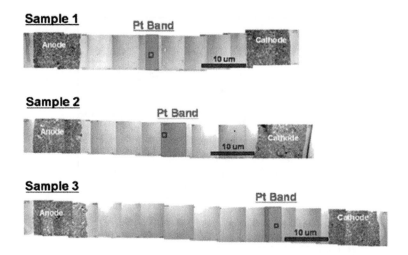

FIGURE 7.5 Transmission electron microscopy images of sample cross-sections [16]. (For color version of this figure, the reader is referred to the online version of this book.)

TABLE 7.4 OCV Hold Test Conditions [16]

	Sample 1	Sample 2	Sample 3
Anode gas	10% H_2	H_2	H_2
Cathode gas	O_2	O_2	O_2
Cell temperature (K)	363	363	363
Backpressure	Ambient	Ambient	Ambient
Anode dew point (K)	334	334	334
Cathode dew point (K)	334	334	334
Anode flow rate (N m^{-3})	0.5×10^{-3}	0.5×10^{-3}	0.5×10^{-3}
Cathode flow rate (N m^{-3})	0.5×10^{-3}	0.5×10^{-3}	0.5×10^{-3}
Duration of OCV hold (h)	110	110	110

Kundu et al. [17] also investigated the degradation of fuel cells during OCV durability testing. OCV durability testing is believed to promote chemical degradation of the electrolyte membrane material via radical attack, as well as degradation of the catalyst layer. Fuel cell degradation during OCV durability testing includes reversible and irreversible processes, as shown in Fig. 7.6. The results indicated that irreversible voltage decay at OCV was caused by

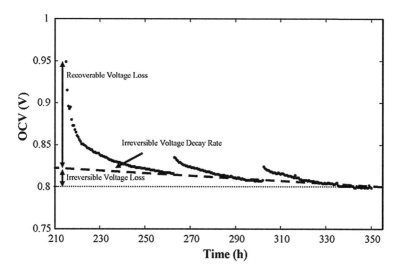

FIGURE 7.6 Reversible and irreversible degradation as well as irreversible degradation rates for a fuel cell. MEA active area: 80.1 cm^2; cell temperature: 90 °C; RH: 100% [17].

irreversible material changes, such as hydrogen crossover and loss of active electrochemical surface area.

7.5. CHAPTER SUMMARY

The OCV of a PEM fuel cell was addressed in this chapter. OCV can be decreased by increasing the operating temperature. At certain operating conditions, the OCV is mainly determined by the hydrogen crossover rate and the mixed potential formed by the electrochemical reactions of Pt surface oxidation and the ORR.

OCV measurement can be a useful diagnostic tool in fuel cell operation or stack lifetime testing. An abnormal OCV usually indicates fuel cell failure, such as a broken membrane, a gasket leak, and/or bipolar plate failure. OCV measurement can also help in obtaining information from within the fuel cell and in localizing the position of broken MEAs.

The OCV hold test can be used in accelerated fuel cell durability tests to understand the failure mechanisms that lead to fuel cell degradation at OCV.

REFERENCES

[1] Amphlett JC, Baumert RM, Mann RF, Peppley BA, Roberge PR, Harris TJ. J Electrochem Soc 1995;142:9–15.
[2] Zhang J, Tang Y, Song C, Zhang J, Wang H. J Power Sources 2006;163:532–7.
[3] Vilekar SA, Datta R. J Power Sources 2010;195:2241–7.

[4] Hoare JP. The electrochemistry of oxygen. New York: Interscience Publishers; 1968.

[5] Bockris JOM, Srinivasan S. Fuel cells: their electrochemistry. New York; 1969.

[6] Laraminie L, Dicks A. Fuel cell systems explained. Chichester, England: John Wiley; 2000.

[7] Sompalli B, Litteer BA, Gu W, Gasteiger HA. J Electrochem Soc 2007;154:B1349–57.

[8] Vielstich W, Lamm A, Gasteiger HA. In: Cleghorn S, Kolde J, Liu W, editors. Handbook of fuel cells-fundamentals, technology and applications. New York: John Wiley & Sons; 2003. p. 566.

[9] Hoare JP. J Electrochem Soc 1962;109:858–65.

[10] Thacker R, Hoare JP. J Electroanal.Chem Inter Electrochem 1971;30:1–14.

[11] Lü W, Liu Z, Wang C, Mao Z, Zhang M. Int J Energy Res 2011;35:24–30.

[12] Stumper J, Rahmani R, Fuss F. J Power Sources 2010;195:4928–34.

[13] Jao T-C, Jung G-B, Ke S-T, Chi P-H, Su A. Int J Energy Res, John Weily & Sons, Ltd 2010, http://dx.doi.org/10.1002/er.

[14] Ralph TR, Barnwell DE, Bouwman PJ, Hodgkinson AJ, Petch MI, Pollington M. J Electrochem Soc 2008;155:B411–22.

[15] Mittal VO, Kunz HR, Fenton JM. J Electrochem Soc 2006;153:A1755–9.

[16] Ohma A, Suga S, Yamamoto S, Shinohara K. J Electrochem Soc 2007;154:B757–60.

[17] Kundu S, Fowler M, Simon LC, Abouatallah R. J Power Sources 2008;182:254–8.

[18] Yoon W, Huang X. J Electrochem Soc 2010;157:B599–606.

Relative Humidity (RH) Effects on PEM Fuel Cells

8.1. INTRODUCTION

Conventional proton exchange membrane (PEM) fuel cells (typically operated at $<90\ °C$) make use of a perfluorosulfonic acid (PFSA) membrane (e.g. Nafion membrane) as the PEM. A certain level of relative humidity (RH), typically near saturation ($>80\%$ RH) is required to achieve high PEM fuel cell

PEM Fuel Cell Testing and Diagnosis. http://dx.doi.org/10.1016/B978-0-444-53688-4.00008-5

performance because the proton conductivity of PFSA membranes depends on their water content. Therefore, the RH is one of the important factors to consider in PEM fuel cell performance. Both modeling and experimental results have shown that the RH can influence PEM fuel cell performance by affecting the proton conductivity of membranes [1–9], the proton activity in the catalyst layers [4,6–8,10], electrode reaction kinetics [5,8,9,11], and the mass transfer process [4–6,8,11].

As a PEM fuel cell is a complicated system, the effect of RH is also related to other operating conditions, such as temperature, pressure, flow field design. This chapter will address the effect of RH on fuel cell performance through theoretical analysis and typical experimental examples.

8.2. DEFINITION OF RELATIVE HUMIDITY

8.2.1. General Concept of Relative Humidity

Humidity is a parameter that helps describe the water content in a gas and water mixture. There are many different methods to express humidity: absolute humidity, RH, dew point temperature, or mixing ratio. In PEM fuel cell technology, RH is commonly used. The RH is a measure of the amount of water vapor in the air relative to the maximum amount of water vapor that air can hold at a specific temperature, and is usually expressed as a percentage value. The RH at temperature T ($RH(T)$) is defined as the ratio of the partial pressure and the saturated vapor pressure of water, and can be expressed as in Eqn (8.1) [12]:

$$RH(T) = \frac{P_{H_2O}(T)}{P^o_{H_2O}(T)} \times 100 \qquad (8.1)$$

Here, $P_{H_2O}(T)$ is the partial pressure of water at temperature T and $P^o_{H_2O}(T)$ is the saturated vapor pressure above water at temperature T. Actually, the saturated vapor pressure is the partial pressure of water when a mixture of air and liquid water is at the equilibrium state, where the rate of evaporation is equal to the rate of condensation. When the air cannot hold any more water vapor, it is said to have reached a saturated state.

Figure 8.1 shows a closed chamber to explain the formation of water vapor. In equilibrium, the number of water particles leaving the surface of the liquid water is equal to that rejoining it, and the number of gaseous water particles remains statistically constant. The partial pressure of the water vapor is equal to $P^o_{H_2O}(T)$, which can be expressed by the Clapeyron Equation:

$$\frac{dP^o_{H_2O}(T)}{dT} = \frac{L_V}{T\Delta V} \qquad (8.2)$$

where L_V is the latent heat, T is the absolute temperature, and ΔV is the volume change of the phase transition. If we assume that the vapor gas is an ideal gas,

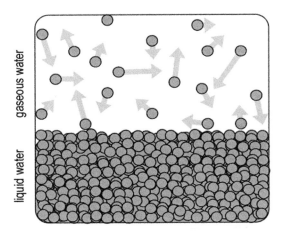

FIGURE 8.1 Explanation of the formation of water vapor in a closed water chamber. At the equilibrium state, the number of water particles (vapor) leaving the surface of liquid water equals the number of particles rejoining it. The gaseous water particles exert pressure against the wall of the chamber [1]. (For color version of this figure, the reader is referred to the online version of this book.)

we can describe $P^o_{H_2O}(T)$ as a function of temperature (t) in $°C$; this expression is known as the Magnus Equation:

$$P^o_{H_2O}(T) = \alpha\exp\left(\frac{\beta \cdot t}{\gamma + t}\right) \tag{8.3}$$

Here, $\alpha = 623.424$ Pa, $\beta = 17.62$, and $\gamma = 243.12\,°C$, and the range of t above water is -45 to $60\,°C$.

It is well known that $P^o_{H_2O}(T)$ (with a unit of atmospheres) increases with increasing temperature (with a unit of Kelvins), and this relationship can be expressed as in Eqn (8.4) [13,14]:

$$\begin{aligned}P^o_{H_2O}(T) = \; & 6.02724 \times 10^{-3} + 4.38484 \times 10^{-4}(T - 273.15) + 1.39844 \\ & \times 10^{-5}(T - 273.15)^2 + 2.71166 \times 10^{-7}(T - 273.15)^3 + 2.57731 \\ & \times 10^{-9}(T - 273.15)^4 + 2.82254 \times 10^{-11}(T - 273.15)^5\end{aligned}$$

$$\tag{8.4}$$

Figure 8.2 clearly shows the relationship between $P^o_{H_2O}(T)$ and temperature. It can be seen that $P^o_{H_2O}(T)$ does not behave linearly with temperature, but rather it increases more and more rapidly at higher temperatures.

8.2.2. Relative Humidity in PEM Fuel Cells

The water balance inside a PEM fuel cell can be strongly affected by the RHs of the inlet gases, such as H_2 and O_2. To gain a clear understanding of the RH effect, one should understand the water balance inside a fuel cell. As shown in Fig. 8.3,

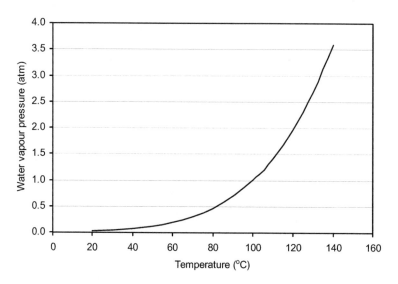

FIGURE 8.2 Calculated saturated water vapor pressure as a function of temperature [13,14]. (For color version of this figure, the reader is referred to the online version of this book.)

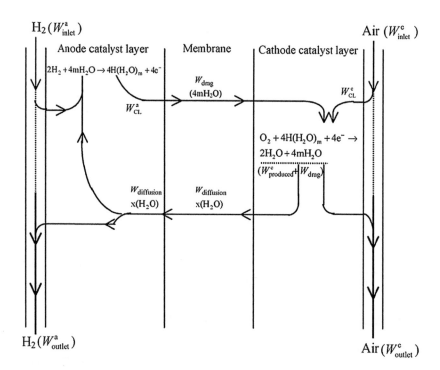

FIGURE 8.3 Schematic of water balance inside an operating PEM fuel cell [14].

in an operating PEM fuel cell, water is usually introduced into the fuel cell at both the anode and the cathode with fuel and oxidant flows, respectively, so that it reaches the anode and cathode catalyst layers through their respective flow fields and gas diffusion layers. At the interface of the anodic catalyst layer and membrane, a part of the anode water is combined with protons and electro-osmotically dragged into the cathode catalyst layer through the membrane. This dragged water can be expressed as W_{drag}. At the cathode catalyst layer, the oxygen reduction reaction (ORR) takes place, and this produces some water and releases W_{drag}. If the RHs at the anode inlet and the cathode inlet are controlled at the same level, some of the water will diffuse back to the anode from the cathode through the membrane because of the water concentration gradient between the anode and the cathode. At the anode, the remaining water (including the backdiffused water and part of the humidified water) is drained out by the exhaust fuel gas (H_2). Similarly, the residual water (including the produced water, the osmotically dragged water, and part of the humidified water) at the cathode side is drained out by the exhaust oxidant gas (O_2 or air). The water balance inside an operating PEM fuel cell is shown in Fig. 8.3, where W_{inlet}^a and W_{inlet}^c represent the water that flows into the fuel cell at the anode and the cathode, respectively; W_{outlet}^a and W_{outlet}^c represent the water that is drained out of the fuel cell at the anode and the cathode, respectively; W_{CL}^a and W_{CL}^c represent the water in the anode and the cathode catalyst layers, respectively; $W_{produced}^c$ stands for the water produced by the ORR at the cathode catalyst layer; W_{drag} indicates the water osmotically dragged from the anode to the cathode; and $W_{diffusion}$ indicates the water that diffused back from the cathode to the anode.

The concept of RH in a PEM fuel cell stream can be defined as follows. The RHs at the cathode (RH_c^{inlet} and RH_c^{outlet}) and the anode (RH_a^{inlet} and RH_a^{outlet}) are the percentages of corresponding water vapor partial pressure in the cathode stream ($P_{H_2O}^{c-inlet}$ and $P_{H_2O}^{c-outlet}$) and anode stream ($P_{H_2O}^{a-inlet}$ and $P_{H_2O}^{a-outlet}$), divided by the saturated water vapor pressure ($P_{H_2O}^o(T)$) at the fuel cell operating temperature T. RH_c^{inlet}, RH_c^{outlet}, RH_a^{inlet}, and RH_a^{outlet} can be expressed using Eqns (8.5)–(8.8), respectively [14]:

$$RH_c^{inlet} = \frac{P_{H_2O}^{c-inlet}}{P_{H_2O}^o(T)} \times 100\% \quad \text{(cathode inlet)} \tag{8.5}$$

$$RH_c^{outlet} = \frac{P_{H_2O}^{c-outlet}}{P_{H_2O}^o(T)} \times 100\% \quad \text{(cathode outlet)} \tag{8.6}$$

$$RH_a^{inlet} = \frac{P_{H_2O}^{a-inlet}}{P_{H_2O}^o(T)} \times 100\% \quad \text{(anode inlet)} \tag{8.7}$$

$$RH_a^{outlet} = \frac{P_{H_2O}^{a-outlet}}{P_{H_2O}^o(T)} \times 100\% \quad \text{(anode outlet)} \tag{8.8}$$

8.2.3. Distributions of RH in PEM Fuel Cells

From Fig. 8.3, it can be deduced that at a steady state, the water balance inside the cathode and the anode of an operating PEM fuel cell can be expressed as in Eqns (8.9) and (8.10), respectively [14]:

$$n_{c-H_2O}^{outlet} = n_{c-H_2O}^{inlet} + n_{c-H_2O}^{p} + n_{H_2O}^{o} - n_{c-H_2O}^{b} \quad \text{(cathode)} \qquad (8.9)$$

$$n_{a-H_2O}^{outlet} = n_{a-H_2O}^{inlet} - n_{H_2O}^{o} + n_{c-H_2O}^{b} \quad \text{(anode)} \qquad (8.10)$$

where $n_{c-H_2O}^{outlet}$ and $n_{a-H_2O}^{outlet}$ are the water draining rates (here the rate is the mole rate with a unit of mol s^{-1}) in the outlet streams at the cathode and the anode; $n_{c-H_2O}^{inlet}$ and $n_{a-H_2O}^{inlet}$ are the water feeding rates in the inlet streams at the cathode and the anode, respectively; $n_{c-H_2O}^{p}$ is the water production rate by the ORR at the cathode; $n_{H_2O}^{o}$ is the water osmotic-dragging rate by protons from the anode to the cathode; and $n_{c-H_2O}^{b}$ is the water backdiffusing rate from the cathode to the anode.

For a PEM fuel cell that is operated under a load of I (current, with a unit of A), the relationship between RH$_c^{inlet}$ and RH$_c^{outlet}$ can be expressed as in Eqn (8.11):

$$\text{RH}_c^{outlet} = \text{RH}_c^{inlet} \left(\frac{V_{c-H_2O}^{inlet}}{V_{c-H_2O}^{outlet}} \right) + \frac{RT}{2FP_{H_2O}^{o}(T)V_{c-H_2O}^{outlet}} \left[(1+2m)I - 2Fn_{c-H_2O}^{b} \right]$$

$$(8.11)$$

Similarly, the relationship between RH$_a^{inlet}$ and RH$_a^{outlet}$ can be expressed as in Eqn (8.12):

$$\text{RH}_a^{outlet} = \text{RH}_a^{inlet} \left(\frac{V_{a-H_2O}^{inlet}}{V_{a-H_2O}^{outlet}} \right) + \frac{RT}{FP_{H_2O}^{o}(T)V_{a-H_2O}^{outlet}} \left(Fn_{c-H_2O}^{b} - mI \right) \qquad (8.12)$$

In Eqns (8.11) and (8.12), $V_{c-H_2O}^{inlet}$ and $V_{c-H_2O}^{outlet}$ are the cathode water flow rates in the inlet and outlet streams, respectively, with a unit of L s^{-1}; $V_{a-H_2O}^{inlet}$ and $V_{a-H_2O}^{outlet}$ are the anode water flow rates in the inlet and outlet streams, respectively, with a unit of L s^{-1}; I is the fuel cell current (with a unit of A); R is the gas constant (0.08,206 L atm K^{-1} mol^{-1}); T is the temperature (with a unit of K); 2 is the electron transfer number for each H_2O molecule produced by a 4-electron ORR; F is the Faraday constant (96,487 A s mol^{-1}); and m is the amount of water osmotically dragged per proton transfer from the anode to the cathode (m is also called the osmotic drag coefficient). For a water-vapor-equilibrated system, the value of m is likely to be a constant number over a broad range of water content values, from a nearly dry membrane (~2H_2O per SO$_3^-$) to a fully hydrated membrane (with water vapor ~14H_2O per SO$_3^-$). The values reported for this constant number vary from 1.0 to 1.4 [15,16]. Equations (8.11) and (8.12) are based on the assumptions that air, H_2, and water vapor obey the ideal gas law, and the water produced by the ORR and osmotically dragged by protons can be totally evaporated and removed by the outlet stream [14].

For an operating PEM fuel cell, $V_{c-H_2O}^{outlet}$ and $V_{a-H_2O}^{outlet}$ can easily be calculated by collecting the water for a given time at the fuel cell cathode and anode outlets. Then, $V_{c-H_2O}^{inlet}$ and $V_{a-H_2O}^{inlet}$ can be determined, as the controlled fuel and oxidant flow rates and their respective RHs (RH_a^{inlet} and RH_c^{inlet}) at the anode and cathode inlets are known. Then, $n_{c-H_2O}^b$ at a certain current can be calculated according to Eqns (8.9) and (8.10). Thus, RH_a^{outlet} and RH_c^{outlet} can be calculated according to Eqns (8.11) and (8.12).

8.3. HUMIDIFICATION METHODS IN PEM FUEL CELLS

As discussed in Chapter 5, PEM fuel cells widely use PFSA membranes, whose proton conductivity strongly depends on their water content. To achieve high membrane proton conductivity and good PEM fuel cell performance, it is necessary to add water to fuel cell systems to maintain a sufficient membrane hydration level. Water is often added externally with the reactant gases at the anode and the cathode. So far, several humidification methods, such as bubble humidification and direct liquid water injection, have been developed for PEM fuel cells.

In a laboratory test system, the reactant gas is usually humidified by passing it through a water bath in a bubble humidifier that is controlled at a desired temperature. This process assumes that the dew point of the air is the same as the temperature of the bubble humidifier. This temperature can be controlled independently of the cell temperature to achieve the desired gas RH. Water vapor is absorbed by the gas, and water uptake is a function of the water–gas interfacial area. This is a conventional humidification method and is widely used in small-scale laboratory fuel cells due to its simplicity and low cost, but it is not very practical to be used in large-scale stacks. The main disadvantages of the method are its limited water transfer capacity and slow response to the changes in the RH level.

Direct liquid water injection [17] is another method for humidifying the reactant gases. In this method, water is injected by a pump into two heated stainless steel coils, where it is preheated to the cell operating temperature. This method involves the use of pumps to pressurize the water, as well as a solenoid valve to open and close the injector, which increases the cost and complexity of the fuel cell system. Another concern is electrode "flooding" by the liquid water.

Direct vapor injection is another easy way to humidify the reactant gases in PEM fuel cell systems. Vapor is directly introduced into the reactant gases. Thermal energy in the vapor can heat the reactant gas to the desired temperature to entrain all of the water vapor. A temperature-controlled cooler is required to lower the gas temperature and condense the excess water to obtain humidified reactant gas for the PEM fuel cell. This method provides high water transfer capacity and is suitable for large-scale PEM fuel cell systems.

Rather than introducing water into the fuel cell system, the membrane can also be humidified directly by internal methods. Wicks [18,19] are usually incorporated into the structure of the gas diffusion layer or flow field plate. These wicks adsorb water and draw it to the membrane. In this method, the heat produced in the fuel cell can be used as an energy source for vaporizing the water. However, the incorporated wicks may cause difficulty in fuel cell sealing. Alternatively, instead of introducing water to humidify the membrane, self-humidifying membranes [20–24] are also used in PEM fuel cells. Catalyst particles are often embedded in the self-humidifying membrane, and they catalyze the reaction between oxygen and crossover hydrogen to produce water to humidify the membrane.

8.4. EFFECT OF RH ON FUEL CELL REACTION KINETICS

In PEM fuel cells, the following parameters are typically used to evaluate the fuel cell reaction kinetics: open circuit voltage (OCV), Tafel slope, and exchange current density. The OCV is related to thermodynamics and the reactant gas partial pressures. The hydrogen crossover and mixed potential caused by the ORR and Pt/Pt–O coupling can reduce fuel cell OCV by 200–300 mV [25,26], as has been discussed in Chapter 7. The Tafel slope is used to measure how much activation polarization is required to reach a given reaction rate. The smaller the Tafel slope, the better the electrode performance. The exchange current density is the rate of the reaction occurring forward and backward at the reversible potential, and it measures the readiness of the electrochemical reaction. The higher the exchange current density, the lower will be the energy barrier limiting the movement of charge from the electrolyte to the catalyst surface, and vice versa. Also, the higher the exchange current density, the smaller will be the activation loss at a given net current density, and this results in a higher fuel cell performance.

8.4.1. Analysis of Fuel Cell Reaction Kinetics

The measured fuel cell voltage (V_{cell}) can be expressed as in Eqn (8.13):

$$V_{cell} = E^{OCV} - \eta_c - \eta_a - \eta_m \qquad (8.13)$$

where η_c and η_a are the overpotentials at the cathode and the anode, respectively, and η_m is the membrane resistance polarization, which dominates the ohmic polarization of PEM fuel cells. η_c, and η_a can be defined by the Butler–Volmer and mass diffusion theories [27]. In general, the reaction kinetics in PEM fuel cells is dominated by the cathode electrochemical reaction, due to the sluggish electrode kinetics of the ORR. Therefore, the value of η_a is much smaller than that of η_c, and can be omitted from Eqn (8.13). In a very low polarization range ($\eta_c < 30$ mV), Eqn (8.13) can be approximately expressed as Eqn (8.14):

$$V_{cell} = E^{OCV} - \frac{RT}{n_{\alpha o}F i^o_{O_2}} I_{cell} - I_{cell}R_m \tag{8.14}$$

In a high polarization range ($\eta_c > 30$ mV), Eqn (8.13) can be approximately expressed as Eqn (8.15):

$$V_{cell} = E^{OCV} + \frac{RT}{\alpha_o n_{\alpha o}F}\ln(i^o_{O_2}) - \frac{RT}{\alpha_o n_{\alpha o}F}\ln(I_{cell}) - \frac{RT}{\alpha_o n_{\alpha o}F}\ln\left(\frac{I_{cell}I^f_{dc}}{I^f_{dc} - I_{cell}}\right)$$
$$- I_{cell}R_m$$
$$\tag{8.15}$$

In Eqns (8.14) and (8.15), α_o is the charge transfer coefficient for the ORR and is related to the temperature and RH; $n_{\alpha o}$ is the electron transfer number in the rate-determining step of the ORR; $i^o_{O_2}$ is the apparent exchange current density (AECD) of the ORR; I^f_{dc} is the apparent diffusion limiting current density for the ORR; and R_m is the measured membrane resistance. In Eqns (8.14) and (8.15), E^{OCV} is a fuel cell thermodynamic term. The second right-hand-side element in Eqn (8.14) and the second and third right-hand-side elements in Eqn (8.15) are kinetic terms, the fourth one in Eqn (8.15) is the term determined by mass transfer, and the last term in both equations is related to the membrane resistance. By differentiating Eqn (8.15) against fuel cell current density (I_{cell}), Eqn (8.16) can be obtained:

$$-\frac{\partial V_{cell}}{\partial I_{cell}} = \frac{RT}{\alpha_o n_{\alpha o}F}\frac{1}{I_{cell}} + \frac{RT}{\alpha_o n_{\alpha o}F}\frac{I^f_{dc}}{I_{cell}(I^f_{dc} - I_{cell})} + R_m \tag{8.16}$$

In Chapter 3, electrochemical impedance spectroscopy (EIS) was introduced as a powerful technique for PEM fuel cell diagnosis. EIS measurement can be conducted at OCV and under load. The AECD of the ORR ($i^o_{O_2}$) can be calculated using Eqn (3.8), based on the simulated $R^{OCV}_{ct-O_2}$ (charge transfer resistance at the OCV for the ORR) from the Nyquist plot obtained by EIS that is shown in Fig. 3.12. The values of the membrane resistance (R_m), charge transfer resistance (R_t) and mass transfer resistance (R_{mt}) in a PEM fuel cell at different current densities can also be simulated using measured EIS, based on the equivalent circuit shown in Fig. 3.11. In Eqn (8.16), $-\dfrac{\partial V_{cell}}{\partial I_{cell}}$ represents the total fuel cell AC impedance. The first right-hand-side term in Eqn (8.16) represents R_t $\left\{R_t = \dfrac{RT}{\alpha_o n_{\alpha o}F}\dfrac{1}{I_{cell}}\right\},$ the   second   term   represents

$$R_{mt} \left\{ R_{mt} = \frac{RT}{\alpha_o n_{\alpha o} F} \frac{I_{dc}^f}{I_{cell}(I_{dc}^f - I_{cell})} \right\}, \text{ and the third term represents the}$$

membrane resistance (R_m).

As an important operating condition that influences fuel cell performance, RH should affect the parameters shown in Eqns (8.14) and (8.15), specifically the values of E^{OCV}, $i_{O_2}^o$, I_{dc}^f, R_m, R_t, and R_{mt}. The following sections will discuss these issues in detail.

8.4.2. Effect of RH on the Partial Pressures of Reactant Gases in the Feed Streams

For an operating PEM fuel cell, the O_2 partial pressure ($P_{O_2}^{c\text{-}inlet}$) in the cathode inlet feed stream can be expressed as in Eqn (8.17):

$$P_{O_2}^{c\text{-}inlet} = \frac{1}{4.76} \left(P_c^{inlet} - P_{H_2O}^{c\text{-}inlet} \right) = \frac{1}{4.76} \left(P_c^{inlet} - P_{H_2O}^o RH_c^{inlet} \right) \quad (8.17)$$

and the H_2 partial pressure ($P_{H_2}^{a\text{-}inlet}$) in the anode inlet feed stream can be expressed as in Eqn (8.18):

$$P_{H_2}^{a\text{-}inlet} = \left(P_a^{inlet} - P_{H_2O}^{a\text{-}inlet} \right) = \left(P_a^{inlet} - P_{H_2O}^o RH_a^{inlet} \right) \quad (8.18)$$

Similarly, the partial pressures of O_2 and H_2 in the cathode and anode outlets can be expressed as in Eqns (8.19) and (8.20), respectively:

$$P_{O_2}^{c\text{-}outlet} = \frac{1}{4.76} \left(P_c^{outlet} - P_{H_2O}^{c\text{-}outlet} \right) = \frac{1}{4.76} \left(P_c^{outlet} - P_{H_2O}^o RH_c^{outlet} \right) \quad (8.19)$$

$$P_{H_2}^{a\text{-}outlet} = \left(P_a^{outlet} - P_{H_2O}^{a\text{-}outlet} \right) = \left(P_a^{outlet} - P_{H_2O}^o RH_a^{outlet} \right) \quad (8.20)$$

where $P_{H_2O}^{c\text{-}inlet}$, $P_{H_2O}^{c\text{-}outlet}$, $P_{H_2O}^{a\text{-}inlet}$, and $P_{H_2O}^{a\text{-}outlet}$ are the water partial pressures in the gas streams of the fuel cell inlet and outlet at the cathode and the anode, respectively, and P_c^{inlet}, P_c^{outlet}, P_a^{inlet}, and P_a^{outlet} are the gas pressures of the fuel cell inlet and outlet at the cathode and the anode, respectively, and can be monitored by their respective pressure sensors. Equations (8.17)–(8.20) clearly show that the reactant partial pressures at the fuel cell inlets are different from those at the fuel cell outlets. Therefore, the average partial pressures, $\overline{P}_{O_2}^{c\text{-}average}$ for O_2 and $\overline{P}_{H_2}^{a\text{-}average}$ for H_2, expressed by Eqns (8.21) and (8.22), respectively, should be more useful in appropriately describing fuel cell partial pressures.

$$\overline{P}_{O_2}^{c\text{-}average} = \frac{P_{O_2}^{c\text{-}inlet} + P_{O_2}^{c\text{-}outlet}}{2} \quad (8.21)$$

$$\overline{P}_{H_2}^{a\text{-}average} = \frac{P_{H_2}^{a\text{-}inlet} + P_{H_2}^{a\text{-}outlet}}{2} \quad (8.22)$$

In practical terms, the inlet partial pressure can be used instead of the average partial pressure to describe the fuel cell reactant partial pressure if the pressure drop from the fuel cell inlet to the outlet is small [14]. This often happens in the study of a small single cell with high reactant gas flow rates.

Figure 8.4 shows the O_2 and H_2 inlet partial pressures as a function of inlet RHs on a Nafion®-112-membrane-based PEM fuel cell operated at 120 °C and 1.0 atm backpressure. It can be seen that the values of both $P_{O_2}^{c-inlet}$ and $P_{H_2}^{a-inlet}$ decrease as the inlet RH increases. This is because $P_{H_2O}^{c-inlet}$ and $P_{H_2O}^{a-inlet}$ increase dramatically when the inlet RH increases. As shown in Fig. 8.4, the inlet water partial pressure increases linearly in the RH range of 25–100%. From Eqns (8.21) and (8.22), we can deduce that the values of both $\overline{P}_{O_2}^{c-average}$ and $\overline{P}_{H_2}^{a-average}$ decrease with increasing RH.

8.4.3. Effect of RH on Fuel Cell Reaction Thermodynamics

The OCV of an H_2/O_2 PEM fuel cell can be thermodynamically expressed by using Eqn (8.23) [14]:

$$E^{OCV} = E^0 + \frac{RT}{2F}\left[\ln\left(\overline{P}_{H_2}^{a-average}\right) + \frac{1}{2}\ln\left(\overline{P}_{O_2}^{c-average}\right)\right] \qquad (8.23)$$

where E^{OCV} is the fuel cell OCV and E^0 is the fuel cell voltage at standard conditions. This equation is based on the assumption that the water activity equals one at the interface between the catalyst layer and the membrane. For

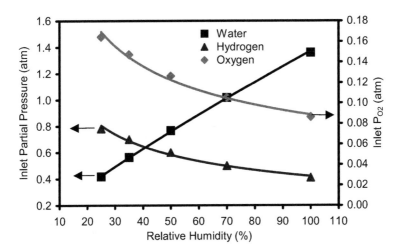

FIGURE 8.4 Inlet partial pressures of O_2 and H_2 in the fuel cell feed streams as a function of inlet relativity humidity at 120 °C and 1.0 atm backpressure. Nafion®-112-membrane-based membrane electrode assembly (MEA) with an active area of 4.4 cm^2 [14]. (For color version of this figure, the reader is referred to the online version of this book.)

a single cell with a small pressure drop between the fuel cell inlet and outlet, $\overline{P}_{O_2}^{c\text{-average}}$ and $\overline{P}_{H_2}^{a\text{-average}}$ can be replaced with $P_{O_2}^{c\text{-inlet}}$ and $P_{H_2}^{a\text{-inlet}}$. According to Eqn (8.23), it can be deduced that the fuel cell OCV decreases when both $\overline{P}_{O_2}^{c\text{-average}}$ and $\overline{P}_{H_2}^{a\text{-average}}$ decrease. For example, when the fuel cell inlet RH increases from 25 to 100% at 120 °C and 1.0 atm backpressure, $\overline{P}_{O_2}^{c\text{-average}}$ will be reduced from 0.164 to 0.086 atm, and $\overline{P}_{H_2}^{a\text{-average}}$ from 0.783 to 0.411 atm. Based on Eqn (8.23), the fuel cell OCV will decrease by approximately 16.4 mV, which will decrease the fuel cell performance.

8.4.4. Effect of RH on the Exchange Current Density of the ORR

The parameter $i_{O_2}^o$ in Eqns (8.14) and (8.15) represents the AECD for the ORR in a PEM fuel cell and is related to the temperature, the RH, $\overline{P}_{O_2}^{c\text{-average}}$, the cathode electrochemical Pt surface area ((EPSA)$_c$, with a unit of $cm^2\,cm^{-2}$, which implies the Pt active surface area per cm^2 of the electrode), and the intrinsic exchange current density ($I_{O_2(Pt)}^o$). The relationship can be described using Eqn (8.24) [14,27]:

$$i_{O_2}^o = (EPSA)_c I_{O_2(Pt)}^o \left(\frac{\overline{P}_{O_2}^{c\text{-average}}}{P_{O_2}^o} \right)^{(0.001552\overline{RH}_c^{average} + 0.000139)T} \tag{8.24}$$

In Eqn (8.24), (EPSA)$_c$ can be measured by means of cyclic voltammetry, as discussed in Section 3.5.1 of Chapter 3; $\overline{P}_{O_2}^{c\text{-average}}$ can be obtained according to Eqn (8.21); and $i_{O_2}^o$ can be experimentally obtained, as stated in Section 8.4.1 of this Chapter. Then, $I_{O_2(Pt)}^o$ can be calculated based on Eqn (8.24); the results are given in Table 8.1.

TABLE 8.1 Measured and Simulated Kinetic Parameters at 120 °C and 1.0 atm Backpressure with Different Inlet RHs. Nafion®-112-Based MEA. $P_{O_2}^{inlet}$ and RH_c^{inlet} were Treated as being Approximately Equivalent to $\overline{P}_{O_2}^{c\text{-average}}$ and $\overline{RH}_c^{inlet}$, and Were Used for Calculation [14]

RH (%)	$i_{O_2}^o$ (A cm^{-2})	EPSA (cm^2 cm^{-2})	$\overline{P}_{O_2}^{c\text{-average}}$ (atm)	$I_{O_2(Pt)}^o$ (A cm^{-2})
100	8.6×10^{-4}	53.9	0.08624	7.97×10^{-5}
70	6.3×10^{-4}	49.8	0.10502	5.51×10^{-5}
50	5.2×10^{-4}	46.4	0.12628	2.22×10^{-5}
35	4.9×10^{-4}	42.9	0.14716	1.93×10^{-5}
25	3.9×10^{-4}	36.0	0.16435	1.59×10^{-5}

As shown in the table, $I^o_{O_2(Pt)}$ increases with increasing RH, which suggests that an increase in the RH will speed up the electrode kinetics of the ORR in PEM fuel cells. This observed trend is consistent with other reported results [4–6,8–11,28,29]. $(EPSA)_c$ also increases with increasing RH (Table 8.1). These results can be attributed to the restructuring of the ionomer surface, as suggested by Uribe et al. [29]. The ionic clusters and channels will shrink at a low RH level, and the ionic channels can even collapse at a very low RH. As a result, some of the hydrophobic components in the ionomer structures come into direct contact with the Pt surface at a low RH level [14]. It has also been reported that a hydrophobic interfacial configuration can apparently lead to poor ORR reactivity [29].

8.4.5. Effect of RH on the Tafel Slope of the ORR

In the low current density range of an operating PEM fuel cell, the Tafel equation is used to describe the voltage loss caused by activation polarization and is expressed as follows:

$$\eta_{act} = a + b\log(i) \tag{8.25}$$

$$a = -2.303\frac{R \cdot T}{\alpha_o \cdot n \cdot F}\log(i^o) \tag{8.26}$$

$$b = 2.303\frac{R \cdot T}{\alpha_o \cdot n \cdot F} \tag{8.27}$$

Here, b is the Tafel slope, with a unit of $V\,(dec)^{-1}$; R is the ideal gas constant $(8.314\,J\,mol^{-1}\,K^{-1})$; T is the absolute temperature (K); F is the Faraday constant; α_o is the charge transfer coefficient for the ORR and is related to temperature and RH; and n is the electron transfer number in the rate-determining step of the ORR. It has been reported [28,30] that the RH of the reactant gas affects the Tafel slope.

Xu et al. [4,28] investigated the effect of RH on electrode kinetics at 120 °C with Nafion®-112-membrane-based PEM fuel cells. Figure 8.5 shows the influences of RH on the IR-corrected cell voltage in a low current density range $(<100\,mA\,cm^{-2})$.

The Tafel slopes for the ORR under different operating conditions can be calculated based on the data shown in Fig. 8.6. The respective calculated Tafel slopes are $131\,mV\,(dec)^{-1}$ at 20% RH, $105\,mV\,(dec)^{-1}$ at 35% RH, $82\,mV\,(dec)^{-1}$ at 50% RH, $75\,mV\,(dec)^{-1}$ at 72% RH, and $72\,mV\,(dec)^{-1}$ at 100% RH. All these values deviate from the theoretical value, calculated based on $2.303RT/F$. The authors attributed the deviation to a decreased catalyst activity and/or the effect of anode overpotential, and even to a change in the ORR mechanism [4,28].

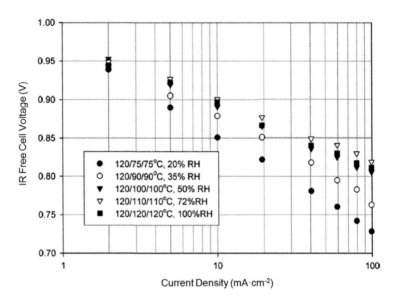

FIGURE 8.5 IR-corrected cell voltage at 120 °C in a low current density range (<100 mA cm^{-2}) at different RHs. Both H_2 and O_2 partial pressures were kept at 1.0 atm. The flow rate was kept constant at 100 ml min^{-1} for both the gases [4,28].

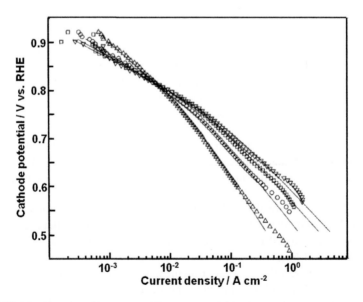

FIGURE 8.6 The effect of RH on ORR kinetics. $RH_a = RH_c = 34\%$ (Δ), $RH_a = RH_c = 60\%$ (O), $RH_a = RH_c = 78\%$ ($\square$), $RH_a = RH_c = 90\%$(∇); $P_a = P_C = 1$ atm; pure H_2 and O_2 [30].

Ihonen et al. [30] established a one-dimensional, steady-state agglomerate model to investigate the effects of gas humidity on the Tafel slope in both low and high current density regions. As shown in Fig. 8.6, two Tafel slopes appear in the entire current density region in the RH range of 34–90%. The obtained Tafel slopes are given in Table 8.2. The value of the kinetic Tafel slope varies from 101 mV (dec)$^{-1}$ at 34% RH to 67 mV (dec)$^{-1}$ at 90% RH. The ratio of the first slope to the second slope is between 1.7 and 1.8. The appearance of two Tafel slopes indicates that the cathode reaction could be limited by the ORR kinetics and by either proton migration or oxygen diffusion in the agglomerates. The value of the second Tafel slope also decreases with increasing RH. Meanwhile, the oxygen diffusion coefficient (Table 8.2) in the Nafion® film increases with increasing RH. It has been concluded that the second Tafel slope might be caused by oxygen diffusion being limited in the agglomerates.

8.4.6. Effect of RH on Charge Transfer Resistance

As discussed in Section 8.4.1, the charge transfer resistance (R_t), expressed by the first right-hand term in Eqn (8.16), can be simulated from the AC impedance spectra based on the equivalent circuit shown in Fig. 3.11 of Chapter 3. Figure 8.7 shows the simulated R_t as a function of current density at 120 °C and 1.0 atm backpressure with different cathode gas inlet RHs. It can be seen that R_t decreases with increasing RH, which indicates that the fuel cell reaction kinetics at higher RHs is faster than at lower RHs. This is consistent with the trend in the dependence of $i_{O_2}^o$ and $I_{O_2(Pt)}^o$ on RH, as observed in Table 8.1.

From Fig. 8.7, it can also be seen that R_t decreases with increasing current density. This could be attributed to the increase in hydrophilic area as more water is produced by electrochemical reaction at higher current densities. Simulating the data in Figure 8.7 according to R_t can give the RH-dependent

TABLE 8.2 Kinetic and Transport Parameters as Determined by Fitting the Agglomerate Model to the Experimental Curves [30]

RH	Kinetic Tafel Slope (mV (dec)$^{-1}$)	Second Tafel Slope (mV (dec)$^{-1}$)	O$_2$ Diffusion Coefficient (mol m^{-3} s^{-1})
34%	101	180	3.0×10^{-14}
60%	79	140	5.7×10^{-14}
78%	70	125	8.0×10^{-14}
90%	67	114	13.1×10^{-14}

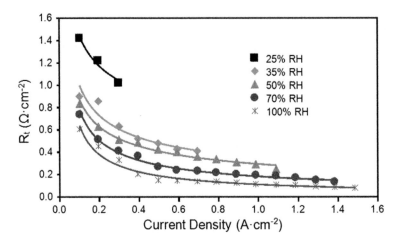

FIGURE 8.7 Charge transfer resistance as a function of current density at 120 °C and 1.0 atm backpressure, with different inlet RHs. Nafion®-112-membrane-based MEA with an active area of 4.4 cm² [14]. (For color version of this figure, the reader is referred to the online version of this book.)

values, and then an empirical equation of α_O as a function of RH at 120 °C can be deduced as in Eqn (8.28):

$$\alpha_O = 0.6099 \text{RH}_c^{\text{inlet}} + 0.0546 \tag{8.28}$$

This equation clearly shows that α_O increases with increasing RH. It is known that α_O is a function of temperature at 100% RH, and can be expressed as Eqn (8.29) [25,27]:

$$\alpha_O = 0.001678T \tag{8.29}$$

By comparing Eqns (8.28) and (8.29), we can express the dependence of α_O on temperature and RH as in Eqn (8.30) [14]:

$$\alpha_O = (0.001552 \text{RH}_c^{\text{inlet}} + 0.000139)T \tag{8.30}$$

Equations (8.30) show that α_O increases with increasing temperature and RH.

8.5. EFFECT OF RH ON MASS TRANSFER

RH affects not only the electrode kinetics of the ORR but also the mass transfer of oxygen within the catalyst layer. As shown in Fig. 8.8, the mass transfer resistance (R_{mt}) in the high current density range (>0.4 A cm^{-2}) decreases when the RH increases from 35 to 100%. This R_{mt} results mainly from oxygen gas transfer inside the catalyst layer [27,31,32]. At low RH levels, the

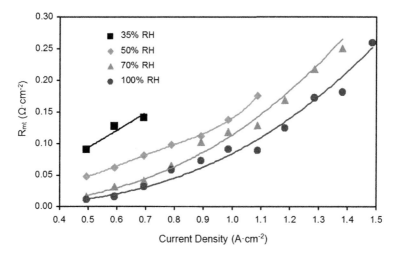

FIGURE 8.8   Mass transfer resistances in the high current density range (>0.4 A cm^{-2}) at 120 °C and 1.0 atm backpressure, with different RHs. Nafion®-112-membrane-based MEA with an active area of 4.4 cm^2 [14]. (For color version of this figure, the reader is referred to the online version of this book.)

hydrophobicity of the catalyst layer is high, which results in a reduced amount of dissolved oxygen [4] and low hydrogen and oxygen permeabilities in the ionomer of the catalyst layer [33]. This is because the gas mainly diffuses through the hydrated ion clusters and channels in the polymer, as suggested by Sakai [34].

Xu et al. [4] investigated the effect of RH on oxygen transport in a PEM fuel cell at 120 °C using a Nafion®-112-membrane-based MEA. Their results revealed that the catalyst layer structure is a crucial factor that affects reactant transport in PEM fuel cells at high temperature and low RH. At a reduced RH, the Nafion® ionomer inside the catalyst layer is dehydrated, and small Nafion® ionic clusters form in the electrode. This leads to a decreased oxygen permeability in the Nafion® ionomer and results in a severe diffusion problem. Figure 8.9 shows the polarization curves in the high current density region under different RHs. The cell voltages are corrected by both the membrane resistance and the charge transfer resistance to better reflect the effect of RH on oxygen mass transport. As shown in Fig. 8.9, the polarization curves are nearly straight lines over almost two decades at high RH levels of 100% and 72%, indicating that the influence of water flooding on oxygen transport can be negligible. As the RH decreases to 50%, 35%, and 20%, the polarization curves bend down steeply with increasing current densities. Especially at 20% and 35% RH, the cell voltage decreases significantly with current density, leading to much smaller limiting current densities at both the RHs. For example, at a current density of 600 mA cm^{-2}, the voltage difference between the

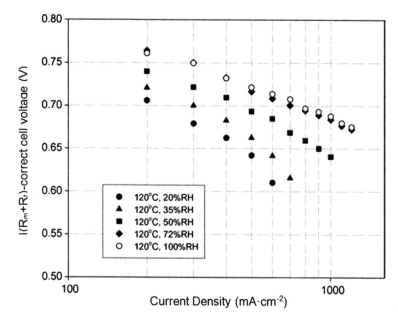

FIGURE 8.9 IR-correct polarization curves in the high current density region under different RHs at 120 °C. Both H_2 and O_2 partial pressures are controlled at 1.0 atm. The stoichiometries of H_2 and O_2 are 3 and 4, respectively [4].

polarization curve at 100% RH and that at 20% RH is about 105 mV. This large loss arises from oxygen transport.

The above two examples are for high-temperature PEM fuel cells (usually operated at >90 °C) without internal "water-flooding" issues. They indicate that the mass transfer resistance decreases with increasing RH. However, for a conventional PEM fuel cell operated at <90 °C, the fuel cell structure and design need to be considered to prevent "water flooding" at high RH levels. When "water flooding" occurs, water droplets block the flow channels on the flow field, or the passages for gas transportation, leading to a sharp increase in mass transfer resistance and a decrease in PEM fuel cell performance. Therefore, for conventional PEM fuel cells, R_{mt} may first decrease with RH, reaching its lowest value at an optimum RH level, and may then increase sharply with further increases in RH, due to "water flooding." At this optimum RH level, the best fuel cell performance will be achieved.

8.6. EFFECT OF RH ON MEMBRANE RESISTANCE

Membrane resistance (R_m) usually represents the proton transfer resistance within the membrane, and can be measured through the high-frequency interception of AC impedance spectroscopy at the real axis of the Nyquist plot, as

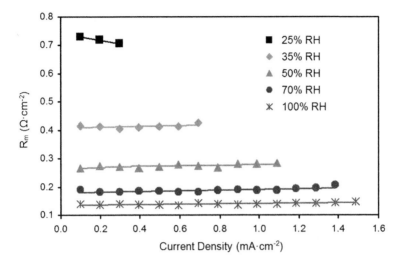

FIGURE 8.10   Membrane resistance as a function of current density at 120 °C and 1.0 atm backpressure with different RHs. Nafion®-112-membrane-based MEA with an active area of 4.4 cm² [14]. (For color version of this figure, the reader is referred to the online version of this book.)

discussed in Chapters 3 and 5. Figure 8.10 shows R_m as a function of current density at 120 °C, ambient pressure, with different RHs. It can be seen that R_m decreases significantly with increasing RH. For example, the membrane resistance at 25% RH is about five times higher than that at 100% RH. The membrane adsorbs more water at higher RHs and makes it well hydrated. Thus, more ionic clusters or channels are filled with water. It has been reported that proton mobility increases with water content [35]. Therefore, protons can get transported easily as free ions through the water-filled clusters or ionic channels inside the polymer membrane networks, resulting in a lower membrane resistance. Figure 8.10 shows that R_m is almost constant across the whole current density range studied, which is consistent with Eqn (8.16), where R_m is independent of I_{cell}.

Based on the data for R_m in Fig. 8.10, the membrane conductivities (σ, S cm^{-1}) at different current densities in the RH range studied can easily be obtained. Figure 8.11 shows the plots of membrane conductivity with RH. It can be seen that σ increases with RH. This is because the increased water content in the membrane results in an increased proton mobility. In the low RH range (20–50%), σ initially increases quickly, then it increases smoothly as RH continues to increase. This trend of σ increasing with RH shown in Fig. 8.11 was also reported by other groups [1–3,36]. Figure 8.11 also shows that the effect of operating current density on σ is negligible, which proves that R_m is independent of I_{cell}, as shown in Eqn (8.16).

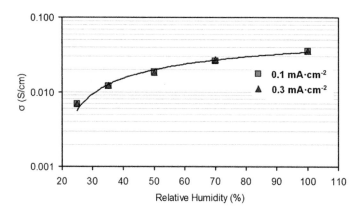

FIGURE 8.11 Membrane conductivity as a function of current density at 120 °C and 1.0 atm backpressure, with different RHs. Nafion®-112-membrane-based MEA with an active area of 4.4 cm². The thickness of the Nafion®-112 membrane was taken to be 50 μm [14]. (For color version of this figure, the reader is referred to the online version of this book.)

8.7. EFFECT OF RH ON PEM FUEL CELL PERFORMANCE

Based on the discussion above, it can be said that RH significantly influences R_m, R_t, R_{mt}, the reactant gas partial pressure, and the electrode kinetics in a PEM fuel cell, and hence can have a large effect on the overall performance of a cell. Figure 8.12 shows the fuel cell performance at 120 °C and ambient backpressure with different inlet RHs. Clearly, the performance decreases

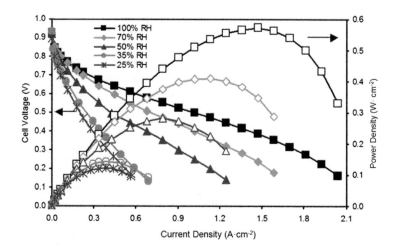

FIGURE 8.12 PEM fuel cell performance at 120 °C and 1.0 atm backpressure with different RHs. Nafion®-112-membrane-based MEA with an active area of 4.4 cm²; hydrogen and air flow rates were kept at 0.75 and 1.0 L min⁻¹, respectively; the anode and cathode inlet RHs were kept constant [14]. (For color version of this figure, the reader is referred to the online version of this book.)

dramatically with decreasing RHs. For example, at a current density of 0.34 A cm^{-2}, the cell voltages are 0.675 V (100% RH), 0.642 V (70% RH), 0.556 V (50% RH), 0.414 V (35% RH), and 0.358 V (25% RH). It can be seen from Fig. 8.12 that the maximum power density of the MEA at 120 °C increases linearly from 0.122 to 0.572 W cm^{-2} when the inlet RH increases from 25 to 100%. The improvement in fuel cell performance at high RH values is a synthetic result of changes in reactant (O_2 and H_2) partial pressures, fuel cell reaction thermodynamics and kinetics, mass transfer, and membrane conductivity, as discussed in earlier sections of this chapter.

Figure 8.13 shows cell voltage as a function of current density at 120 °C under five different RHs, from 20 to 100%. For each condition, the hydrogen and oxygen pressures were fixed at 1.0 atm, and hence, the total pressure at each side varied with the change in RH. It can be observed that the effect of RH on fuel cell performance is distinguishable in different polarization regions (low, medium, and high current density regions). The overall performance significantly decreases when the hydration level in a single cell is reduced. The cell voltage at a current density of 600 mA cm^{-2} is selected to evaluate the cell performance. As shown in Fig. 8.13, the fuel cell voltages are 0.617 V, 0.593 V, and 0.512 V at the RHs of 100%, 72% and 50%, respectively. When the RH is reduced to 20%, the cell voltage drops to 0.226 V. The polarization curves do not show a bending tendency until 400 mA cm^{-2}, and the limiting current

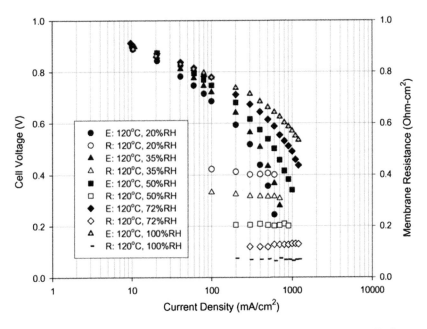

FIGURE 8.13 I–V polarization curves at 120 °C under different RHs. Pure oxygen and hydrogen with a constant pressure of 1.0 atm [4].

densities are >1000 mA cm^{-2} for 72% and 100% RH. At 20%, however, the polarization curve shows a severe mass transfer loss even at a current density of 400 mA cm^{-2}.

The above two examples are for high-temperature PEM fuel cells, which usually experience no "water flooding," which indicates that the overall fuel cell performance increases with RH. However, the effect of RH on fuel cell performance is complicated at low temperatures. Below 90 °C, "water flooding," may occur inside the fuel cell catalyst layer, flow field channel, and gas diffusion layer at high RH levels, resulting in a decreased overall PEM fuel cell performance. Thus, there is an optimum RH level for a conventional operating PEM fuel cell, depending on the fuel cell structure and design, plus other operating conditions. The fuel cell performance will be compromised if the fuel cell is operated above or below this optimum value.

8.8. CHAPTER SUMMARY

RH is one of the key parameters that affect fuel cell performance and kinetics. In this chapter, the definition and general concept of RH were presented, and RH distribution in PEM fuel cells was described. The effects of RH on PEM fuel cell performance were then addressed in detail.

The RH affects fuel cell performance by influencing the cell thermodynamics, reaction kinetics (e.g. Tafel slope, exchange current density), OCV, membrane conductivity, and mass transfer. With increasing RH, the partial pressures of the reactant gases will drop due to the resulting increase in water partial pressure. Thus, the fuel cell OCV will decrease thermodynamically. A reduction in the RH will cause an increase in the Tafel slope and a decrease in the exchange current density.

In this chapter, the effects of RH on R_m, R_t, and R_{mt} are also addressed thoroughly. Both the R_m and R_t increase with decreasing RH. The proton conductivity is increased for a well-hydrated membrane as the proton mobility is increased with RH. At high temperatures (e.g. 120 °C) and in the absence of "water-flooding," inside the fuel cell, R_{mt} increases with RH due to the increase in the oxygen and hydrogen permeability in the ionomer of the catalyst layer. These effects can be reflected by an overall fuel cell performance.

At high temperatures, fuel cell performance increases with RH. However, the effect of RH on fuel cell performance becomes complicated at low temperatures. For conventional PEM fuel cells, depending on the cell design and structure, an increase in the RH may cause "water flooding" and result in a decrease in the overall PEM fuel cell performance.

REFERENCES

[1] Sumner JJ, Creager SE, Ma JJ, DesMarteau DD. J Electrochem Soc 1998;145:107–10.
[2] Sone Y, Ekdunge P, Simonsson D. J Electrochem Soc 1996;143:1254–9.
[3] Anantaraman AV, Gardner CL. J Electroanal Chem 1996;414:115–20.

[4] Xu H, Kunz HR, Fenton JM. Electrochim Acta 2007;52:3525–33.

[5] Saleh MM, Okajima T, Hayase M, Kitamura F, Ohsaka T. J Power Sources 2007;164:503–9.

[6] Jiang R, Kunz HR, Fenton JM. J Power Sources 2005;150:120–8.

[7] Knights SD, Colbow KM, St-Pierre J, Wilkinson DP. J Power Sources 2004;127:127–34.

[8] Jiang R, Russell Kunz H, Fenton JM. Electrochim Acta 2006;51:5596–605.

[9] Abe T, Shima H, Watanabe K, Ito Y. J Electrochem Soc. 2004;151:A101–5.

[10] Ciureanu M. J Appl Electrochem 2004;34:705–14.

[11] Jang J-H, Yan W-M, Li H-Y, Chou Y-C. J Power Sources 2006;159:468–77.

[12] http://www.faqs.org/faqs/meteorology/temp-dewpoint/.

[13] http://hyperphysics.phy-astr.gsu.edu/hbase/kinetic/watvap.html.

[14] Zhang J, Tang Y, Song C, Xia Z, Li H, Wang H, et al. Electrochim Acta 2008;53:5315–21.

[15] Zawodzinski TA, Davey J, Valerio J, Gottesfeld S. Electrochim Acta 1995;40:297–302.

[16] Fuller TF, Newman J. J Electrochem Soc 1992;139:1332–7.

[17] Wood Iii DL, Yi JS, Nguyen TV. Electrochim Acta 1998;43:3795–809.

[18] Ge S, Li X, Hsing IM. Electrochim Acta 2005;50:1909–16.

[19] Ge S-H, Li X-G, Hsing IM. J Electrochem Soc 2004;151:B523–8.

[20] Watanabe M, Uchida H, Seki Y, Emori M, Stonehart P. J Electrochem Soc 1996;
     143:3847–52.

[21] Uchida H, Ueno Y, Hagihara H, Watanabe M. J Electrochem Soc 2003;150:A57–62.

[22] Liu F, Yi B, Xing D, Yu J, Hou Z, Fu Y. J Power Sources 2003;124:81–9.

[23] Kwak S-H, Yang T-H, Kim C-S, Yoon KH. J Power Sources 2003;118:200–4.

[24] Watanabe M, Uchida H, Emori M. J Phys Chem B 1998;102:3129–37.

[25] Zhang J, Tang Y, Song C, Zhang J, Wang H. J Power Sources 2006;163:532–7.

[26] Amphlett JC, Baumert RM, Mann RF, Peppley BA, Roberge PR, Harris TJ. J Electrochem
     Soc 1995;142:9–15.

[27] Zhang J, Tang Y, Song C, Zhang J. J Power Sources 2007;172:163–71.

[28] Xu H, Song Y, Kunz HR, Fenton JM. J Electrochem Soc 2005;152:A1828–36.

[29] Uribe FA, Springer TE, Gottesfeld S. J Electrochem Soc 1992;139:765–73.

[30] Ihonen J, Jaouen F, Lindbergh G, Lundblad A, Sundholm G. J Electrochem Soc
     2002;149:A448–54.

[31] Tang Y, Zhang J, Song C, Liu H, Zhang J, Wang H, et al. J Electrochem Soc
     2006;153:A2036–43.

[32] Jalani NH, Ramani M, Ohlsson K, Buelte S, Pacifico G, Pollard R, et al. J Power Sources
     2006;160:1096–103.

[33] Broka K, Ekdunge P. J Appl Electrochem 1997;27:117–23.

[34] Sakai T, Takenaka H, Torikai E. J Electrochem Soc 1986;133:88–92.

[35] Zawodzinski Jr TA, Springer TE, Uribe F, Gottesfeld S. Solid State Ionics 1993;60:199–211.

[36] David Ofer BN. Electrochem. Soc. Proceedings, vol. 2003–31, 2005.

Pressure Effects on PEM Fuel Cell Performance

9.1. INTRODUCTION

As one of the key operating conditions of proton exchange membrane (PEM) fuel cells, the operating pressure plays a significant role in determining PEM fuel cell performance. A fuel cell can be operated under a wide range of pressures, from ambient to 5 atm [1]. Operating pressure can influence the fuel cell performance by affecting the fuel cell open circuit voltage (OCV), the partial pressures of the reactant gases, hydrogen crossover, exchange current densities, and mass transfer in the electrode reactions [1–3]. Usually, the performance of a PEM fuel cell in terms of voltage and power density can be improved by increasing the operating pressure [1,4–6]. However, pressurization of the PEM fuel cell system can bring about increased gas permeation (e.g. hydrogen or oxygen crossover), water management issues [1,7], increased cost, size, and weight, and parasitic energy loss. Kazim [8], for example, conducted a comprehensive exergoeconomic analysis of a 10-kW PEM fuel cell stack and concluded that the energy cost could be increased by operating at higher

PEM Fuel Cell Testing and Diagnosis. http://dx.doi.org/10.1016/B978-0-444-53688-4.00009-7

pressures. Boyer et al. [9] indicated that operating a fuel cell under higher pressure could help overcome the oxygen transfer limitation but would decrease the fuel efficiency and increase the complexity and cost of the fuel cell system. Al-Baghdadi et al. [10] showed that a higher operating pressure could generate a more even distribution of local current density due to high oxygen concentration in the catalyst layer. However, Sun et al. [6] concluded that higher pressure could result in a nonuniform current distribution.

Thus, the operation of a PEM fuel cell at high pressures has both advantages and disadvantages. This chapter examines in detail, the effects of operating backpressure on fuel cell performance.

9.2. OPERATING PRESSURE IN PEM FUEL CELLS

In PEM fuel cells, the distribution of reactant gases is not uniform across the flow field from inlet to outlet. Thus, a pressure (or reactant concentration) gradient is generated between the inlet and the outlet, which is indicated by a pressure drop (ΔP) across the flow field. Hence, the inlet pressure (P^{inlet}) and the outlet pressure (P^{outlet}) in a PEM fuel cell are different, and their relationship can be expressed by using Eqn (9.1):

$$P^{inlet} = P^{outlet} + \Delta P \tag{9.1}$$

In a fuel cell, the backpressure ($P^{backpressure}$) is the same as the P^{outlet}, so Eqn (9.1) can also be expressed as Eqn (9.2):

$$P^{inlet} = P^{backpressure} + \Delta P \tag{9.2}$$

During fuel cell operation, the system pressure can be determined by controlling P^{inlet} or $P^{backpressure}$. Backpressure control is used more often in practical fuel cell operation than inlet pressure control.

Because of the pressure gradient along the flow field, the average fuel cell pressure ($\overline{P}$) can be more convenient for accurately describing a fuel cell's operating pressure; $\overline{P}$ can be expressed as in Eqns (9.3) or (9.4):

$$\overline{P} = \frac{1}{2}(P^{inlet} + P^{backpressure}) \tag{9.3}$$

$$\overline{P} = \frac{1}{2}(2P^{backpressure} + \Delta P) \tag{9.4}$$

The average pressures at the anode and the cathode can then be expressed as in Eqns (9.5) and (9.6), respectively:

$$\overline{P_{anode}} = \frac{1}{2}(2P^{backpressure}_{anode} + \Delta P_{anode}) \tag{9.5}$$

$$\overline{P_{\text{cathode}}} = \frac{1}{2}(2P_{\text{cathode}}^{\text{backpressure}} + \Delta P_{\text{cathode}}) \tag{9.6}$$

where $\overline{P_{\text{anode}}}$, $P_{\text{anode}}^{\text{backpressure}}$, and ΔP_{anode} are the average pressure at the anode, the backpressure at the anode, and the pressure drop at the anode; $\overline{P_{\text{cathode}}}$, $P_{\text{cathode}}^{\text{backpressure}}$, and $\Delta P_{\text{cathode}}$ are the average pressure at the cathode, the backpressure at the cathode, and the pressure drop at the cathode.

The total pressure $(P_{\text{anode}}^{\text{total}})$ at the anode side is the sum of the hydrogen pressure (P_{H_2}) and the water pressure $(P_{\text{anode}}^{\text{H}_2\text{O}})$ at the anode, and it can be expressed as in Eqn (9.7):

$$P_{\text{anode}}^{\text{total}} = P_{\text{H}_2} + P_{\text{anode}}^{\text{H}_2\text{O}} \tag{9.7}$$

So, the average anode pressure is the sum of the average H_2 partial pressure $(\overline{P_{\text{H}_2}})$ and the average anode water vapor partial pressure $(\overline{P_{\text{anode}}^{\text{H}_2\text{O}}})$. According to Eqns (9.5) and (9.7), $\overline{P_{\text{H}_2}}$ can be expressed as in Eqn (9.8):

$$\overline{P_{\text{H}_2}} = \frac{1}{2}(2P_{\text{anode}}^{\text{backpressure}} + \Delta P_{\text{anode}}) - \overline{P_{\text{anode}}^{\text{H}_2\text{O}}} \tag{9.8}$$

It is well known that saturated water vapor pressure $(P_{\text{H}_2\text{O}}^0(T))$ is a function of temperature [11,12] and scarcely changes with pressure. As discussed in Chapter 8, the relative humidity (%RH) can be expressed as the ratio of water vapor pressure $(\overline{P_{\text{H}_2\text{O}}})$ to saturated water vapor pressure at temperature T

$\left(\%\text{RH} = \dfrac{\overline{P_{\text{H}_2\text{O}}}}{P_{\text{H}_2\text{O}}^0(T)}\right)$. Thus, for the anode, Eqn (9.8) can be written as Eqn (9.9):

$$\overline{P_{\text{H}_2}} = \frac{1}{2}\left(2P_{\text{anode}}^{\text{backpressure}} + \Delta P_{\text{anode}}\right) - P_{\text{H}_2\text{O}}^0(T) \times \%\text{RH}_{\text{anode}} \tag{9.9}$$

If the anodic RH is controlled at 100%, then Eqn (9.9) should be rewritten as Eqn (9.10):

$$\overline{P_{\text{H}_2}} = \frac{1}{2}\left(2P_{\text{anode}}^{\text{backpressure}} + \Delta P_{\text{anode}}\right) - P_{\text{H}_2\text{O}}^0(T) \tag{9.10}$$

Because the water pressure does not change at a constant operating temperature and RH, the average hydrogen partial pressure should proportionally increase with increasing backpressure.

Similarly, the average cathode pressure is the sum of the average O_2 partial pressure $(\overline{P_{\text{O}_2}})$, average N_2 partial pressure $(\overline{P_{\text{N}_2}})$, and average water vapor partial pressure at the cathode $(\overline{P_{\text{cathode}}^{\text{H}_2\text{O}}})$, assuming that the average air pressure is the sum of $\overline{P_{\text{O}_2}}$ and $\overline{P_{\text{N}_2}}$. Similar to Eqn (9.8), the average O_2 partial pressure $(\overline{P_{\text{O}_2}})$ can be expressed as in Eqn (9.11):

$$\overline{P_{\text{O}_2}} = \frac{1}{2}\left(2P_{\text{cathode}}^{\text{backpressure}} + \Delta P_{\text{cathode}}\right) - \overline{P_{\text{N}_2}} - \overline{P_{\text{cathode}}^{\text{H}_2\text{O}}} \tag{9.11}$$

Assuming that the volume ratio of oxygen in air is 21%, Eqn (9.11) can be rewritten as Eqn (9.12):

$$\overline{P_{O_2}} = \left[\frac{1}{2} \left(2P_{\text{cathode}}^{\text{backpressure}} + \Delta P_{\text{cathode}} \right) - P_{H_2O}^0(T) \times \%RH_{\text{cathode}} \right] \times 0.21$$

$$(9.12)$$

At an RH of 100%, the above equation should be rewritten as follows:

$$\overline{P_{O_2}} = \left[\frac{1}{2} \left(2P_{\text{cathode}}^{\text{backpressure}} + \Delta P_{\text{cathode}} \right) - P_{H_2O}^0(T) \right] \times 0.21 \qquad (9.13)$$

Obviously, the average oxygen partial pressure increases with back-pressure when the fuel cell is operated at a constant temperature and RH. In our previous work [1], the average oxygen partial pressures at 70 °C and 100% RH with different backpressures and current densities were calculated based on Eqn (9.13), the controlled cathode backpressures, and the measured cathode pressure drops between fuel cell inlet and outlet. As shown in Fig. 9.1, the average oxygen partial pressure increases as the operating backpressure increases from 1.0 to 3.04 atm across the entire current density range of 0–2.2 A cm^{-2}.

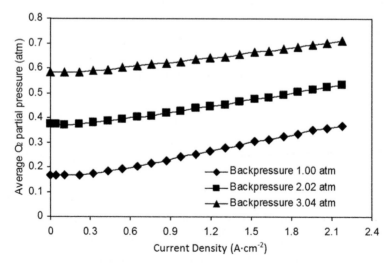

FIGURE 9.1 Oxygen partial pressure as a function of current density at 70 °C and 100% RH with different backpressures. Gore®-membrane-based membrane electrode assembly (MEA) area: 46 cm^2. Gas diffusion layer (GDL): 25-DC. Stoichiometries of H$_2$ and air: 1.2 and 2.5, respectively. Single serpentine flow channel with both a width and depth of 1.0 mm [1].

9.3. THEORETICAL AND SEMIEMPIRICAL ANALYSIS OF BACKPRESSURE EFFECTS ON FUEL CELL PERFORMANCE

As discussed in Chapter 1, when an H_2/air (O_2) PEM fuel cell is operated at a current density of I_{cell}, the cell voltage (V_{cell}) can be expressed as follows:

$$V_{cell} = E_{cell}^{OCV} + \frac{2.303RT}{(1 - \alpha_{2O})n_{\alpha 2O}F}\log(i_{O_2/H_2O}^{o}) + \frac{2.303RT}{\alpha_H n_{\alpha H}F}\log(i_{H_2/H^+}^{o})$$

$$- \frac{2.303RT}{(1 - \alpha_{2O})n_{\alpha 2O}F}\log\left(\frac{I_{cell}\overrightarrow{I}_{O_2/H_2O}^{l}}{\overrightarrow{I}_{O_2/H_2O}^{l} - I_{cell}}\right) - \frac{2.303RT}{\alpha_H n_{\alpha H}F}\log\left(\frac{\overrightarrow{I}_{H_2/H^+}^{l}I_{cell}}{\overrightarrow{I}_{H_2/H^+}^{l} - I_{cell}}\right) - I_{cell}R_m$$

$$(9.14)$$

In Eqn (9.14), E_{cell}^{OCV} is the measured fuel cell OCV; i_{H_2/H^+}^{o} and i_{O_2/H_2O}^{o} are the apparent exchange current densities of the hydrogen oxidation reaction (HOR) and oxygen reduction reaction (ORR); α_H and α_{2O} are, respectively, the electron transfer coefficients from H_2 to the anode catalyst surface and from the cathode catalyst surface to O_2; $n_{\alpha H}$ and $n_{\alpha 2O}$ are the corresponding electron transfer numbers in the determining steps of the HOR and ORR reactions; and $\overrightarrow{I}_{O_2/H_2O}^{l}$ and $\overrightarrow{I}_{H_2/H^+}^{l}$ are the mass transfer limited current densities. The R_m in Eqn (9.14) is the membrane resistance, whose value is dependent on temperature, humidity, backpressure, and current density.

As discussed in our previous publications [1], pressure has four major effects: (1) on the thermodynamics, by affecting the fuel cell OCV, (2) on the electrode kinetics, by affecting the exchange current densities (i_{H_2/H^+}^{o} and i_{O_2/H_2O}^{o}), (3) on mass transfer, by affecting the mass transfer limited current densities ($\overrightarrow{I}_{O_2/H_2O}^{l}$ and $\overrightarrow{I}_{H_2/H^+}^{l}$) in Eqn (9.14), and (4) on membrane resistance. Therefore, E_{cell}^{OCV}, i_{H_2/H^+}^{o}, i_{O_2/H_2O}^{o}, $\overrightarrow{I}_{O_2/H_2O}^{l}$, $\overrightarrow{I}_{H_2/H^+}^{l}$, and R_m are all related to the operating pressure. To achieve a fundamental understanding of the effects of operating pressure on a fuel cell, the relationships of these parameters to the operating pressure should be known.

In a semiempirical approach, the relationships of i_{H_2/H^+}^{o}, i_{O_2/H_2O}^{o}, $\overrightarrow{I}_{O_2/H_2O}^{l}$, and $\overrightarrow{I}_{H_2/H^+}^{l}$ to both temperature and average partial pressure can be obtained from experimental data in the temperature range of 23–100 °C, as reported in a work on fuel cell reaction kinetics [13]. The values of i_{H_2/H^+}^{o}, i_{O_2/H_2O}^{o}, $\overrightarrow{I}_{O_2/H_2O}^{l}$, and $\overrightarrow{I}_{H_2/H^+}^{l}$ can be expressed as in Eqns (9.15) to (9.18), respectively [14]:

$$i_{H_2/H^+}^{o} = A_H(\overline{P_{H_2}})^{0.5} \qquad (9.15)$$

$$i_{O_2/H_2O}^{o} = A_O(P_{O_2})^{0.001678T} \qquad (9.16)$$

$$\overrightarrow{I}^{1}_{H_2/H^+} = A_{H(L)}\overline{P_{H_2}} \tag{9.17}$$

$$\overrightarrow{I}^{1}_{O_2/H_2O} = A_{O(L)}\overline{P_{O_2}} \tag{9.18}$$

where A_H is related to the catalyst active surface area and the intrinsic exchange current density of the HOR; A_O is related to the catalyst active surface area and the intrinsic exchange current density of the ORR; $A_{H(L)}$ is related to the diffusion coefficient and Henry's constant of H_2; and $A_{O(L)}$ is related to the diffusion coefficient and to Henry's constant of O_2. All of these can be expressed as a function of temperature, according to Arrhenius's theory. For example, if the data listed in Table 9.1 in the temperature range of 23–100 °C, measured by Song et al. [13], can be simulated to obtain the expressions of A_H, A_O, $A_{H(L)}$, and $A_{O(L)}$, then $i^0_{H_2/H^+}$, $i^0_{O_2/H_2O}$, $\overrightarrow{I}^{1}_{O_2/H_2O}$, and $\overrightarrow{I}^{1}_{H_2/H^+}$ can be expressed as in Eqns (9.19)–(9.22), in their respective numerical forms:

$$i^0_{H_2/H^+} = 2.252 \times 10^2 \exp\left(-\frac{2.347 \times 10^3}{T}\right)\left(\overline{P_{H_2}}\right)^{0.5} \tag{9.19}$$

$$i^0_{O_2/H_2O} = 3.695 \times 10^{-2} \exp\left(-\frac{1.525 \times 10^3}{T}\right)\left(\overline{P_{O_2}}\right)^{1.678 \times 10^{-3}T} \tag{9.20}$$

$$\overrightarrow{I}^{1}_{H_2/H^+} = 1.241 \times 10^2 \exp\left(-\frac{1.438 \times 10^3}{T}\right)\overline{P_{H_2}} \tag{9.21}$$

TABLE 9.1 Measured and Simulated Kinetic Parameters at 3.0 atm (Absolute) Backpressure, 100% RH, and Different Temperatures. The Values of α_H, α_{2O}, n_{aH}, and n_{a2O} were Taken to be 0.001,678T, 2.0, 0.5, and 1.0, Respectively. The Cathode Catalyst Surfaces were Pt/PtO in the Low Current Density Range and Pure Pt in the High Current Density Range, and the Anode Catalyst Surface was Pure Pt in Both Ranges [13]

Temperature, (°C)	23	40	60	80	100
P_{H_2}, (atm)	3.02	2.97	2.84	2.57	2.04
P_{O_2}, (atm)	0.634	0.623	0.597	0.540	0.489
$i^0_{H_2}$, (A cm^{-2})	0.134	0.198	0.344	0.607	0.604
$\overrightarrow{I}^{1}_{H_2/H^+}$, (A cm^{-2})	3.01	3.71	4.39	5.46	5.54
$i^0_{O_2}$, (A cm^{-2})	1.22×10^{-4}	2.43×10^{-4}	3.92×10^{-4}	4.60×10^{-4}	3.43×10^{-4}
$\overrightarrow{I}^{1}_{O_2/H_2O}$, (A cm^{-2})	1.32	1.48	1.59	1.82	1.95

$$\overrightarrow{I}^{1}_{O_2/H_2O} = 4.707 \times 10^1 \exp\left(-\frac{9.332 \times 10^2}{T}\right)\overline{P_{O_2}} \qquad (9.22)$$

If these four equations are substituted into Eqns (9.14), (9.9), and (9.12), a semiempirical equation can be obtained, from which the fuel cell polarization curve can be calculated at a desired temperature, backpressure (obtained using gas partial pressure, according to Eqn (9.9) or (9.12)), and RH. Figure 9.2 shows the calculated polarization curves at two different backpressures (2.0 and 3.0 atm absolute backpressure, respectively). Evidently, the backpressure has an effect on performance. Note that during calculation, the values of both E_{cell}^{OCV} and R_m were measured from their corresponding experiments.

9.3.1. Backpressure Effect on Fuel Cell (OCV)

As reported in the literature [1,2], due to O_2 reacting with the Pt catalyst and H_2 crossover, the measured OCV in a fuel cell is normally lower than the thermodynamic OCV. Equation (9.14) describes the experimentally measured OCV; however, the OCV can be expressed semiempirically as in Eqn (9.23) [2,14]:

$$E_{cell}^{OCV} = E_{theory}^{OCV} - \Delta E_{O_2-Pt}^{OCV} - \Delta E_{H_2-xover}^{OCV} \qquad (9.23)$$

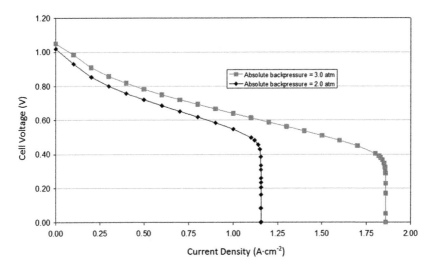

FIGURE 9.2   Calculated fuel cell polarization curves at backpressures of 2.0 atm and 3.0 atm. In the calculation, $P_{cathode}^{backpressure} = P_{anode}^{backpressure}$; $\Delta P_{cathode} = \Delta P_{anode} = 0.2$ atm; $P_{H_2O}^o(80\,°C) = 0.455$ atm; $T = 80\,°C$; α_H, α_O, $n_{\alpha H}$, and $n_{\alpha O} = 0.5$, 0.001,678T, 1.0, and 2.0, respectively; and $R_m = 0.130\,\Omega\,cm^2$ are assumed [14]. (For color version of this figure, the reader is referred to the online version of this book.)

where E_{theory}^{OCV} is the theoretical OCV, $\Delta E_{O_2-Pt}^{OCV}$ is the OCV drop caused by a mixed reaction between the Pt surface and O_2, and $\Delta E_{H_2-x-over}^{OCV}$ is the OCV drop caused by H_2 crossover from the anode side to the cathode side.

Differentiating Eqn (9.23) with respect to backpressure $P^{backpressure}$ (assuming that both anode and cathode backpressures are controlled at the same value, i.e. $P^{backpressure} = P_{anode}^{backpressure} + P_{cathode}^{backpressure}$), the effect of backpressure on the OCV can be expressed as follows:

$$
\left(\frac{\partial E_{cell}^{OCV}}{\partial P^{backpressure}} \right)_{T,I_{cell}=0} = \left(\frac{\partial E_{theory}^{OCV}}{\partial P^{backpressure}} \right)_{T,I_{cell}=0} - \left(\frac{\partial (\Delta E_{O_2-Pt}^{OCV})}{\partial P^{backpressure}} \right)_{T,I_{cell}=0}
$$
$$
- \left(\frac{\partial (\Delta E_{H_2-x-over}^{OCV})}{\partial P^{backpressure}} \right)_{T,I_{cell}=0}
$$

(9.24)

where $\left(\dfrac{\partial E_{theory}^{OCV}}{\partial P^{backpressure}} \right)_{T,I_{cell}=0}$ is the effect of backpressure on the fuel cell theoretical OCV. This theoretical OCV can be expressed as follows:

$$
E_{theory}^{OCV} = E_{O_2/H_2O}^o - E_{H_2/H^+}^o + \frac{RT}{2F} \ln \left(\frac{P_{H_2} P_{O_2}^{-\frac{1}{2}}}{P_{H_2O}^o} \right)
$$
(9.25)

where E_{H_2/H^+}^o is the electrode potential of the H_2/H^+ redox couple at standard conditions (1.0 atm, T), defined as zero at any temperature; E_{O_2/H_2O}^o is the electrode potential of the O_2/H_2O redox couple at standard conditions (1.0 atm, T), which has a temperature dependency of approximately 0.846 mV deg^{-1} (e.g. at 25 °C, the value is 1.229 V (vs. standard hydrogen electrode), whereas at 80 °C, its value is 1.182 V); F is Faraday's constant; and R and T have their usual values.

Figure 9.3 shows how OCV changes with backpressure, with OCV calculated according to Eqn (9.24); for a more detailed discussion of this figure, please see Ref. [14].

9.3.2. Backpressure Effect on Fuel Cell Kinetics (Electrode Kinetics and Mass Transfer Process)

The fuel cell kinetics is expressed by the terms on the right-hand side in Eqn (9.14), except for E_{cell}^{OCV}, which has been dealt within Eqn (9.24), and $I_{cell}R_m$, which will be discussed later. In actuality, fuel cell kinetics should include both the electrode kinetics and the mass transfer process. If Eqn (9.14) is

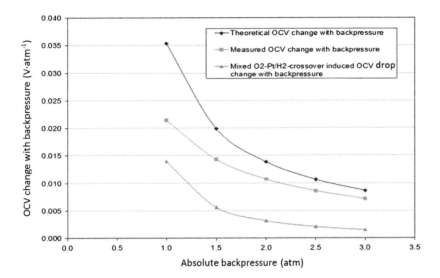

FIGURE 9.3 Calculated theoretical OCV, measured OCV, and mixed O_2–Pt/H_2-crossover induced OCV drop changes, as a function of backpressure. The calculation assumes $P_{cathode}^{backpressure} = P_{anode}^{backpressure}$, $\Delta P_{cathode} = \Delta P_{anode} = 2.0$ atm, $P_{H_2O}^o(80\ °C) = 0.455$ atm, and $T = 80\ °C$ [14]. (For color version of this figure, the reader is referred to the online version of this book.)

differentiated with respect to backpressure, the backpressure effect on fuel cell kinetics can be obtained:

$$
\begin{aligned}
\left(\frac{\partial V_{cell}}{\partial P^{backpressure}}\right)_{T,I_{cell}} &= \frac{RT}{2\alpha_H n_{\alpha H}F}\frac{1}{0.5(2P_{anode}^{backpressure}+\Delta P_{anode})-\%RH_{anode}P_{H_2O}^o} \\
&+0.001678\frac{RT^2}{(1-\alpha_{2O})n_{\alpha2O}F}\frac{1}{0.5(2P_{cathode}^{backpressure}+\Delta P_{cathode})-\%RH_{cathode}P_{H_2O}^o} \\
&+I_{cell}\frac{RT}{\alpha_H n_{\alpha H}F}\left\{\frac{1}{\begin{array}{c}(0.5(2P_{anode}^{backpressure}+\Delta P_{anode})-\%RH_{anode}P_{H_2O}^o)\\(A_{H(L)}(0.5(2P_{anode}^{backpressure}+\Delta P_{anode})-\%RH_{anode}P_{H_2O}^o)-I_{cell})\end{array}}\right\} \\
&+I_{cell}\frac{RT}{(1-\alpha_{2O})n_{\alpha2O}F}\left\{\frac{1}{\begin{array}{c}(0.5(2P_{cathode}^{backpressure}+\Delta P_{cathode})-\%RH_{cathode}P_{H_2O}^o)\\(0.21A_{O(L)}(0.5(2P_{cathode}^{backpressure}+\Delta P_{cathode})-\%RH_{cathode}P_{H_2O}^o)-I_{cell})\end{array}}\right\}
\end{aligned}
$$

$$(9.26)$$

According to Eqn (9.26), the anode, cathode, and total cell voltage changes, induced by electrode kinetics, which occur with backpressure can be calculated as a function of current density. Figure 9.4 shows the calculated results at a 3.0-atm backpressure. It can be seen that in the current density range controlled by electrode kinetics ($< \sim0.8$ A cm^{-2}), the anode voltage change with increasing or decreasing backpressure is more sensitive than the cathode voltage change,

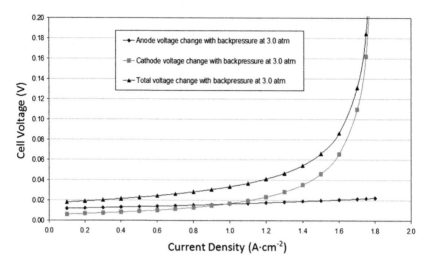

FIGURE 9.4 Calculated anode, cathode, and total cell voltage changes with backpressure as a function of current density. In the calculation, $P_{\text{cathode}}^{\text{backpressure}} = P_{\text{anode}}^{\text{backpressure}} = 3.0$ atm; $\Delta P_{\text{cathode}} = \Delta P_{\text{anode}} = 0.2$ atm; $P_{\text{H}_2\text{O}}^{\text{o}}(80\ °C) = 0.455$ atm; $T = 80\ °C$; α_{H}, α_{O}, $n_{\alpha\text{H}}$, and $n_{\alpha\text{O}}$ equal to 0.5, 0.001,6787, 1.0, and 2.0, respectively; A_{H}^{o}, A_{O}^{o}, $A_{\text{H}(\text{L})}^{\text{o}}$, and $A_{\text{O}(\text{L})}^{\text{o}}$ are, respectively, $2.252 \times 10^2\ \text{A cm}^{-2}\cdot(\text{atm})^{-0.5}$, $3.695 \times 10^{-2}\ \text{A cm}^{-2}(\text{atm})^{-0.001,678\text{K}}$, 1.241×10^2 A cm^{-2} atm, and $4.707 \times 10^1\ \text{A cm}^{-2}$ atm [14]. (For color version of this figure, the reader is referred to the online version of this book.)

and the change increases monotonically with increasing current density. However, in the mass transfer controlled range ($>1.2\ \text{A cm}^{-2}$), the voltage change at the cathode with increasing or decreasing backpressure is much more sensitive than that at the anode, because the mass transfer limiting current density of the cathode is smaller than that of the anode.

In fact, the magnitude of the cell voltage change with increasing or decreasing backpressure should be dependent on the operating backpressure, which can be seen from Eqn (9.26). The magnitude of the kinetic cell voltage change with backpressure is mainly determined by four parameters—A_{H}, A_{O}, $A_{\text{H}(\text{L})}$, and $A_{\text{O}(\text{L})}$—as shown in Eqns (9.15)–(9.18), because they are determined by fuel cell reaction exchange current densities and mass transfer limiting current densities. In general, different kinds of MEAs and different operating conditions can yield different performance levels, based on which these four parameters (A_{H}, A_{O}, $A_{\text{H}(\text{L})}$, and $A_{\text{O}(\text{L})}$) can be simulated using I–V polarizations. If these four parameters are known, Eqns (9.19), (9.23), and (9.26) can predict the effects of backpressure or partial pressure on the fuel cell performance.

Exchange current densities can be evaluated using the charge transfer resistance values ($R_{\text{ct}}^{\text{OCV}}$) of the electrode reactions at OCV, and $R_{\text{ct}}^{\text{OCV}}$ can be obtained by simulating the AC impedance spectra using a suitable equivalent circuit [2]; in principle, the larger this resistance value, the slower the electrode kinetics. In actuality, AC impedance spectroscopy can also be used to determine the electrode kinetics under fuel cell load conditions.

The overall impedance of the cell can be obtained by differentiating Eqn (9.14) with respect to the change in current density:

$$R_{cell} = -\left(\frac{\partial V_{cell}}{\partial I_{cell}}\right)_{T,p^{backpressure}} = \overrightarrow{I}^{1}_{H_2/H^+}\frac{RT}{\alpha_H n_{\alpha H}F}\frac{1}{I_{cell}\left(\overrightarrow{I}^{1}_{H_2/H^+} - I_{cell}\right)}$$

$$+ \overrightarrow{I}^{1}_{O_2/H_2O}\frac{RT}{\alpha_O n_{\alpha O}F}\frac{1}{I_{cell}\left(\overrightarrow{I}^{1}_{O_2/H_2O} - I_{cell}\right)} + R_m$$

(9.27)

Because the cell voltage decreases as the current density increases, $(\partial V_{cell}/\partial I_{cell})_{T,p^{backpressure}}$ has a negative value. To ensure a positive value, a negative sign is put before it. In the low current density range (electrode kinetic range), both $\overrightarrow{I}^{1}_{H_2/H^+}$ and $\overrightarrow{I}^{1}_{O_2/H_2O}$ are much larger than the current density, and Eqn (9.23) can be simplified as follows:

$$R_{cell} = \frac{RT}{\alpha_H n_{\alpha H}F}\frac{1}{I_{cell}} + \frac{RT}{\alpha_O n_{\alpha O}F}\frac{1}{I_{cell}} + R_m$$

(9.28)

Because the current density in Eqns (9.28) is related to the cell voltage, OCV, exchange current densities, and mass transfer limiting current densities, the cell resistance should be a function of these parameters. Since these parameters are related to the partial pressures of a gas, the cell resistance should also be a function of the partial pressures or backpressures.

Figure 9.5 [1] shows charge transfer resistances simulated according to AC impedance spectra collected at three different backpressures—1.00, 2.02, and

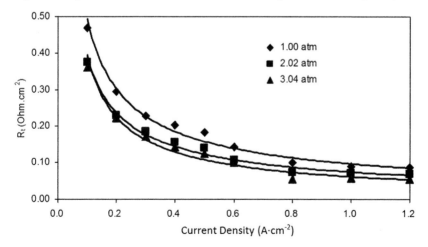

FIGURE 9.5   Charge transfer resistance as a function of current density at 70 °C, 100% RH, and three different backpressures from 1.00 to 3.04 atm. Gore®-membrane-based MEA area: 46 cm². GDL: 25-DC. Stoichiometries of H_2 and air were 1.2 and 2.5, respectively. Single serpentine flow channel with a width and depth of 1.0 mm [1].

3.04 atm—in the current density range of 0–1.2 A cm^{-2}. Clearly, the charge transfer resistance of the electrode reaction decreases with increasing back-pressure, indicating a faster electrode kinetics at higher backpressures, which suggests that a better fuel cell performance will be obtained at a higher back-pressure due to the acceleration of the electrode kinetics. This can be explained by the difference in the partial pressures of the reactant gases. Because the operating temperature and RH were kept at the same level, the partial pressures of both hydrogen and oxygen were higher at a higher backpressure than at a lower one. According to the theoretical analysis given above, an increase in the partial pressures of the reactant gases will speed up the electrode kinetics by increasing the exchange current densities and the mass transfer limiting current densities.

Regarding the mass transfer process, the transportation of reactant gases through the GDL to the catalyst layer is mainly via the diffusion process and partially through convection. Concentration polarization occurs when the reactant is quickly consumed within the catalyst layer by the electrochemical reaction, and the reactant transportation rate is not high enough to compensate for the reactant loss. Thus, a reactant concentration gradient will form along the gas diffusion and catalyst layers, particularly in the high current density range. This voltage loss is usually called concentration loss (it is also called mass transfer polarization, as it is caused by ineffective mass transfer of the reactant gases). Several processes may contribute to concentration polarization: in the gas phase, slow diffusion in the electrode pores; solution/dissolution of reactants and products into and out of the electrolyte; or diffusion of reactants and products through the electrolyte to and from the electrochemical reaction sites. At high current densities, the slow transport of reactants and products to or from the electrochemical reaction site is a major contributor to the concentration polarization, which is expressed as a mass transfer limiting current density, as shown in Eqn (9.17) and (9.18). Here, the mass transfer limiting current densities are directly related to the partial pressures of a gas, and can be alternatively expressed as a function of backpressure:

$$\overrightarrow{I}^{l}_{H_2/H^+} = A_{H(L)}\overline{P_{H_2}} = A_{H(L)}(0.5(2P^{backpressure}_{anode} + \Delta P_{anode}) - \%RH_{anode}P^o_{H_2O})$$

$$(9.29)$$

$$\overrightarrow{I}^{l}_{O_2/H_2O} = A_{O(L)}\overline{P_{O_2}}$$
$$= A_{O(L)}(0.21(0.5(2P^{backpressure}_{cathode} + \Delta P_{cathode}) - \%RH_{cathode}P^o_{H_2O}))$$

$$(9.30)$$

Both the equations indicate that when the backpressure is increased, the mass transfer limiting current densities will increase. This is mainly due to the enhancement of mass transfer from the GDL to the catalyst layer, and also to

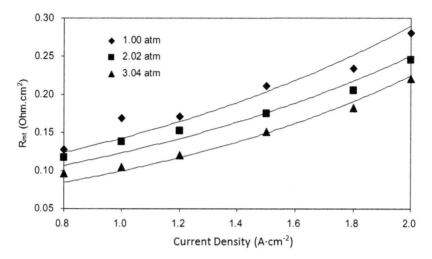

FIGURE 9.6   Measured and simulated mass transfer resistance for fuel cell reactions, as a function of current density at 70 °C, 100% RH, and backpressures from 1.00 to 3.04 atm. Gore®-membrane-based MEA area: 46 cm²; GDL: 25-DC. Stoichiometries of H_2 and air were 1.2 and 2.5, respectively. Single serpentine flow channel with a channel width and depth of 1.0 mm [1].

the increase in reactant concentration in the effective reaction sites [15]. Figure 9.6 shows the mass transfer resistances as measured using AC impedance spectroscopy at three different backpressures of 1.00, 2.02, and 3.04 atm [1]. It can be seen that as the backpressure increases, the mass transfer resistance decreases, suggesting improved PEM fuel cell performance.

9.3.3. Backpressure Effect on Membrane Resistance

It is well known that the proton conductivity of a perfluorosulfonic acid membrane (e.g. a Nafion® membrane) is a function of temperature and water content. Although pressure has an insignificant effect on membrane conductivity, one recent study [1] showed that the membrane resistance measured by AC impedance spectroscopy when operating Gore®-membrane-based fuel cells was slightly decreased when the backpressure was increased from 1.0 to 3.04 atm. This can be explained in terms of the water content on the membrane surface at different backpressures. Because the mass flow rates of the reactant gases are fixed, the volume gas flow rate at a low backpressure is faster than at a higher backpressure. Therefore, the water purging effect at a lower backpressure is higher than at a higher one. At a lower backpressure, more of the water generated by the electrochemical reaction at the interface of the membrane and the catalyst layer is purged by the gas flow. Thus, the water content on the surface of the membrane is lower, yielding relatively lower proton conductivity compared with the conductivity at a higher backpressure.

9.3.4. Backpressure Effect on Hydrogen Crossover

Due to the porous property of the membrane and the H_2 concentration gradient between the anode and the cathode, hydrogen crossover from the anode to the cathode is inevitable in PEM fuel cells, as discussed in Chapter 6. An increase in the anode backpressure can effectively increase the partial pressure of H_2 at the anode, which leads to a higher H_2 concentration gradient across the membrane. According to Chapter 6, an alternative hydrogen crossover rate $(J_{H_2}^{x-over})$ can be expressed as a function of backpressure:

$$J_{H_2}^{x-over} = \left(\frac{\psi_{H_2}^{PEM}}{l_{PEM}}\right)\overline{P_{H_2}} = \left(\frac{\psi_{H_2}^{PEM}}{l_{PEM}}\right)\left(0.5(2P_{anode}^{backpressure} + \Delta P_{anode})\right) \quad (9.31)$$

where $\psi_{H_2}^{PEM}$ is the permeability coefficient of H_2 in the PEM, and l_{PEM} is the thickness of the membrane. This equation clearly indicates that an increase in the anode backpressure can effectively increase the H_2 crossover.

As described in Chapter 6, the hydrogen crossover rate can be measured using the electrochemical method [1–3,12,16,17], whereby the current generated by the electrochemical oxidation of permeated hydrogen at the cathode can be measured and the hydrogen crossover rate can then be determined [1–3,12,16,17].

As an example, Figure 9.7 shows the linear sweep voltammetry (LSV) recorded under different backpressures from 0.05 to 0.2 MPa at 70 °C and 100% RH. The current density generated by electrochemical oxidation of the crossed over hydrogen at cathode $(I_{H_2-x-over}$, represented by the limiting

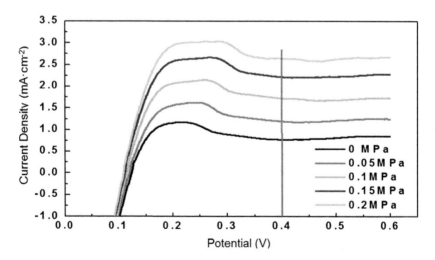

FIGURE 9.7 Current density as a function of operating backpressure, measured using the LSV technique in a single cell with Nafion® 212 as the membrane. $T = 70$ °C; RH $= 100\%$. (For color version of this figure, the reader is referred to the online version of this book.)

current density at ~0.4 V in Fig. 9.7) increases from 0.764 to 2.63 mA cm^{-2} when the anode operating backpressure of the fuel cell increases from 0 to 0.2 MPa, demonstrating that an increase in the operating backpressure can cause an increase in the hydrogen crossover rate. The same trend has been observed in our recent studies [1,3]. The higher the hydrogen crossover rate, the less efficient the fuel cell will be, particularly when the fuel cell is operated at OCV or in a low current density range.

9.3.5. Backpressure Effect on Overall Fuel Cell Performance [18–21]

As discussed in the preceding sections, the partial pressures of the reactant gases increase as the operating backpressure increases, leading to enhanced electrode kinetics and improved mass transfer in PEM fuel cells. These are the positive effects of operating pressure on fuel cell performance. However, the increased pressure can also cause an increased hydrogen crossover rate; this is a negative effect that increasing the operating backpressure has on fuel cell performance, because the hydrogen crossover rate is normally low and its contribution to fuel cell performance may be insignificant, especially when the fuel cell is operated at high current densities. Therefore, the positive effect of increasing the operating pressure should be much larger than the negative effect. Even at OCV, the fuel cell theoretical OCV can be increased by increasing the operating backpressure, as discussed above. Therefore, when we consider only the fuel cell, its performance can be improved by increasing the operating backpressure.

As an example, Fig. 9.8 presents fuel cell performance at several backpressures. Evidently, the OCV can be improved from 0.936 V for the H$_2$/air PEM fuel cell operating at 0 MPa to 0.97 V at 0.2 MPa, and the overall cell voltage can be improved significantly when the backpressure is increased.

Normally, a PEM fuel cell can be operated at pressures of 0–5 atm. It has been demonstrated that a higher operating pressure yields a higher cell performance. However, high gas pressures introduce some difficulties for fuel cell seals and cause durability issues. Moreover, for a PEM fuel cell system, higher operating pressures will bring about negative effects such as more parasitic power loss, higher cost for compression, increased volume of the fuel cell system, and fast degradation of the air pump. Thus, high-pressure operation does not always yield benefits for fuel cell systems [21]. An optimized operating backpressure is needed, according to the requirements of the entire fuel cell system.

9.4. CHAPTER SUMMARY

In this chapter, the effect of operating pressure on PEM fuel cell performance was analyzed theoretically and experimentally. An increase in the fuel cell operating pressure can affect the reversible thermodynamic potential, such as

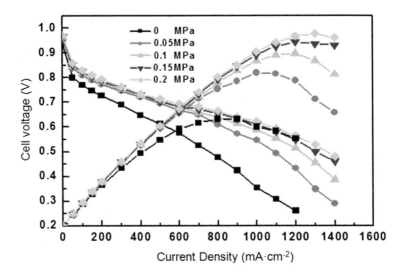

FIGURE 9.8 Polarization curves of Nafion®-212 membrane-based PEM fuel cell at different backpressures from 0 to 0.2 MPa. $T = 70\,°C$; $RH = 100\%$; H_2 and O_2 stoichiometries are 1.5 and 2.0, respectively. (For color version of this figure, the reader is referred to the online version of this book.)

the OCV, the exchange current densities of the electrode reactions, the membrane conductivity, and the mass transfer properties. Overall, incremental increases in fuel cell pressure will result in performance enhancement. However, negative effects, such as high crossover, sealing problems, parasitic power loss, higher cost for compression, and enlarged fuel cell system volume can occur with increasing operating pressure. Thus, it is necessary to optimize the fuel cell operating pressure to balance these positive and negative effects.

REFERENCES

[1] Zhang J, Li H, Zhang J. ECS Trans 2009;19:65–76.
[2] Zhang J, Tang Y, Song C, Zhang J, Wang H. J Power Sources 2006;163:532–7.
[3] Cheng X, Zhang J, Tang Y, Song C, Shen J, Song D, et al. J Power Sources 2007;167:25–31.
[4] Wang L, Liu H. J Power Sources 2004;134:185–96.
[5] Wang L, Husar A, Zhou T, Liu H. Int J Hydrogen Energy 2003;28:1263–72.
[6] Sun H, Zhang G, Guo L-J, Liu H. J Power Sources 2006;158:326–32.
[7] Alzate V, Fatih K, Wang H. J Power Sources 2011;196:10625–31.
[8] Kazim A. Energy Convers Manage 2005;46:1073–81.
[9] Boyer C, Gamburzev S, Appleby AJ. J Appl Electrochem 1999;29:1095–102.
[10] Al-Baghdadi MARS, Al-Janabi HAKS. Int J Hydrogen Energy 2007;32:4510–22.
[11] hyperphysics. http://hyperphysics.phy-astr.gsu.edu/hbase/kinetic/watvap.html.
[12] Zhang J, Tang Y, Song C, Xia Z, Li H, Wang H, et al. Electrochim Acta 2008;53:5315–21.
[13] Song C, Tang Y, Zhang JL, Zhang J, Wang H, Shen J, et al. Electrochim Acta 2007;52: 2552–61.

[14] Zhang J, Song C, Zhang J, Baker R, Zhang L. J Electroanal Chem., in press, corrected proof, 2012, http://dx.doi.org.pass.cisti-icist.nrc-cnrc.gc.ca/10.1016/j.jelechem.2012.09.033.

[15] Kong CS, Kim D-Y, Lee H-K, Shul Y-G, Lee T-H. J Power Sources 2002;108:185–91.

[16] Song Y, Fenton JM, Kunz HR, Bonville LJ, Williams MV. J Electrochem Soc 2005;152:A539–44.

[17] Inaba M, Kinumoto T, Kiriake M, Umebayashi R, Tasaka A, Ogumi Z. Electrochim Acta 2006;51:5746–53.

[18] Tobias CW. In: Meredith RE, Tobias CW, editors. Advances in electrochemical science and engineering. New York: Interscience; 1962.

[19] Amirinejad M, Rowshanzamir S, Eikani MH. J Power Sources 2006;161:872–5.

[20] Kazim A, Forges P, Liu HT. Int J Energy Res 2003;27:401–14.

[21] Yan W-M, Wang X-D, Mei S-S, Peng X-F, Guo Y-F, Su A. J Power Sources 2008;185: 1040–8.

High-Temperature PEM Fuel Cells

PEM Fuel Cell Testing and Diagnosis. http://dx.doi.org/10.1016/B978-0-444-53688-4.00010-3

10.1. INTRODUCTION

High-temperature proton exchange membrane fuel cells (HT-PEM fuel cells), which use modified perfluorosulfonic acid (PFSA) polymers [1–3] or acid–base polymers as membranes [4–8], usually operate at temperatures from 90 to 200 °C with low or no humidity. The development of HT-PEM fuel cells has been pursued worldwide to solve some of the problems associated with current low-temperature PEM fuel cells (LT-PEM fuel cells, usually operated at <90 °C); these include sluggish electrode kinetics, low tolerance for contaminants (e.g. carbon monoxide (CO)), and complicated water and heat management [4,5]. However, operating a PEM fuel cell at >90 °C also accelerates degradation of the fuel cell components, especially the membranes and electrocatalysts [8].

In this chapter, we will first discuss in detail the benefits of HT-PEM fuel cells. Then, we will review the recent research in developing membranes and catalysts for these fuel cells, followed by the descriptions of their design, testing, and diagnosis. Finally, we will assess the challenges that HT-PEM fuel cells present, and make some suggestions for future research directions.

10.2. BENEFITS OF HT-PEM FUEL CELLS

10.2.1. Improved Electrode Kinetics

As discussed in Chapter 1, the exchange current density for the oxygen reduction reaction (ORR) on Pt-based catalysts is in the range of 10^{-9}–10^{-8} A cm^{-2}, whereas the exchange current density for the hydrogen oxidation reaction (HOR) is in the range of 10^{-4}–10^{-3} A cm^{-2}. This significant difference suggests that the overall electrochemical kinetics of PEM fuel cells is limited by the relatively slow ORR.

For both the HOR and ORR, the exchange current densities are given by the following equations:

$$i^o_{O_2} = I^o_{O_2} e^{-(E^o_c/RT)} \tag{10.1}$$

$$i^o_{H_2} = I^o_{H_2} e^{-(E^o_a/RT)} \tag{10.2}$$

where $I^o_{O_2}$ and $I^o_{H_2}$ are, respectively, the cathodic and anodic exchange current densities at infinite temperature; E^o_c and E^o_a are the activation energies for the

ORR and HOR reactions, respectively; while R and T have their usual significance. As the operating temperature increases, the cathodic and anodic exchange current densities will subsequently increase, and therefore, both the electrode's kinetics will be improved.

10.2.2. Improved Contaminant Tolerance

Fuel cell feed streams, such as air for the cathode and hydrogen for the anode, contain some undesirable impurities or contaminants. For the air stream, the main contaminants are NO_x (NO/NO_2), SO_x (SO_2/SO_3), CO_x (CO_2/CO_3), and other volatile organic compounds (VOCs), while for the hydrogen stream, the main contaminants are CO_x, H_2S, SO_x, and other sulfur-containing trace organic compounds. All these contaminants can poison fuel cell catalysts, which can lead to significant degradation in the fuel cell performance.

Let us take anode CO contamination as an example to discuss how high-temperature operation can improve the contamination tolerance of the catalysts. As we know, PEM fuel cells generally perform best with pure hydrogen as the fuel. However, with the current technology, hydrogen is generated by steam reforming of various organic fuels, such as methanol, natural gas, gasoline, which contain trace amounts of CO in the range of 0.1–2%, and other trace impurities. For LT-PEM fuel cells, usually operated around 80 °C, fuel CO content as low as 10–20 ppm can result in a significant loss in cell performance due to the poisoning of the anodic Pt catalysts [9]. To get the CO level <10 ppm, additional purification of the reformate gas is necessary, which increases the complexity of the fuel processing system.

CO poisons Pt catalysts due to its strong adsorption on the Pt surface; it occupies the Pt active sites for the HOR and hence decreases the anodic catalyst activity. Figure 10.1 portrays CO adsorption on a Pt surface [10], which suggests two typical bonding modes. The linear adsorption of CO occupies one Pt site, whereas the bridge-bonded adsorption of CO occupies two adjacent Pt sites.

To reduce the CO poisoning of Pt catalysts, a few strategies have been developed, and these include (1) creating CO-tolerant electrocatalysts (e.g. PtRu/C [11,12], PtSn/C [11,13], $Pt/Ti_{0.7}W_{0.3}O_2$ [14]); (2) feeding a small

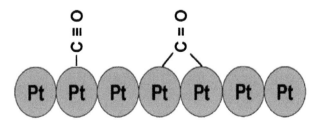

FIGURE 10.1   Two types of CO adsorption on a Pt surface [10]. (For color version of this figure, the reader is referred to the online version of this book.)

amount of oxidant (e.g. O_2 or H_2O_2) into the fuel (called air bleeding) [15,16]; and (3) adding another fuel processing system. The first two strategies are based on the oxidation of the CO in the fuel stream, either chemically or electrochemically. The third strategy is based on purification of the fuel. However, all these strategies have significant drawbacks.

Although considerable efforts have been made to develop CO-tolerant electrocatalysts, currently their effectiveness is far from satisfactory. PtRu alloys seem to be the most attractive CO-tolerant electrocatalysts. It is believed that the presence of Ru in the alloys can promote water dissociation, and the resulting oxygen-containing species facilitate CO oxidation—a so-called bifunctional mechanism. The CO-tolerant activity and long-term durability of PtRu catalysts are still high enough for practical applications. However, the strategy of adding an oxidant into the fuel stream could decrease fuel utilization and lead to safety issues. Additional fuel processing could also increase the total system cost, as well as considerably increase the system complexity and the time for startup and transient response.

As noted, the strong adsorption of CO on Pt causes poisoning. An effective approach to mitigate CO poisoning is to reduce its adsorption through high-temperature operation. It is well known that the adsorption of CO on Pt exhibits high negative entropy, and this indicates that CO adsorption is strongly favored at low temperatures [17,18]. Figure 10.2 shows the competitive adsorption of CO and H_2 on a smooth Pt surface as a function of temperature [2]. These adsorption plots are calculated based on the adsorption equilibrium constants for CO and H_2 on Pt(111) surfaces under conditions of 100 ppm CO in 1.0 bar of H_2 gas. It can be seen that the H_2 coverage increases from 0.02 monolayers at 350 K (77 °C) to 0.39 monolayers at 450 K (127 °C). Therefore, if the fuel cell is operated at higher temperatures, the CO adsorption can be significantly reduced, leading to better CO tolerance. Because the current density is

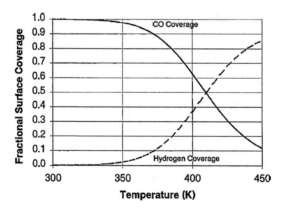

FIGURE 10.2 Langmuir-type adsorption of hydrogen and CO on a smooth platinum surface as a function of the temperature [2].

proportional to the H_2 coverage at the anode, a 50 °C temperature increase from 80 to 130 °C can increase the current density by a factor of 20. With respect to the effect of temperature on CO adsorption, high-temperature operation of a PEM fuel cell (>120 °C) has been reported in the literature. For example, Lakshmanan et al. [19] used poly(ether ether ketone) (PEEK) membranes and PtRu catalysts to fabricate their MEA and then tested it at a high temperature; an increase in CO tolerance from 50 to 1300 ppm was observed as the operating temperature was increased from 70 to 120 °C. Li et al. [20] also investigated the CO poisoning effect on a Pt/C catalyst by using a HT-PEM fuel cell that contained phosphoric-acid-doped polybenzimidazole (PBI) membrane over a temperature range of 125–200 °C. As shown in Fig. 10.3, by defining the CO tolerance as a voltage loss <10 mV, a tolerance of 1% CO can be obtained at current densities up to 0.3, 1.2, and 1.3 A cm^{-2} as the operating temperature is increased from 125 to 150 and 200 °C, respectively. At 125 °C, a tolerance of 0.1% CO was obtained at a current density <0.3 A cm^{-2}. However, at 80 °C, even for H_2 fuel containing as little as 25 ppm CO, tolerance can only be maintained at a current density of 0.2 A cm^{-2}. As an approximate estimate, operation of a PEM fuel cell at temperatures >130 °C results in sufficient CO tolerance to use H_2 directly from a simple reformer.

10.2.3. Simplified Management of Water and Heat

Water management is absolutely critical in PEM fuel cells, because it affects the overall power and efficiency of a system. For a PFSA-membrane-based PEM fuel cell operated at <90 °C, complicated humidifier subsystems are usually needed to humidify the feeding gases (H_2 and O_2 or air) to maintain

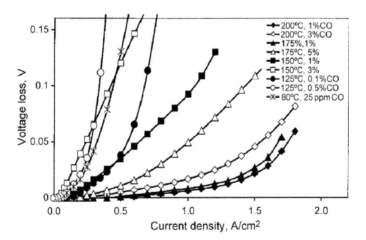

FIGURE 10.3   Loss in voltage as a function of current density at different temperatures and different CO concentrations [20].

a high water content for ionic conduction in the membrane. However, excessive water in the electrode often causes "flooding," especially in the cathode, where water is generated by an electrochemical reaction. The excess water in the electrode fills the pores in the catalyst layer and gas diffusion layer, or blocks the channels on the flow fields, thus hindering gas transport and significantly decreasing cell performance. Operation of a PEM fuel cell at high temperatures could mitigate the potential issues associated with water flooding. Unfortunately, because water is necessary in a PFSA-based membrane, eliminating the humidifier in a PEM fuel cell is not an option, so the benefit of high-temperature operation is lessened. For example, an increase in the operating temperature from 80 to 200 °C could lead to a saturated water vapor pressure as high as 15 atm, which would greatly complicate the system if high relative humidity was to be achieved during operation. Therefore, PFSA-based membranes are not the right choice for high-temperature operation.

Fortunately, for acid–base polymer membranes, such as the PBI membrane, proton transfer is conducted by a doped acid, H_3PO_4, through a solid PBI matrix, whose conductivity is less dependent on the water content than in a PFSA-membrane system. A H_3PO_4–PBI-membrane-based PEM fuel cell can be operated without humidification at a temperature as high as 200 °C; in such a fuel cell system, flooding issues are avoidable, and the system design can be significantly less complicated because humidification is not necessary. Also, because there is no liquid water in the electrode, water management in this system becomes much easier. Further, due to the lack of liquid water in the electrode, two other benefits emerge from high-temperature operation: (1) much less mass transport limitation in the catalyst layer and gas diffusion layer and (2) simplified flow-field design, as there is only one-phase flow in the flow channels.

Another benefit of high-temperature operation is rapid heat removal, which reduces the load on the cooling system. A PEM fuel cell operating at a low temperature, such as 80 °C, with an efficiency of 40–50% can produce a large amount of heat that must be removed to keep the fuel cell stack at a controllable temperature. Therefore, a cooling subsystem is necessary to maintain reliable operation, good performance, high efficiency, and long durability in the stack. However, if the stack is operated at a high temperature, such as 200 °C, its rate of heat dispersal will be much faster than at a low temperature; this will lead to a significant reduction in the heat exchanger size of the cooling subsystem. For example, when the operating temperature increases from 80 to 200 °C, the front area of the radiators may be reduced 3- to 4-fold. This reduction will simplify the cooling subsystem and increase the mass-specific and volume-specific power density of the fuel cell system. In addition, operating PEM fuel cells at high temperatures can yield useful high-temperature water, which can be used for other purposes and result in an improved overall efficiency of the whole PEM fuel cell system.

10.2.4. Other Benefits

When residual water produced during fuel cell operation remains in the electrodes after the stack is shut down, problems can arise, particularly when the environmental temperature is <0 °C. When the stack is exposed to subzero conditions, the residual water will freeze, so the volume of the electrodes (in particular, the catalysts layers) will expand due to ice formation, which will lead to structural damage and decreased electrochemical active surface area. This has been reported as an additional degradation mechanism in PEM fuel cells [21]. However, if the PEM fuel cell is operated at high temperatures, less liquid water will remain in the electrode and thus decrease the impact of fuel cell structure failure caused by frozen water.

In addition, if we operate a PEM fuel cell at around 200 °C, which is close to the temperature for methanol reforming, it is possible to integrate the fuel cell with a methanol reformer. Such integration is expected to increase the overall power system efficiency and resolve the problems associated with hydrogen storage [22].

Unfortunately, operation of PEM fuel cells at high temperatures also poses challenges, such as rapid degradation of both catalyst and membrane. This issue will be discussed in a later section of this chapter.

10.3. MEMBRANE DEVELOPMENT FOR HT-PEM FUEL CELLS

As discussed above, some of the shortcomings associated with PFSA-membrane-based LT-PEM fuel cells could be solved or avoided by operating PEM fuel cells at high temperatures [23]. High-temperature operation also yields some other benefits, as discussed above. To facilitate high-temperature operation, a key technology to develop is high-temperature proton-conducting membranes.

We have established that at high temperatures and/or at low humidity, conventional PFSA membranes will dehydrate and will result in low proton conductivity. In the past several years, great efforts have been made to develop polymer membranes that are capable of retaining high proton conductivity in anhydrous environments, while still possessing chemical and electrochemical stability at high temperatures. These membranes can be classified into three groups: (1) modified PFSA membranes, (2) alternative sulfonated polymers and their composite membranes, and (3) acid–base polymer membranes.

10.3.1. Modified PFSA Membranes

Traditionally, PEM fuel cells have been developed based on the use of PFSA polymer membranes. These polymers contain a perfluorinated backbone similar to that of polytetrafluoroethylene (PTFE, Teflon®), which has pendent perfluorinated chains linked to the main chain by ether bonds, and a sulfonated group at the end of each chain. Among the PFSA membranes, the most widely used and studied are the Nafion® membranes. The structure of Nafion® is shown in Fig. 10.4. One intrinsic drawback of Nafion® and other PFSA polymer membranes is that their proton conductivity is dependent on water content; they therefore require a high relative humidity and an operating temperature below approximately 90 °C to avoid water loss.

To achieve high-temperature operation, three approaches have been developed to modify PFSA membranes: (1) swelling the membrane with nonaqueous and low-volatility solvents; (2) reducing the membrane thickness; and (3) impregnating the membrane with hygroscopic oxide nanoparticles or solid inorganic proton conductors.

For example, Savinell et al. [24] attempted to incorporate phosphoric acid (PA) in a Nafion® membrane and achieved a proton conductivity of 0.05 S cm^{-1} at 150 °C. The low-volatility PA acts as a Brønsted base and solvates the protons from the sulfonic acid group in the same way as water does. Due to the low volatility of PA, the operating temperature can even be increased to 200 °C. By following this strategy, other acids or ionic liquids, such as phosphotungstic acid [25], 1-butyl, 3-methylimidazolium triflate, BMI tetrafluoroborate [26], and heterocycles [27], have been used to impregnate Nafion® membranes. The conductivities of these modified membranes were better than those of the Nafion® membranes at high temperatures. However, almost no fuel cell tests that made use of these modified membranes have been reported, due to two major difficulties: (1) immobilization of the acid and/or ionic liquid, especially in the presence of water and (2) catalyst poisoning due to solvent adsorption.

Reducing the membrane thickness can facilitate the backdiffusion of water from the cathode to the anode, which will keep water in the membranes and lead to improved conductivity at high temperature and low relative humidity [28]. One challenge in developing thin membranes is the consequent reduction in the mechanical strength; to improve mechanical strength, PTFE-reinforced PFSA membranes have been well investigated. For example, Lin et al. [29] achieved a performance of 250 mA cm^{-2} at 0.6 V by using a porous PTFE–PFSA membrane with a thickness of 25 μm for operation at 120 °C. This performance is better than that with thick membranes under similar conditions, but is much lower than 700 mA cm^{-2} at 0.6 V achieved at 80 °C with the same relative humidity.

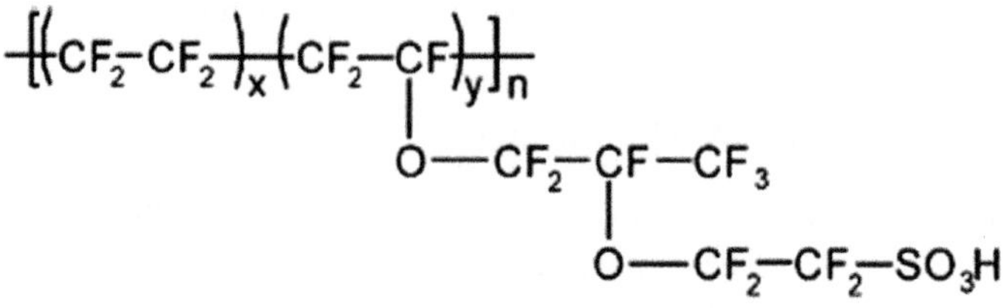

FIGURE 10.4 Structure of Nafion®.

Another way to achieve low-humidity, high-temperature operation of PFSA-membrane-based PEM fuel cells is to incorporate hygroscopic oxides, such as SiO_2 and TiO_2, into the hydrophilic domains inside the membrane to improve the composite membrane water retention, thus increasing the proton conductivity at high temperatures. Two approaches have been developed to prepare these composite membranes: (1) direct addition of hygroscopic oxide particles into a Nafion® solution, followed by casting and (2) impregnation of membranes with solutions of inorganic precursors, followed by in situ sol–gel reaction to produce metal oxides inside the membranes. For this latter method, Watanabe et al. [30] reported that the water uptake of the oxide-containing membrane was higher than that of a pristine Nafion® membrane. For recast Nafion® membranes, the water uptake was 17 wt.%, whereas for recast membranes containing 3 wt.% SiO_2, the water uptake was as high as 43 wt.% [31]. Due to water adsorption on the oxide surface, backdiffusion of the product water from the cathode to the anode can be enhanced, and the electro-osmotic drag of water from the anode to the cathode can be reduced. Consequently, much more water remains in the membrane, which thus facilitates the fuel cell operation at high temperature and low relative humidity.

Figure 10.5 compares the cell performance of different MEAs when using a pristine Nafion®-115-membrane, recast Nafion® membrane, and recast

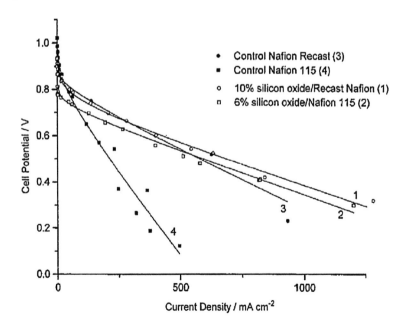

FIGURE 10.5   Cell potential vs. current density of (●) recast Nafion® control, (■) Nafion® 115 control, (○) 10% silicon oxide/recast Nafion®, and (□) 6% silicon oxide/Nafion®-115 membranes at a pressure of 3.0 atm. H_2 and O_2 humidifier temperature: 130 °C; cell temperature: 130 °C [2].

Nafion® composite membranes that contain SiO$_2$ particles at an operating temperature of 130 °C [2]. These results confirm that SiO$_2$-containing membranes can perform better than a pristine Nafion® 115 membrane, irrespective of whether the SiO$_2$ is impregnated into the Nafion® membrane or is directly doped into a Nafion® solution, followed by recasting. Figure 10.6 shows the stability tests of MEAs at 130 °C when using SiO$_2$-containing Nafion® 115 and pristine Nafion® 115 [2]. The Nafion®-115 membrane modified with 6 wt.% SiO$_2$ demonstrates a much better stability during 50 h of testing; the pristine Nafion®-115 membrane, in contrast, degrades dramatically and fails in an hour.

In summary, impregnation of the nanopores of Nafion® with SiO$_2$ can result in the membrane having better water retention and less susceptibility to high-temperature damage, making it possible for PEM fuel cells to be operated at high temperatures.

10.3.2. Sulfonated Polymers and Their Composite Membranes

Originally, sulfonated polymer membranes were used as an alternative to PFSA membranes for low-temperature operation, due to the former's lower cost. Recently, however, some—especially sulfonated hydrocarbon membranes—were found to be less dependent on humidity than were PFSA membranes if their associated H$_2$O/SO$_3^-$ ratios were low. This allowed them to have good conductivity at high temperatures, which is promising for high-temperature operation.

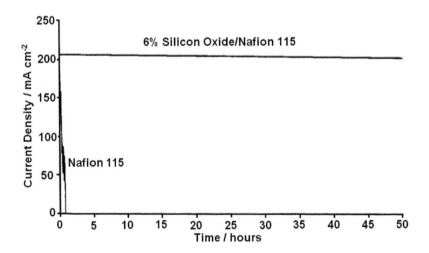

FIGURE 10.6   Time performance test of Nafion® 115 and Nafion® 115/silicon oxide at a pressure of 3.0 atm. H$_2$ and O$_2$ humidifier temperature: 130 °C; cell temperature: 130 °C; potential: 0.65 V [2].

In the literature, the most widely investigated sulfonated polymer membrane systems include sulfonated PEEK (SPEEK) [32,33], polysulfones [34], poly-imides [35], and polyphenylenes [36]. In general, the microstructures of sulfonated aromatic polymers are different from those of PFSA membranes. As shown in Fig. 10.7, the perfluorinated polymer backbones of PFSA membranes are hydrophobic, and their terminal sulfonic acid groups are hydrophilic [37]. Therefore, when water is present, only the hydrophilic domain of the nano-structure is hydrated; this maintains proton conductivity, while the hydrophobic domains provide mechanical strength. Thus, water uptake by PFSA membranes is very high and also very sensitive to the relative humidity. In the case of sulfonated hydrocarbon polymers, their hydrocarbon backbones are less hydro-phobic, and their sulfonic acid functional groups are less hydrophilic. Hence, the water molecules are completely dispersed in the polymer nanostructure. The advantage of this hydrocarbon nanostructure is that conductivity is less dependent on humidity, which allows for high proton conductivity at high temperatures.

When these sulfonated polymer membranes are used in PEM fuel cells, especially for high-temperature operation, the membrane stability is a large concern. Their thermal stability is primarily due to the desulfonation of the sulfonic acid side chains [38]. For Nafion® membranes, the sulfonic acid group is stable up to 280 °C in air [39]. For most of the sulfonated hydrocarbons, the

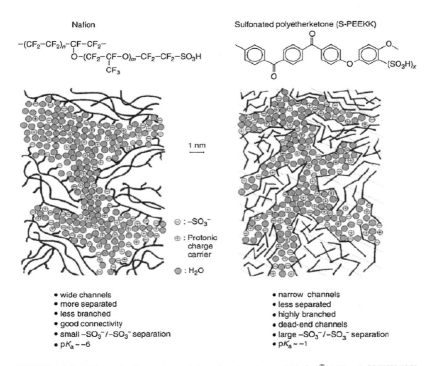

FIGURE 10.7   Schematic illustrations of the microstructures of Nafion® 117 and SPEEK [37].

sulfonic acid group is usually stable between 240 and 330 °C [33,40], which suggests that this kind of membrane has good thermal stability. However, the chemical stability of sulfonated polymer membranes is of more concern for the lifetime of a membrane in a PEM fuel cell. The H_2O_2 and the HO· and HO_2· radicals produced by the incomplete reduction of oxygen on the cathode are believed to be more aggressive in attacking the hydrogen-containing terminal bonds in hydrocarbon polymers; this is assumed to be the principal degradation mechanism of sulfonated hydrocarbon polymers [4]. To date, limited information is available about the long-term stability of sulfonated hydrocarbon polymers, especially at high temperatures. Another concern for high-temperature operation is the polymers' mechanical strength, especially when the degree of sulfonation is high, which is necessary to achieve high conductivity.

To improve the mechanical strength, thermal stability, and proton conductivity of membranes, especially at high temperatures, sulfonated hydrocarbon polymers are widely used as a matrix for preparing inorganic–organic composite membranes. As with the modification of PFSA membranes, solid inorganic compounds such as SiO_2 [6] and TiO_2 [41]; inorganic proton conductors such as ZrP [42], SiWA [39], and PWA [43]; and SiO_2-supported inorganic proton conductors [44] are incorporated into sulfonated hydrocarbon polymers to develop inorganic–organic composite membranes that yield a better performance under higher operating temperatures. As reported in the literature, some of these composite membranes exhibited promising conductivities at a temperature of >100 °C. However, most of these composite membranes have not been tested in real PEM fuel cells at high temperatures.

10.3.3. Acid–Base Polymer Membranes

Although many efforts have been made to develop modified PFSA and sulfonated polymer membranes that will function well during high-temperature operation, the proton-conducting nature of these membranes still requires them to be fully hydrated to maintain high proton conductivity. Further, the maximum operating temperatures of these membranes are still relatively low. In this regard, acid–base polymer membranes represent an effective approach to achieve high-temperature operation with high proton conductivity. In general, polymers have basic sites such as ether, alcohol, imine, amide, or imide groups, which can react with strong acids such as PA or sulfuric acid. Many basic polymers have been investigated for fabricating acid–base polymer membranes, such as poly(ethylene oxide) [45], polyvinyl alcohol [46], polyethylenimine [47], and polyacrylamide [48]. Most of these acid–base polymer membranes exhibit proton conductivity at $<10^{-3}$ S cm^{-1} at room temperature. To enhance conductivity, high acid content is usually needed. Unfortunately, the high acid content in these polymers can decrease their mechanical strength. Even worse, the oxidative stability of the polymers that have been studied inhibits their application in PEM fuel cells, especially at high temperatures.

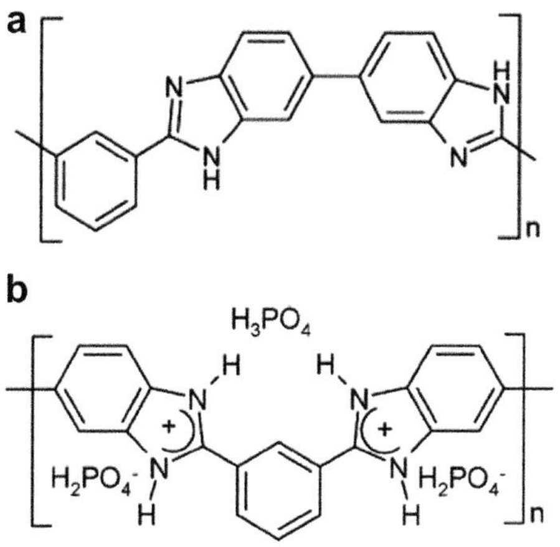

FIGURE 10.8 Structures of PBI (a) and phosphoric-acid-doped PBI (b) [4,8].

Among the acid–base polymer membranes, an exception is phosphoric acid (PA)-doped PBI, which has received the most attention and has been used in PEM fuel cells at temperatures as high as 200 °C without humidity. Figure 10.8a shows the structure of PBI, which is an amorphous thermoplastic polymer with a glass transition temperature of 425–436 °C; this may be the reason for its high thermal stability. PBI also has good chemical resistance and excellent mechanical strength. When it is doped with PA, the first two PA molecules absorbed by the PBI membrane can form a salt by protonation of the imine N group at the imidazole ring. The structure of PA-doped PBI is shown in Fig. 10.8b. Excess PA molecules are present as free or unbounded acid—also very important for improving the conductivity [4]. It is known that the conductivity of PA-doped PBI follows the Arrhenius law with a "hopping" conduction mechanism. In addition, the conductivity is strongly dependent on the acid-doping level, temperature, and humidity. In the conduction mechanism, the proton hopping that occurs from one N–H site to another in the PBI membrane contributes little to the conductivity, but the proton hopping from one N–H site to the PA anions contributes significantly to the conductivity.

10.4. CATALYST DEVELOPMENT FOR HT-PEM FUEL CELLS

As discussed in the previous sections, an increase in PEM fuel cell operating temperatures to >90 °C not only resolves or at least alleviates some LT-PEM fuel cell problems but it also provides other benefits. One of the key components when operating PEM fuel cells at high temperatures is the proton exchange membrane; another is the electrocatalyst. However, the literature contains limited information about HT-PEM fuel cell catalysts. It is expected that high-temperature operation will pose challenges for catalyst stability, but both the electrode kinetics and the catalyst's contamination tolerance improve with increasing temperature.

10.4.1. Pt/C Stability in HT-PEM Fuel Cells

The stability of carbon-supported Pt catalyst (Pt/C)—to date, the most practical catalyst for PEM fuel cell technology—in LT-PEM fuel cells has been well studied in the literature [49,50]. Three degradation modes during LT-PEM fuel cell operation have been proposed for this catalyst: (1) Pt particle agglomeration, either through coalescence growth or the Ostwald ripening process; (2) Pt dissolution and detachment; and (3) carbon corrosion. Figure 10.9 shows a schematic for the catalyst agglomeration mechanism [51].

Only a few publications have reported the stability of Pt/C catalysts for HT-PEM fuel cells, especially PA-doped PBI -membrane-based HT-PEM fuel cells. For example, Liu et al. [52] conducted a 600-hour lifetime test of a PA-doped PBI -membrane-based HT-PEM fuel cell by using commercially available Pt/C as both anode and cathode catalysts. The Pt particle sizes before and after the lifetime test were evaluated by transmission electron microscopy (TEM) measurements. Figure 10.10 shows the TEM images and Pt particle size distribution histograms of the Pt/C catalysts before and after the test. The TEM results reveal that the Pt particle agglomeration occurred at both the anode and the cathode, but more severely on the latter. For the fresh Pt/C catalyst, the Pt particle size distribution was relatively narrow, with a range of 2–5 nm and an

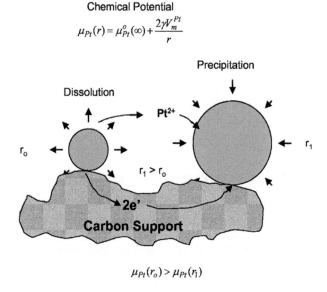

Chemical Potential

$$\mu_{Pt}(r) = \mu^o_{Pt}(\infty) + \frac{2\gamma V^{Pt}_m}{r}$$

$$\mu_{Pt}(r_o) > \mu_{Pt}(r_1)$$

FIGURE 10.9 Schematic of Pt agglomeration [51].

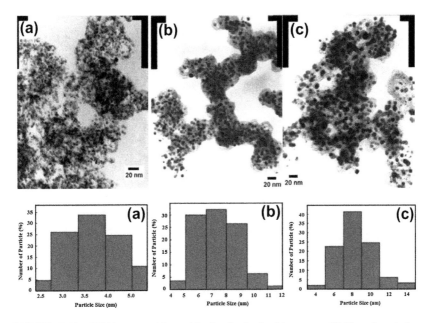

FIGURE 10.10   TEM images and Pt particle size distribution histograms of Pt/C catalysts before and after lifetime testing [52]. (For color version of this figure, the reader is referred to the online version of this book.)

average particle size of 3.7 nm. After the lifetime test, the dispersion became very broad, from 4 to 14 nm, with an average size of 8.4 nm. Zhai et al. [53] also observed Pt agglomeration during stability testing of Pt/C catalysts in a PA-doped PBI -membrane-based HT-PEM fuel cell. After the Pt particle size distribution was analyzed using a method developed by Ascarelli et al. [54], the Pt/C catalyst agglomeration was found to follow the coalescence mechanism, in which the small Pt particles move together to form large particles.

10.4.2. New Catalysts for HT-PEM Fuel Cells

Because commercially available Pt/C catalysts show significant degradation during the long-term operation of PA-doped PBI -membrane-based HT-PEM fuel cells, it is necessary to develop highly durable catalysts for these conditions.

The agglomeration of small Pt particles during the operation of a HT-PEM fuel cell is very severe, and this makes it one of the main factors to consider in catalyst degradation. The literature presents some efforts to decrease this agglomeration. For example, Liu et al. [55] introduced stable metal oxides such as ZrO_2 into Pt/C catalysts to decrease Pt agglomeration. Their Pt_4ZrO_2/C catalyst was prepared by depositing ZrO_2 on carbon, followed by deposition of Pt on the ZrO_2/C composite support. A potential sweep test at 150 °C between 0.6 and 1.2 V was conducted to evaluate the stability of the newly developed

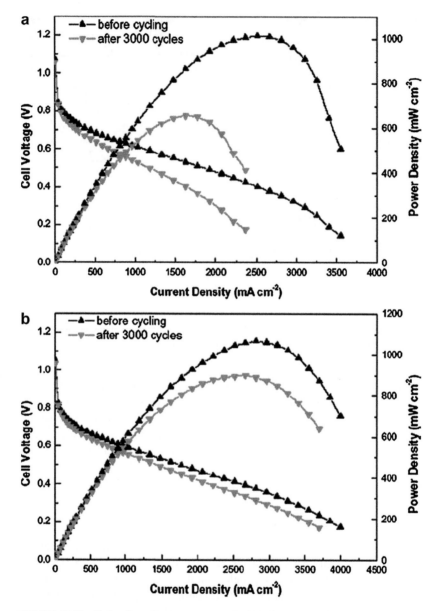

FIGURE 10.11 Fuel cell performance curves of (a) TKK 46.6% Pt/C-based MEA and (b) Pt_4ZrO_2/C-based MEA before and after a potential sweep test for 3000 cycles between 0.6 and 1.2 V vs. RHE. Anode: commercial TKK 46.6% Pt/C catalyst with Pt loading of 1.0 mg cm^{-2}. Cathode Pt loading: 1.0 mg cm^{-2}. Cell temperature: 150 °C. $H_2/O_2 = 0.1$ MPa/0.1 MPa. Pure hydrogen and oxygen were fed directly into the cell without humidification [55]. (For color version of this figure, the reader is referred to the online version of this book.)

catalyst. For comparison, the stability of commercially available Pt/C catalyst was also evaluated using the same accelerated aging test. Figure 10.11 shows the fuel cell performances of Pt/C-based and Pt_4ZrO_2/C-based MEAs before and after a potential sweep test of 3000 cycles. Evidently, the performance degradation of the Pt_4ZrO_2/C-based MEA is less significant than that of the Pt/C-based MEA. For example, the cell voltage loss of the Pt_4ZrO_2/C-based MEA at $100\ \text{mA cm}^{-2}$ is 18 mV, while the voltage loss of the Pt/C-based MEA is 27 mV. Similarly, at $1000\ \text{mA cm}^{-2}$, the cell voltage loss of the Pt_4ZrO_2/C-based MEA is 37 mV, whereas that of the Pt/C-based MEA is as high as 83 mV. The Pt particle size distributions of the Pt_4ZrO_2/C and Pt/C catalysts before and after the potential tests are shown in Figs 10.12 and 10.13. It is clear that the particle size of the Pt_4ZrO_2/C catalyst increases from approximately 4 to 6 nm, while the particle size of the Pt/C catalyst increases from approximately 3 to12 nm, indicating that Pt nanoparticles on ZrO_2/C are more resistant to agglomeration or sintering. The authors explained that ZrO_2 on the carbon support acts an anchor or barrier to the adjacent Pt nanoparticles, which thereby inhibits their agglomeration.

Another catalyst degradation mode is catalyst carbon support corrosion or oxidation. Figure 10.14 shows a schematic of the carbon corrosion mechanism [56].

As discussed above, catalyst sintering or agglomeration reduces PEM fuel cell durability, which can then reduce the electroactive surface area of the Pt catalyst, reduce Pt use, and degrade its catalytic activity. One of the major causes of this agglomeration has been identified as oxidation of the carbon catalyst support. This carbon oxidation or corrosion effect is more pronounced when PEM fuel cells are operated at higher temperatures. For example, one recent study showed that at 300 °C, Pt particle size increased from 4.2 to 8.7 nm after just a few hours of operation [56]. In the quest to eliminate carbon support oxidation, development of noncarbon materials to replace carbon as the catalyst support has become an active research area in recent years. Such materials must meet the following requirements: high thermal stability, high electrical conductivity, high electrochemical stability, low solubility in acid, high surface area, and high mass activity and durability once loaded with Pt catalysts. Among noncarbon support materials, TiO_2 meets some of the support material requirements, except for its low electrical conductivity and low mass activity. However, with modifications, such as doping and compositing with carbon, catalysts using TiO_2-based support materials have proven to be feasible in PEM fuel cell applications. Aside from Pt/C composited with ZrO_2 [55], it has also been reported that introduction of TiO_2 into a Pt/C catalyst can form Pt/TiO_2/C, which helps to reduce Pt agglomeration and at the same time to improve catalytic activity [57–59]. Huang et al. [60] developed a TiO_2-supported Pt catalyst that exhibited high stability along with activity comparable to a state-of-the-art, carbon-based Pt catalyst. Bauer et al. [61] reported a Pt/TiO_2 nanofiber catalyst that showed high durability. However, it is difficult to achieve high

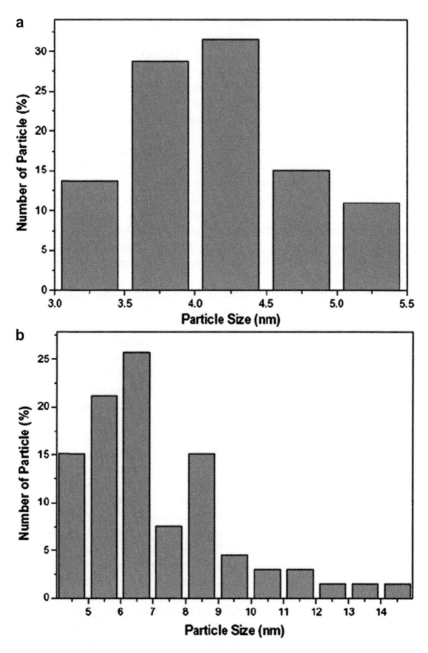

FIGURE 10.12 Histograms of the particle size distribution of the Pt$_4$ZrO$_2$/C catalyst before (a) and after (b) the potential sweep test [55]. (For color version of this figure, the reader is referred to the online version of this book.)

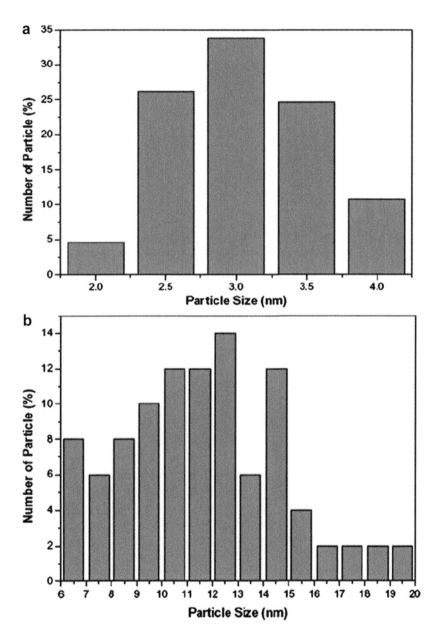

FIGURE 10.13 Histograms of the particle size distribution of the Pt/C catalyst before (a) and after (b) the potential sweep test [55]. (For color version of this figure, the reader is referred to the online version of this book.)

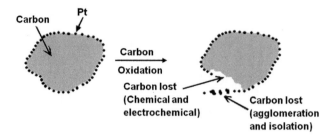

FIGURE 10.14   Schematic representations of carbon corrosion [56].

mass activity with a TiO_2-supported catalyst due to its low conductivity. Doping other conductive elements into the bulk material seems to be effective in further improving the mass activity of TiO_2-supported Pt catalysts, as this improves their conductivity. It has been shown that Nb-doped TiO_2 (Nb-TiO_2) materials can have improved conductivity. For example, Do et al. [62] reported using Nb-TiO_2 as a support and achieving an ORR mass activity of 160 mA $mgPt^{-1}$. Huang et al. [63]. also reported a highly conductive Nb-TiO_2-supported catalyst with improved ORR mass activity and durability. However, in comparison with the targets set by the DOE, the mass activity of Nb-TiO_2 is still far off, as is its durability (this target being 4 times the mass activity). New approaches are therefore required to further improve both the ORR mass activity and the durability of noncarbon-supported catalysts for PEM fuel cell applications.

10.5. DESIGN OF HT-PEM FUEL CELLS

10.5.1. Catalyst Design

To address the problem of low catalyst stability during HT-PEM fuel cell operation, the catalysts should have suitable resistance to sintering and corrosion. The general opinion among researchers is that development of corrosion-resistant catalyst supports and proper anchoring of the metal phase on these supports are both important for improving catalyst stability.

The application of graphitized rather than amorphous carbon is one way to increase the corrosion resistance of the catalyst support [64]. However, the low surface area and the hydrophobicity of graphitized carbon prevent the uniform dispersion of Pt nanoparticles on this type of support. Thus, the surface of the graphitized carbon should be functionalized to control its hydrophobic properties before loading the Pt nanoparticles [65]. Different methods have been developed to increase the hydrophilicity of graphitized carbon, including chemically treating it in HNO_3 or H_2SO_4/HNO_3 acid solutions, and non-covalent functionalization of graphitized carbon via π–π interaction. However, the chemical oxidation method leads to many defects on the carbon surface, and this results in increased electrochemical carbon corrosion [66]. The noncovalent functionalization of graphitized carbon by 1-pyrenecarboxylic acid, 1-aminopyrene, and benzyl mercaptan can preserve the intrinsic properties of graphitized carbon as well as increase its hydrophilicity [65,67]. Thus, noncovalent functionalization is highly recommended for obtaining uniform dispersion of Pt on graphitized carbon.

Although graphitized carbon performs better than does amorphous carbon in terms of oxidation and corrosion resistance, oxidation of graphitized carbon still occurs during high-potential, high-temperature operation and under fuel starvation conditions. Development of noncarbon materials to replace carbon as the catalyst support is the ultimate way to eliminate carbon-support oxidation. In the literature, conducting [68] and nonconducting [69] polymers, metal oxides [60,61], and heteroatom-doped metal oxides [62,63] have been used as fuel cell catalyst supports. Higher support stability than carbon's has been achieved, as expected. However, the low electronic conductivity of noncarbon supports limits the activities of their supported catalysts. To increase the conductivity of non-carbon-supported catalysts, higher metal loading, and continuous metal films on noncarbon supports have also been developed [70]. However, use of precious metals is still very low with higher metal loadings or when continuous metal films are deposited on noncarbon supports. New technology to develop thin-film metal layers on noncarbon supports is urgently needed.

Besides improved stability, catalysts developed for HT-PEM fuel cells should demonstrate higher activity than Pt/C catalysts do. Based on the developments of catalysts for PA fuel cells and LT-PEM fuel cells, shape- and composition-controlled Pt- and Pt-alloy-supported catalysts on noncarbon or composite supports are the future for developing high-activity fuel cell catalysts [71]. In addition, anchoring Pt and Pt-alloys using a stable phase, such as stable metal oxides [55,57–59], can prevent the agglomeration and sintering of Pt and Pt-alloy particles, thereby increasing the stability of such catalysts in HT-PEM fuel cells.

10.5.2. Membrane Design

For HT-PEM fuel cell operation, membranes should have high proton conductivity at high temperatures and/or low humidity, high chemical/electrochemical and thermal stability at high temperatures, and low cost [8,72]. Currently, three types of high-temperature membranes have been developed: (1) modified PFSA membranes; (2) alternative sulfonated polymers and their composite membranes; and (3) acid–base polymer membranes. These show better performance than do commercially available Nafion® membranes under high-temperature operation. However, they still cannot fulfill all the requirements for high performance, high stability, and low cost.

The "pore-filling membrane" concept can be applied to develop high-temperature membranes by using different substrates and polymer fill materials that have a high thermal and chemical/electrochemical durability (Fig. 10.15) [72]. An inorganic substrate can also be applied during membrane synthesis to

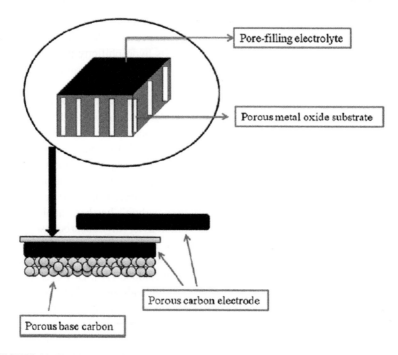

FIGURE 10.15 An electrode–electrolyte membrane integrated system that uses a pore-filling membrane with inorganic substrates [72]. (For color version of this figure, the reader is referred to the online version of this book.)

further enhance the thermal stability. This concept combines the advantages of substrates and polymer fillers and should provide the direction for designing membranes with all the properties required for high-temperature operation.

10.5.3. Gas Diffusion Electrodes, Catalyst Layers, and MEA Designs

The gas diffusion electrode components and electrode structures in HT-PEM fuel cells should differ from those in LT-PEM fuel cells. Because water flooding is not a problem in HT-PEM fuel cells, the requirements for mass transport of the fuel and oxidant within the bipolar plate channels and gas diffusion electrodes may not be as critical as for LT-PEM fuel cells.

Pan et al. [73] investigated the effect of electrode porosity on the performance of PA-doped PBI-membrane-based HT-PEM fuel cells. They optimized the porosity of the gas diffusion layers by applying different amounts of PTFE, and the porosity of the catalyst layer by adding ammonium oxalate into the catalyst layers, followed by heating to remove the additive and make the pores. They observed that an increase in the porosity of the electrode improved the mass transfer, leading to better cell performance (Fig. 10.16).

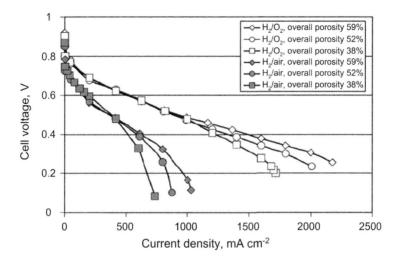

FIGURE 10.16 Polarization curves of PA-doped PBI-membrane-based HT-PEM fuel cells using electrodes with different porosities, and operated with dry gases at atmospheric pressure. The Pt loading of the electrode was 0.5 mg Pt cm^{-2} and the active area of the electrode was 25 cm^2. Hydrogen and oxygen flow rates were 400 ml min^{-1} and the air flow rate was 800 ml min^{-1} [73].

In addition to optimizing the PTFE content in the gas diffusion layers, it is also necessary to optimize the amounts of impregnated PA and PBI ionomer in the catalyst layers to achieve high performance in HT-PEM fuel cells. Wannek et al. [74] found that 40% PTFE in the catalyst layer and 20 mg cm^{-2} H$_3$PO$_4$ per electrode were optimal in their experiments (Figs 10.17 and 10.18). Kim et al. [75] fixed the amount of H$_3$PO$_4$ in the PBI ionomer at 6 H$_3$PO$_4$ molecules per PBI repeating unit, and they used this ionomer to study the catalyst layer performance for a range of 5–40 wt.% H$_3$PO$_4$-doped PBI ionomer. Their results showed that an MEA with 20 wt.% H$_3$PO$_4$-doped PBI ionomer in the catalyst layer performed the best (Fig. 10.19).

Figure 10.20 shows a schematic of the MEA structure in an HT-PEM fuel cell. Most of the components are similar to those in LT-PEM fuel cells [76]; the unique component is the subgasket, which ensures that the MEA maintains proton conductivity and long-term durability.

10.5.4. Single-Cell and Stack Design

Figure 10.21 shows the structure of the hardware of a single cell. Research by Zhang's group [77,78] revealed that when the cell was operated at a temperature <200 °C, the composite graphite material SGL BBP4 could be used for flow-field fabrication, and silicone rubber (Fuel Cell Store 590,363) could be used as the sealing material. But when the cell temperature was 200–300 °C, stainless steel 430 was more suitable as the flow-field material, while silicone rubber only

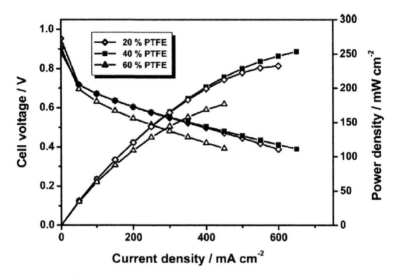

FIGURE 10.17   Performances of HT-PEM fuel cells with different amounts of PTFE in the 20% Pt/C-based catalyst layer ($\sim$1.0 mg Pt cm^{-2} and 20 mg H$_3$PO$_4$ cm^{-2} per electrode) [74].

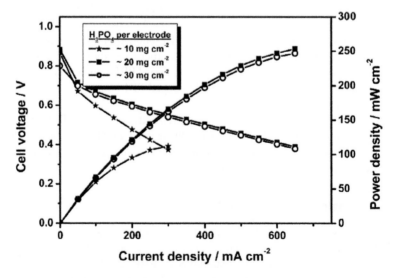

FIGURE 10.18   Influence of acid impregnation amounts in the catalyst layers of gas diffusion electrodes (GDEs; 20% Pt/C, 40% PTFE) on the performance of MEAs, after assembling these GDEs with undoped poly(2,5-benzimidazole) membranes [74].

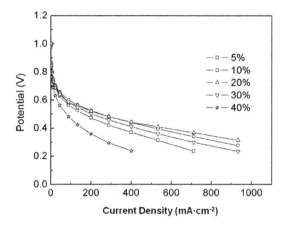

FIGURE 10.19   Polarization curves with various weight percentages of H_3PO_4-doped PBI ionomer in the cathode catalyst layer, at 150 °C using dry H_2/O_2 feed [75].

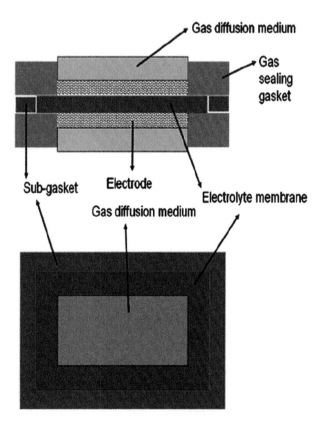

FIGURE 10.20   Schematic of an MEA structure for HT-PEM fuel cells [76]. (For color version of this figure, the reader is referred to the online version of this book.)

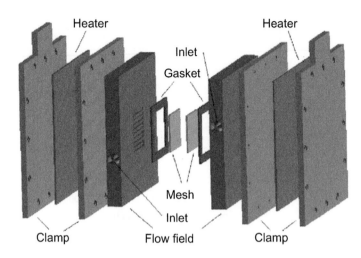

FIGURE 10.21 Schematic of high-temperature fuel cell hardware [77]. (For color version of this figure, the reader is referred to the online version of this book.)

worked up to 160 °C for H_3PO_4-doped PBI-based PEM fuel cells, so they had to change the sealing material. Their research revealed that the use of Teflon$^®$ as the gasket material worked fine only at a temperature <260 °C, whereas Garlock graphite fiber G9900 gaskets remained stable even up to 540 °C.

Because a stack contains many single cells, thermal management is a major challenge for stack design, so the temperature control strategies applied during start-up and normal operation are very important for obtaining high stack performance [79]. Research studies have found that direct electrical and hot air heating are not an efficient means of temperature control for HT-PEM fuel cell stacks [80]. Instead, the use of a cathode air circuit to heat the stack from the inside proved to be more efficient.

To cool stacks, air cooling and liquid cooling are both used. When the operating temperature is high, air cooling is suitable even at extreme outside temperatures, but the startup heating process is prolonged. Liquid cooling involves water or oil [81]. Water cooling using latent heat yields uniform temperature distribution in the stack, but more attention is required to maintain stable stack operation. Oil cooling using sensible heat results in a more stable transient characteristic than water cooling does, but the parabolic temperature distribution varies by about 30 °C.

10.5.5. Direct Use of Methanol Reformate and Integration with Fuel Processor

Because a reformer consumes heat and water, and a fuel cell stack produces heat and water, integration of the stack and the reformer could be expected to improve system efficiency and simplify system construction. This possibility was demonstrated by Pan et al. [22], who integrated a methanol reformer with a PA-doped PBI-membrane-based HT-PEM fuel cell. In their research, they found that methanol reforming took place around 200 °C, with nearly 100% conversion and only 0.2 vol.% CO impurity, which could be tolerated by a PA-doped PBI-based HT-PEM fuel cell operated at that temperature. They also reported 91% hydrogen use.

10.6. TESTING AND DIAGNOSIS OF HT-PEM FUEL CELLS

10.6.1. MEA Preparation

Due to the limited information available on high-temperature MEA preparation, here, we only discuss a typical high-temperature MEA using H_3PO_4-doped PBI membrane as the proton exchange membrane.

Figure 10.22 shows a schematic of the fabrication procedure for this MEA [82]. Typically, a microporous layer, compounded by 1 mg cm^{-2} Ketjen Black 300 carbon (Azko Nobel, UK) and Teflon® (40 wt.%) as binder, is first deposited on Toray Graphite Paper (TGPH-090, 20% wet-proofed) by air brushing, using isopropanol as the solvent to make this layer hydrophobic for gas diffusion.

Catalyst inks are prepared by mixing Pt/C catalyst (20% Pt on Vulcan XC-72, E-TEK Inc.), PBI solution (5 wt.% in N,N-dimethylacetamide, DMAc), and DMAc solvent. After sufficient mixing, the inks are sprayed onto the microporous layer by air brushing. Then, after drying and sintering at 190 °C for an hour inside an inert ventilated oven, the electrodes are doped with 2 M PA and left to dry

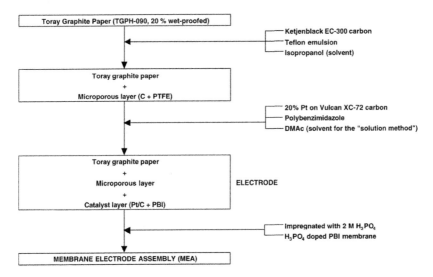

FIGURE 10.22 Schematic of the fabrication procedure for an MEA used in HT-PEM fuel cells [82]. (For color version of this figure, the reader is referred to the online version of this book.)

overnight. The MEA is then fabricated by hot pressing the PA-doped PBI membrane with an electrode on either side at 150 °C and 0.1 t cm^{-2} for 10 min.

10.6.2. Single-Cell and Stack Assembly

Typically, a single cell is fabricated by fixing the assembled MEA between two graphite bipolar plates machined with parallel or serpentine flow fields. Two metal end plates are placed at the outer sides of the two graphite bipolar plates to hold them in place. Rod heaters are inserted into the end plates to control the cell temperature. Figure 10.23 shows a photograph of a typical 50 cm^2 HT-PEM fuel cell [83].

To assemble a stack, a few single cells are arranged between two metal end plates, fixed with threaded tie rods. When the stack has numerous single cells, additional cooling plates are usually inserted for efficient heat integration. Figure 10.24 shows a photograph of a 28-cell stack and an inserted cooling plate [83].

10.6.3. Test Station Modification

HT-PEM fuel cells usually operate with low or no humidification. When the MEA contains a PA-doped PBI membrane, no humidification is needed in the

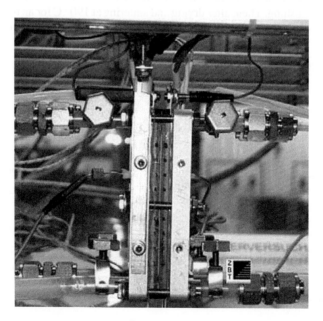

FIGURE 10.23 Photograph of a 50 cm^2 single cell [83]. (For color version of this figure, the reader is referred to the online version of this book.)

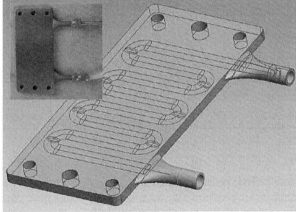

FIGURE 10.24   Photograph of a 28-cell fuel cell stack with integrated cooling plates (inset), and diagram of laser sinter cooling plate (3D design) [83]. (For color version of this figure, the reader is referred to the online version of this book.)

test station, which is the most obvious difference from an LT-PEM fuel cell test station. Figure 10.25 shows a schematic representation of an HT-PEM fuel cell test station [79].

10.6.4. Performance Testing and Diagnosis

Several issues are associated with the testing and diagnosis of HT-PEM fuel cells, including humidity control, thermal management, material corrosion, performance degradation, and lifetime. However, major research activities in this area are currently focused on the development of high-temperature membranes. There has also been a limited amount of independent work on technologies for the testing and diagnosis of HT-PEM fuel cells.

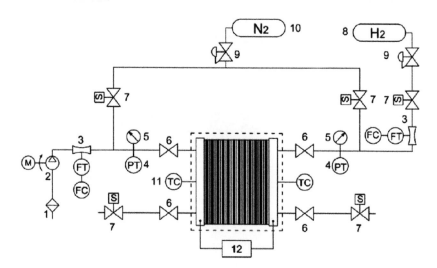

FIGURE 10.25 Schematic representation of a test station (1: air filter, 2: air compressor, 3: flow meter, 4: pressure transmitter, 5: pressure gauge, 6: valve, 7: solenoid valve, 8: hydrogen cylinder, 9: pressure regulator, 10: nitrogen cylinder, 11: thermocouple, 12: electronic load) [79].

10.6.4.1. Testing of HT-PEM Fuel Cells

10.6.4.1.1. Humidity Control

Relative humidity (RH) is determined by the ratio of the vapor pressure, $P(T)$, to the saturated vapor pressure (P_{sat}), multiplied by 100 [84]. The saturated vapor pressure is empirically described by the following equation:

$$\ln P_{sat} = 21.564 - 5420/T \tag{10.3}$$

The saturated vapor pressure increases exponentially with temperature, as shown in Fig. 10.26 [85]. At 180 °C, P_{sat} reaches 10 atm. To maintain 100% RH at 180 °C, it requires a total pressure >10 atm (leaving aside the partial pressure of the fuel and oxidant gases). With a partial pressure of 0.5 atm for the reactant gases, a water-saturated feed stream design with 90% RH at 150 °C still requires a total pressurization of >8 atm.

Two conventional methods are used to humidify the fuel cell reactants and the fuel cell membrane: external and internal [86]. The external method is simple and involves passing of the gases through a temperature-controlled, water-filled container. For conventional PEM fuel cells, the typical pressure limitation is usually <4 atm; below this pressure, the boiling point of water is about 145 °C, which means that internal humidification is not sufficient to provide a high RH at a temperature >150 °C. Internal humidification requires injection of water directly into the gas line leading to the fuel cell. Because less energy is needed for vaporization at high temperatures, injected water is more

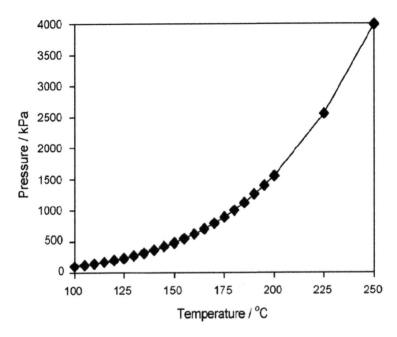

FIGURE 10.26   Saturated water vapor pressure vs. temperature [85].

readily vaporized in the gas line. Internal humidification may be the preferable method for HT-PEM fuel cells.

If the proton conductivity of a high-temperature membrane does not rely on water, humidification is unnecessary; this simplifies the fuel cell system and lowers its cost. One such system is the PBI-based HT-PEM fuel cell, in which proton conductivity is based on doped H_3PO_4.

10.6.4.1.2. Thermal Management

External heating tubes and elements are generally used to raise and maintain the temperature in HT-PEM fuel cells. Residual heat can easily be removed due to the large temperature difference between the fuel cell and the ambient environment. Alternatively, residual heat can be recovered for energy cogeneration. Thermal management in HT-PEM fuel cells is simpler than in LT-PEM fuel cells [10].

10.6.4.2. Diagnosis of HT-PEM Fuel Cells

Diagnostic technologies for LT-PEM fuel cells are the basis for HT-PEM fuel cell diagnosis. Many different investigative tools, including electrochemical and physicochemical methods, are available to elucidate the changes an HT-PEM fuel cell undergoes during operation.

10.6.4.2.1. Cyclic Voltammetry and Linear Sweep Voltammetry

Both cyclic voltammetry (CV) and linear sweep voltammetry (LSV) techniques have been discussed in detail in Chapter 3; for more information, please refer to this chapter.

In fuel cell testing and diagnosis, CV is used to determine the electrochemical Pt surface area (EPSA) of a catalyst layer, and LSV is used to evaluate and monitor fuel crossover. For these methods, pure H_2 and pure inert N_2 or He are passed over the anode and the cathode, respectively. The anode is used as both the reference and counter electrode, whereas the cathode is used as the working electrode.

The EPSA of a fuel cell cathode can be calculated based on the relationship between the Pt surface area and the charge associated with hydrogen adsorption on the electrode, as determined by CV. For CV, the cell is usually cycled between 0.05 and 0.6 V with a high sweep rate of 20 or 50 mV s^{-1}. On a smooth Pt electrode, the hydrogen adsorption charge is 210 µC cm^{-2}. The EPSA is calculated according to the following equation:

$$S_{EPSA} = \frac{Q_H}{G \times 210} \qquad (10.4)$$

where Q_H is the charge quantity, calculated from the integration of CV for hydrogen adsorption–desorption (in µC); G represents the Pt metal loading (mg cm^{-2}) in the catalyst layer; and 210 µC cm^{-2} is the charge required to oxidize a monolayer of hydrogen on the Pt catalyst. CV is an efficient method for measuring the EPSA of Pt catalysts in the catalyst layer and for checking the degradation of Pt catalysts.

LSV is an effective method for measuring the hydrogen crossover through the membrane at elevated temperatures. The cell is usually cycled between 0 and 0.5 V with a low sweep rate of <5 mV s^{-1}. Hydrogen that passes through the membrane is measured as a mass transport "limiting current" at 0.35–0.5 V. Figure 10.27 shows the LSV measurements of a cell at different temperatures [87]. Hydrogen crossover can be determined by using the plateau current density at high voltage, where the current obtained is limited by the hydrogen transport rate through the tested membrane. Some hydrogen oxidation pseudocapacitance is apparent at potentials below approximately 0.35 V, due to the sweep rate used. The effect of carbon double-layer capacitance on the apparent hydrogen limiting current is small at higher potentials.

10.6.4.2.2. Electrochemical Impedance Spectroscopy

Electrochemical impedance spectroscopy (EIS) has also been discussed in Chapter 3. EIS is generally used to diagnose the performance limitations of fuel cells. There are three fundamental sources of voltage loss in fuel cells: kinetic losses (charge-transfer activation), ohmic losses (ion and electron transport), and mass transfer losses (concentration). EIS can be used to distinguish and

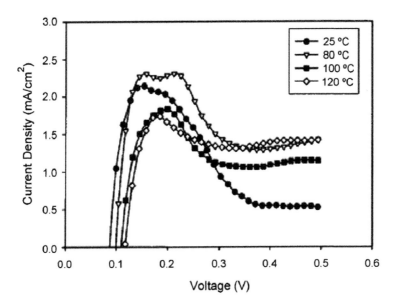

FIGURE 10.27   LSV of a single cell (4 mV s^{-1}, 0.01–0.50 V), 200 ml min^{-1} pure H$_2$ on the anode and 200 ml min^{-1} N$_2$ at the cathode; four conditions are compared: 25/100/100, 80/100/75, 100/70/70, and 120/35/35 [87].

quantify these losses. Figure 10.28 shows a typical EIS spectrum of a fuel cell during a frequency sweep [88]. In the higher frequency range of the spectrum (typically >10 kHz), the high-frequency intercept on the real axis in Fig. 10.28 corresponds to the ohmic resistance, which is dominated by the membrane resistance. The overall kinetic resistance (the sum of the anodic and cathodic charge-transfer resistance) is obtained from the difference between the high-frequency real Z-axis intercept and the next lower frequency real Z-axis intercept. It is evident that the overall charge-transfer resistance is dominated by the sluggish ORR kinetics. The lower frequency part of the spectrum (typically <1 Hz) represents the sum of a capacitive loop in the anode spectrum and an inductive loop in the cathode spectrum, which appear in a similar frequency range. The capacitive arc in the low-frequency range is attributable to a finite diffusion process, which is due to mass transport in the gas diffusion layers and the electrodes.

10.6.4.2.3. Physical Characterization

Different physical characterization techniques, such as X-ray diffraction (XRD), TEM, scanning electron microscopy (SEM), X-ray photoelectron spectroscopy (XPS), are usually applied to measure changes in catalysts, membranes, and MEA structures after HT-PEM fuel cell operations are performed.

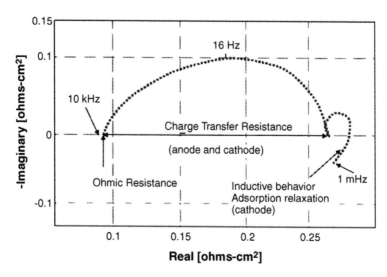

FIGURE 10.28 Correlation between different parts of an EIS spectrum and limiting processes
[88]. (For color version of this figure, the reader is referred to the online version of this book.)

10.7. CHALLENGES OF HT-PEM FUEL CELLS

Although operation of PEM fuel cells at high temperatures has many attractive
benefits, various cell components experience severe challenges, the most
significant problems being membrane dehydration and subsequent decreased
conductivity. In addition, Pt particle agglomeration, carbon support corrosion,
and degradation of other MEA components—such as gaskets, seals, and bipolar
plates—are accelerated at high temperatures.

10.7.1. Membrane Dehydration and Conductivity Loss

The conventional PFSA membranes currently used for PEM fuel cells, such
as Nafion® or other PFSA polymers, show significant conductivity loss at
high temperatures due to membrane dehydration [89]. Moreover, in the
effort to increase proton conductivity, the proton exchange membranes
used in PEM fuel cells are becoming progressively thinner, which can
worsen dehydration. Even more of a problem is that when the membrane is
thinner, gas crossover becomes an issue, as it not only generates mixed
potential—thereby decreasing cell performance—but it also compromises
safety.

The thermal stability of the membrane is another challenge at elevated
temperatures. Usually, the operating temperature of a fuel cell should be well
below the glass transition temperature (T_g) of the membrane. The T_g of
hydrated membranes is dependent on the chemical structure of the polymer

and the water content (due to its plasticizing effect). Nafion® has a T_g between 130 and 160 °C for a dry membrane, and between 80 and 100 °C for a hydrated membrane. It has been recognized that hydration can also decrease the thermal stability of a Nafion® membrane during high-temperature operation.

The chemical stability of membranes at high temperatures is another issue. It is believed that attack by HOO· and HO· radicals is a major reason for membrane degradation [90]. Oxygen crossover provides a means for the formation of peroxide and hydroperoxide radicals, which then attack the membrane and in turn lead to its deterioration. This degradation process is believed to accelerate with increasing temperature. Therefore, researchers who develop membranes for high-temperature operation should make membrane stability the primary consideration, particularly ways to decrease fuel crossover and maintain high conductivity.

10.7.2. Catalyst Agglomeration and Carbon Corrosion

The long-term stability of catalysts, even for LT-PEM fuel cells, is a major challenge. In general, an increase in the operating temperature decreases the catalyst stability, mainly due to the accelerated degradation of platinum and of carbon supports.

As discussed in earlier in this chapter, Pt agglomeration and particle growth is the dominant mechanism for Pt catalyst degradation, and is exacerbated at elevated temperatures. Two mechanisms have been proposed to explain Pt agglomeration: "Pt coalescence" and "Ostwald ripening." In general, the optimum Pt particle size for the ORR is 3–5 nm. Because Pt particle size increases during long-term operation or accelerated stress tests, ORR activity and Pt utilization gradually decrease.

As depicted in Fig. 10.9, Pt dissolution and redistribution are other major reasons for catalyst degradation during long-term operation and high-potential cycling tests [91,92]. In the case of Pt dissolution, two parallel paths have been discerned: (1) the electrochemical dissolution of Pt to Pt^{2+}, according to the reaction $Pt \rightarrow Pt^{2+} + 2e^-$ and (2) the formation of Pt oxide, followed by dissolution according to Reactions (10.I) and (10.II):

$$Pt + H_2O \rightarrow PtO + 2H^+ + 2e^- \tag{10.I}$$

$$PtO + 2H^+ \rightarrow Pt^{2+} + 2H_2O \tag{10.II}$$

The dissolved Pt (Pt^{2+}) either redeposits on Pt particles to form large particles or migrates out of the MEA into the membrane. Figure 10.29 shows Pt particles deposited within the membrane and near the catalyst layer/membrane interface after degradation [93]. These Pt particles originate from the dissolved Pt species at the cathode, which diffuse in the ionomer phase and subsequently

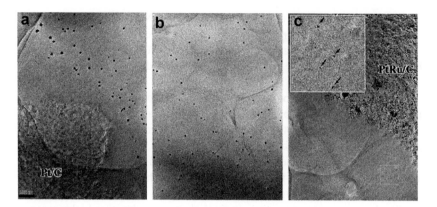

FIGURE 10.29 Cross-sectional TEM images of an MEA after 1.0 V potential holding for 87 h, (a) near the interface of the cathode catalyst layer and membrane, (b) inside the membrane, 10 μm away from the cathode catalyst layer, and (c) near the interface of the anode catalyst layer and the membrane [93].

precipitate in the ionomer phase of the electrode or in the membrane, through the reduction of Pt ions by hydrogen that crosses over from the anode. These Pt particle sites form degradation centers, which causes pinholes and lead to membrane failure.

Carbon can be chemically stable due to its low oxidation kinetics but electrochemically unstable due to Reaction (10.III):

$$C + 2H_2O \rightarrow CO_2 + 4H^+ + 4e^-, \quad E_{298K} = +0.207 \ V \ vs. \ RHE \quad (10.III)$$

The carbon oxidation reaction can be affected by temperature, interfacial electrode potential, and water vapor pressure; electrode potential is the most aggressive of these factors [94]. The presence of Pt can also catalyze carbon oxidation at lower potentials [95]. Corrosion of the catalyst's carbon support can occur at both the cathode and the anode. If the cathode is held at relatively high potentials, its catalyst carbon support will be oxidized, whereas if the anode is fuel starved, oxidation of the anode catalyst's carbon support may occur. For example, when the fuel level is insufficient to provide the expected current for PEM fuel cells, the potential value of the anode can increase to >0.21 V, or even to >1.23 V, at which point water electrolysis and carbon oxidation at the anode will occur to provide the required protons and electrons for the ORR at the cathode.

Another cause of fuel cell degradation is poisoning of Pt catalysts by contaminants [96]. As described earlier in this chapter, contaminants generally come from two sources: the fuel stream and the air stream. The contaminants are CO, CH_4, H_2S, NH_3, NO_x, SO_x, VOCs, and trace amounts of metallic ions or silicon from system components. In HT-PEM fuel cells, the gas poisoning effect is usually less aggressive than in LT-PEM fuel cells.

10.7.3. Degradation of Other Components

Aside from MEA components (membranes and catalysts), other materials in a fuel cell system—such as bipolar plates, seals, and gaskets—are also subject to degradation or other oxidation processes during operation. At elevated temperatures, degradation rates can be higher than for LT-PEM fuel cells [10].

The bipolar plates, usually made from carbon materials, have multiple functions in the fuel cell system, including current collection, gas distribution, water transport, thermal management, and humidification. In fuel cell stacks, the bipolar plates are in contact with the fuel on one side and the oxidant on the other, and are therefore exposed to both reducing and oxidizing conditions [10]. The need for good electrical conductivity and good corrosion resistance make the choice of suitable bipolar plate materials critical. The corrosion rate of a bipolar plate is affected not only by the applied potential but also by the temperature and pressure. For metal bipolar plates, metal dissolution can be enhanced at higher operating temperatures. The dissolved metal ions can then contaminate both the membrane and the catalysts, leading to fuel cell performance degradation. Although a protective coating may inhibit metal corrosion, it will significantly increase the contact resistance between the bipolar plate and the electrodes.

The sealing of the bipolar plates in a fuel cell stack is very important for avoiding fuel and oxidant leaks. O-rings are usually used in fuel cell stacks. For LT-PEM fuel cells, there are many choices of sealing materials, but the majority cannot be used at higher temperatures. For example, if the fuel cell is intended to operate at 180 °C, only those materials with a tolerance >200 °C can be used—silicon rubber, tetrafluoroethylene-propylene, perfluoroelastomer, and so on [97].

In summary, although HT-PEM fuel cells have several advantages over LT-PEM fuel cells, component degradation under high-temperature operation is the largest challenge. Therefore, material selection and degradation mitigation are the two most important considerations in moving forward with this technology.

10.8. CHAPTER SUMMARY

HT-PEM fuel cells have several benefits over LT-PEM fuel cells, and these include improved electrode kinetics, enhanced contaminant tolerance, and facile water and heat management. However, the degradation of key fuel cell components is a major challenge when operating at high temperatures. Currently, research on developing HT-PEM fuel cells is mainly focused on high-temperature membranes, while the fabrication of suitable catalyst layers with improved activity and stability is still in its early stages. To achieve breakthroughs in HT-PEM fuel cell technology, more efforts are needed, especially in the development of novel high-temperature membranes, catalysts, and electrode structures.

REFERENCES

[1] Zhang J, editor. PEM fuel cell electrocatalysts and catalyst layers—fundamentals and applications. London: Springer; 2008. p. 965–1002.

[2] Adjemian K, Lee S, Srinivasan S, Benziger J, Bocarsly A. J Electrochem Soc 2002;149(3): A256–61.

[3] Liu Y, Nguyen T, Kristian N, Yu Y, Wang X. J Memb Sci 2009;330:357–62.

[4] Li Q, He R, Jensen J, Bjerrum N. Chem Mater 2003;15:4896–915.

[5] Asensio J, Sanchez E, Gomez-Romero P. Chem Soc Rev 2010;39:3210–39.

[6] Yu S, Benicewicz B. Macromolecules 2009;42(22):8640–8.

[7] Mader J, Benicewicz B. Macromolecules 2010;43(16):6706–15.

[8] Li Q, Jensen J, Savinell R, Bjerrum N. Prog Polym Sci 2009;34:449–77.

[9] Wilkinson D, Thompsett D. In: Savadogo O, Roberge P, editors. Proceedings of the second international symposium on new materials for fuel-cell and modern battery systems; 1997. p. 266. Publisher:École polytechnique, Montréal, Canada.

[10] Zhang J, Xie Z, Zhang J, Tang Y, Song C, Navessin T, et al. J Power Sources 2006;160: 872–91.

[11] Wee J, Lee K. J Power Sources 2006;157:128–35.

[12] Yamanaka T, Takeguchi T, Wang G, Muhamad E, Ueda W. J Power Sources 2010;195:6398– 404.

[13] Lee D, Hwang S, Lee I. J Power Sources 2005;145:147–53.

[14] Wang D, Subban C, Wang H, Rus E, Disalvo F. J Am Chem Soc 2010;132:10218–20.

[15] Gottesfeld S, Pafford J. J Electrochem Soc 1988;135:2651–2.

[16] Schmidt V, Oetjen H, Divisek J. J Electrochem Soc 1997;144:L237–8.

[17] Dhar H, Christner L, Kush A. J Electrochem Soc 1987;134:3021–6.

[18] Vogel W, Lundquist L, Ross P, Stonehart P. Electrochim Acta 1975;20:79–93.

[19] Lakshmanan B, Huang W, Olmeijer D, Weidner J. Electrochem Solid State Lett 2003;6(12): A282–5.

[20] Li Q, He R, Gao J, Jensen J, Bjerrum N. J Electrochem Soc 2003;150:A1599–605.

[21] Kim S, Ahn B, Mench M. J Power Sources 2008;179:140–6.

[22] Pan C, He R, Li Q, Jensen J, Bjerrum N, Hjulmand H, et al. J Power Sources 2005;145: 392–8.

[23] Li Q, Hjuler H, Hasiotis C, Kallitsis J, Kontoyannis C, Bjerrum N. Electrochem Solid State Lett 2002;5:A125–8.

[24] Savinell R, Yeager E, Tryk D, Landau U, Wainright J, Weng D, et al. J Electrochem Soc 1994;141:L46–8.

[25] Malhotra S, Datta R. J Electrochem Soc 1997;144:L23–6.

[26] Fuller J, Breda AC, Carlin RT. J Electrochem Soc 1997;144:L67–70.

[27] Kreuer KD, Fuchs A, Ise M, Spaeth M, Maier J. Electrochim Acta 1998;43:1281–8.

[28] Dhar HP. U.S. Patent 5,242,764, 1993.

[29] Lin J, Jnuz H, Fenton J. In: Vielstich W, Lamm A, Gasteiger H, editors. vol. 3. New York: John Wiley & Sons Ltd; 2003. p. 457.

[30] Watanabe M, Uchida H, Seki Y, Emori M. The Electrochem Soc. Meeting, PV94-2, Abstract No. 606, Electrochem Soc. Pennington, NJ; 1994. p. 946–947.

[31] Watanabe M, Uchida H, Seki Y, Emori M, Stonehart P. J Electrochem Soc 1996;143:3847–52.

[32] Bailly C, Williams DJ, Karasz FE, MacKnight WJ. Polymers 1987;28:1009–16.

[33] Zaidi SMJ, Mikhailenko SD, Robertson GP, Guiver MD, Kaliaguine S. J Memb Sci 2000;173:17–34.

[34] Kerres J, Cui W, Reichle S. J Polym Sci A Polym Chem 1996;34:2421–38.

[35] Guo X, Fang J, Watari T, Tanaka K, Kita H, Okamoto K-i. Macromolecules 2002;35: 6707–13.

[36] Child AD, Reynolds JR. Macromolecules 1994;27:1975–7.

[37] Kreuer KD. J Memb Sci 2001;185:29–39.

[38] Kopitzke RW, Linkous CA, Nelson GL. Polym Degrad Stab 2000;67:335–44.

[39] Samms SR, Wasmus S, Savinell RF. J Electrochem Soc 1996;143:1498–504.

[40] Luo Y, Huo R, Jin X, Karasz FE. J Anal Appl Pyrolysis 1995;34:229–42.

[41] Wang Z, Tang H, Pan M. J Memb Sci 2011;369:250–7.

[42] Hill ML, Kim YS, Einsla BR, McGrath JE. J Memb Sci 2006;283:102–8.

[43] Honma I, Nakajima H, Nishikawa O, Sugimoto T, Nomura S. J Electrochem Soc 2002;149: A1389–92.

[44] Xing D, Zhang H, Wang L, Zhai Y, Yi B. J Memb Sci 2007;296:9–14.

[45] Donoso P, Gorecki W, Berthier C, Defendini F, Poinsignon C, Armand MB. Solid State Ionics 1988;28–30(Part 2):969–74.

[46] Petty-Weeks S, Zupancic JJ, Swedo JR. Solid State Ionics 1988;31:117–25.

[47] Tanaka R, Yamamoto H, Shono A, Kubo K, Sakurai M. Electrochim Acta 2000;45: 1385–9.

[48] Stevens JR, Wieczorek W, Raducha D, Jeffrey KR. Solid State Ionics 1997;97:347–58.

[49] Yu X, Ye S. J Power Sources 2007;172:145–54.

[50] Zhang S, Yuan X-Z, Hin JNC, Wang H, Friedrich KA, Schulze M. J Power Sources 2009;194:588–600.

[51] Virkar AV, Zhou Y. J Electrochem Soc 2007;154:B540–7.

[52] Liu G, Zhang H, Hu J, Zhai Y, Xu D, Shao Z-g. J Power Sources 2006;162:547–52.

[53] Zhai Y, Zhang H, Xing D, Shao Z-G. J Power Sources 2007;164:126–33.

[54] Ascarelli P, Contini V, Giorgi R. J Appl Phys 2002;91:4556–61.

[55] Liu G, Zhang H, Zhai Y, Zhang Y, Xu D, Shao Z-g. Electrochem Commun 2007;9:135–41.

[56] Zhang J, Song C, Zhang J. J Fuel Cell Sci Tech 2011;8:051006.1-051006.5, http://dx.doi.org/ 10.1115/1.4003977.

[57] Shim J, Lee C-R, Lee H-K, Lee J-S, Cairns EJ. J Power Sources 2001;102:172–7.

[58] Xiong L, Manthiram A. Electrochim Acta 2004;49:4163–70.

[59] Chen J-M, Sarma LS, Chen C-H, Cheng M-Y, Shih S-C, Wang G-R, et al. J Power Sources 2006;159:29–33.

[60] Huang S-Y, Ganesan P, Park S, Popov BN. J Am Chem Soc 2009;131:13898–9.

[61] Bauer A, Lee K, Song C, Xie Y, Zhang J, Hui R. J Power Sources 2010;195:3105–10.

[62] Do TB, Cai M, Ruthkosky MS, Moylan TE. Electrochim Acta 2010;55:8013–7.

[63] Huang S-Y, Ganesan P, Popov BN. Appl Catal B 2010;96:224–31.

[64] Wang C, Waje M, Wang X, Tang JM, Haddon RC, Yan Y. Nano Lett 2004;4:345–8.

[65] Oh H, Kim H. J Electrochem Sci Tech 2010;1:92–6.

[66] Oh H-S, Kim K, Ko Y-J, Kim H. Int J Hydrogen Energy 2010;35:701–8.

[67] Simmons TJ, Bult J, Hashim DP, Linhardt RJ, Ajayan PM. ACS Nano 2009;3:865–70.

[68] Huang S-Y, Ganesan P, Popov BN. Appl Catal B 2009;93:75–81.

[69] Bonakdarpour A, Stevens K, Vernstrom GD, Atanasoski R, Schmoeckel AK, Debe MK, et al. Electrochim Acta 2007;53:688–94.

[70] Debe MK, Schmoeckel AK, Vernstrom GD, Atanasoski R. J Power Sources 2006;161: 1002–11.

[71] Wu J, Gross A, Yang H. Nano Lett 2011;11:798–802.

[72] Bose S, Kuila T, Nguyen TXH, Kim NH, Lau K-t, Lee JH. Prog Polym Sci 2011;36:813–43.

[73] Pan C, Li Q, Jensen JO, He R, Cleemann LN, Nilsson MS, et al. J Power Sources 2007;172:278–86.
[74] Wannek C, Lehnert W, Mergel J. J Power Sources 2009;192:258–66.
[75] Kim J-H, Kim H-J, Lim T-H, Lee H-I. J Power Sources 2007;170:275–80.
[76] Lee H-J, Kim BG, Lee DH, Park SJ, Kim Y, Lee JW, et al. Int J Hydrogen Energy 2011;36:5521–6.
[77] Tang Y, Zhang J, Song C, Zhang J. Electrochem Solid State Lett 2007;10:B142–6.
[78] Tang Y, Zhang J, Song C, Liu H, Zhang J, Wang H, et al. J Electrochem Soc 2006;153:A2036–43.
[79] Radu R, Zuliani N, Taccani R. J Fuel Cell Sci Tech 2011;8:051007.1-051007.5, http://dx.doi.org/10.1115/1.4003753.
[80] Andreasen SJ, Kar SK. Int J Hydrogen Energy 2008;33:4655–64.
[81] Scholta J, Messerschmidt M, Jörissen L, Hartnig C. J Power Sources 2009;190:83–5.
[82] Lobato J, Rodrigo MA, Linares JJ, Scott K. J Power Sources 2006;157:284–92.
[83] http://juwel.fz-juelich.de:8080/dspace/bitstream/2128/3858/1/FC2_7_Beckhaus.pdf.
[84] http://www.faqs.org/faqs/meteorology/temp-dewpoint/.
[85] http://www.engineeringtoolbox.com/24 162.html.
[86] Evans J. Experimental evaluation of the effect of inlet gas humidification on fuel cell performance, electronic thesis, mechanical engineering. Virginia Polytechnic and State University; 2003.
[87] Song Y, Fenton JM, Kunz HR, Bonville LJ, Williams MV. J Electrochem Soc 2005;152:A539–44.
[88] Jalani NH, Ramani M, Ohlsson K, Buelte S, Pacifico G, Pollard R, et al. J Power Sources 2006;160:1096–103.
[89] Yang C, Srinivasan S, Bocarsly AB, Tulyani S, Benziger JB. J Memb Sci 2004;237:145–61.
[90] Wang H, Capuano GA. J Electrochem Soc 1998;145:780–4.
[91] Kawahara S, Mitsushima S, Ota K, Kamiya N. ECS Trans 2006;3:625–31.
[92] Wang X, Kumar R, Myers DJ. Electrochem Solid State Lett 2006;9:A225–7.
[93] Akita T, Taniguchi A, Maekawa J, Siroma Z, Tanaka K, Kohyama M, et al. J Power Sources 2006;159:461–7.
[94] Kangasniemi KH, Condit DA, Jarvi TD. J Electrochem Soc. 2004;151:E125–32.
[95] Roen LM, Paik CH, Jarvi TD. Electrochem Solid State Lett 2004;7:A19–22.
[96] Cheng X, Shi Z, Glass N, Zhang L, Zhang J, Song D, et al. J Power Sources 2007;165:739–56.
[97] Parker O-ring handbook, section II, basic O-ring elastomers, P2-11. Cleveland, OH: Parker Hannifin Corporation, http://www.parker.com/o-ring/Literature/ORD5700.pdf; 1999.

Fuel Cell Degradation and Failure Analysis

11.1. INTRODUCTION

Through tremendous research efforts, significant progress has been achieved in polymer electrolyte membrane (PEM) fuel cell technologies over the past decade, especially in the areas of increasing volumetric and/or gravimetric specific power densities as well as more effective use of materials. However, technical challenges still remain for the onboard storage of hydrogen fuel and the infrastructure for its widespread distribution, as well as for the fuel cell system itself. With regard to the fuel cell system, there are two major challenges: high cost and insufficient durability.

Durability, defined as the maximum lifetime of a fuel cell system with not >10% loss in efficiency at the end of life, is one of the most stringent requirements for PEM fuel cells to be accepted as practical power sources. The requirements for fuel cell lifetime vary significantly, depending on the application. The fuel cell industry has set standards that include the following: (1)

PEM Fuel Cell Testing and Diagnosis. http://dx.doi.org/10.1016/B978-0-444-53688-4.00011-5

for stationary applications, a lifetime durability of >40,000 h, with 8000 h of uninterrupted service at over 80% power; (2) for buses and cars, a former value of 20,000 and a latter of 6000 operating hours. Although performance degradation is unavoidable, the degradation rate can be minimized through a comprehensive understanding of degradation and failure mechanisms.

11.2. FAILURE MODES INDUCED BY FUEL CELL OPERATION

So far, various studies have focused on the degradation mechanisms of either the fuel cell system or its components under steady or accelerated operating conditions. The major failure modes of different components of PEM fuel cells are listed in Table 11.1.

Traditional lifetime data analysis in engineering involves analyzing times-to-failure data obtained under normal operating conditions to quantify the lifetime characteristics of the components and systems. For fuel cells, the time-to-failure data are always very difficult to obtain due to the prolonged test periods required and to the high costs involved. For example, almost seven years of uninterrupted testing is needed to reach the 60,000-hour lifetime requirement for a stationary fuel cell system. To test a fuel cell bus system (275 kW) for 20,000 h, the fuel expense alone would be approximately 2 million US dollars (3.8 billion liters of hydrogen at 5.3 US dollars per cubic meter). To increase sample throughput and reduce the experimental time, different strategies to accelerate PEM fuel cell and component degradation have been suggested. The general accelerated stress test (AST) methods for PEM fuel cell lifetime analysis are summarized in Table 11.2.

Degradation and durability of a PEM fuel cell or stack can be affected by many internal and external factors, including fuel cell design and assembly, operating conditions (e.g. humidification, temperature, cell voltage), impurities or contaminants in the feeds, environmental conditions (e.g. subfreezing or cold start), and operation modes (e.g. startup, shutdown, potential cycling).

The design and assembly of PEM fuel cell components, such as flow fields and manifolds, can have a significant influence on water management and feed flows, which will in turn affect the durability of fuel cell components. For example, an improper design of the flow fields can result in water blockage, and improper manifold design can induce poor cell-to-cell flow distribution, both of which may cause localized fuel starvation. This localized fuel starvation can then induce an increased local anode potential to levels at which carbon oxidation or even water electrolysis may occur to provide the required protons and electrons for the oxygen reduction reaction (ORR) at the cathode. These reactions will induce corrosion of the carbon support and will result in a permanent loss of electrochemically active area at the anode.

Startup and shutdown can have a profound impact on fuel cell durability. Under conditions of prolonged shutdown, air will eventually cross over from the cathode to the anode, filling the anode flow channels with air. In this case,

TABLE 11.1 Major Failure Modes of Different Components in PEM Fuel Cells

Component	Failure Modes	Causes
Membrane	Mechanical degradation	Mechanical stress due to nonuniform press pressure; inadequate humidification or penetration of the catalyst particles, which results in pinholes and gas transfer through the membrane
	Thermal degradation	Thermal stress; membrane drying
	Chemical/electrochemical degradation	Trace metal contamination (foreign cations, such as Ca^{2+}, Fe^{3+}, Cu^{2+}, Na^+, K^+, and Mg^{2+}); radical attack (e.g. peroxy and hydroperoxy). Peroxide radical attack causes membrane polymer chain decomposition and fluorine loss; this results in membrane thinning, pinholes, and gas crossover. Note: these peroxide radicals are generated by both the fuel cell reaction and the chemical reaction between O_2 and H_2 within the membrane.
Catalyst/catalyst layer (CL)	Loss of activation	Sintering or dealloying of electrocatalyst
	Conductivity loss	Corrosion of electrocatalyst support
	Decrease in mass transport rate of reactants	Mechanical stress
	Loss of reformate tolerance	Contamination
	Decrease in water management control	Change in hydrophobicity of materials due to Nafion or PTFE dissolution
Gas diffusion layer (GDL)	Decrease in mass transport	Degradation of backing material
	Conductivity loss	Mechanical stress (e.g. freeze/thaw cycle)
	Decrease in water management control	Corrosion; change in hydrophobicity of materials
Sealing gasket	Mechanical failure	Corrosion; mechanical stress

TABLE 11.2 General Accelerated Stress Test Methods in PEM Fuel Cell Lifetime Analysis

Component	Methods
Fuel cell/stack	Open circuit voltage (OCV); dynamic load cycling; thermal cycling; reduced/variable humidity; fuel or oxidant contaminants; fuel or oxidant starvation
Membrane	OCV operation at reduced humidity for chemical stability; relative humidity (RH) and temperature cycling for mechanical degradation
Catalyst/CL	Potential cycles, acid washing, elevated temperature, fuel or oxidant contaminants
GDL	Chemical oxidation in H_2O_2, elevated potential, low humidity
Bipolar plates	Press stress, acid treatment, potential cycling, temperature cycling
Sealing gasket	Temperature, acid treatment, deformation/press stress

fuel cell startup will create a transient condition in which there is fuel at the inlet, but the outlet is fuel-starved at the anode side. When this air/fuel boundary is formed at the anode, the local cathode potential may increase to >1.8 V vs. NHE, and as a consequence, the carbon oxidation reaction and water electrolysis reaction will occur through "reverse current" mechanisms on the corresponding cathode side. This causes serious deterioration in fuel cell performance and durability.

Another likely cause of severe fuel cell degradation is impurities. In general, impurities can be categorized into two groups, based on the sources: the first source is the contaminants from the fuel reforming process (such as CO, H_2S) or pollutants in the air intake (e.g. NO_x, SO_x, or volatile organic compounds); and the second source includes system-derived impurities, such as trace amounts of metallic ions or silicon from the system components (e.g. catalysts, bipolar plates, membranes, and sealing gaskets). These impurities have been known to adversely affect the fuel cell performance and durability by several means: kinetic effects, caused by poisoning of the anode and cathode catalyst active sites; the conductivity effect, due to increased resistance in the membrane and ionomer; and the mass transfer effect, caused by changes in the structure and hydrophobicity of the CLs and/or GDLs.

Survivability, durability, operation, and rapid startup under subfreezing temperatures are still current barriers that should be addressed before mass-market penetration of hydrogen-powered PEM fuel cell vehicles. When a PEM fuel cell is subjected to subfreezing temperatures, there is a significant decrease in the electrochemical surface area (ESCA) of the electrodes (normally attributed to ice formation and porosity changes in the CL), and this ultimately causes delamination of the CL from the membrane. The effect of freeze/thaw thermal cycles on the durability of fuel cell components has also been extensively investigated, with experiments showing that the water content/state in the CLs, the air permeability of the GDLs, and the conductivity of the membrane will dramatically change as a result of freeze/thaw cycling.

The fuel cell performance shows degradation over operating time, which is dependent on materials, fabrication, and operating conditions. The changes in temperature and RH associated with transitions between low and high power can have adverse effects on the component properties and the integrity of the fuel cell system. For example, transient automotive operating conditions, specifically power (or voltage) cycling, can exacerbate a fuel cell's degradation and reduce its components' durability/reliability as well.

11.3. MAJOR FAILURE MODES OF DIFFERENT COMPONENTS OF PEM FUEL CELLS

11.3.1. Membrane Failure

The membrane separates the anode from the cathode, and at the same time acts as a proton conductor and an electron insulator. Therefore, the requirements for a qualified membrane are manifold and stringent, including high protonic conductivity, low flow reactant gas permeability, and high thermal and chemical stability. The most commonly used and promising membranes for PEM fuel cells are perfluorosulfonic acid (PFSA) membranes, such as Nafion® (Dupont™) and Gore-Select® (Gore™) as well as Aciplex® and Flemion® (Asahi™). Extensive studies have been carried out on the mechanisms of membrane degradation and failure in the fuel cell environment. At present, however, unsatisfactory membrane durability/reliability is still one of the critical issues impeding the commercialization of PEM fuel cells.

11.3.1.1. Membrane Degradation Mechanisms

11.3.1.1.1. Mechanical Degradation of the Membrane

Membrane degradation can be classified into three groups: mechanical, thermal, and chemical/electrochemical degradation. Among these, mechanical degradation can cause early life failure due to perforations, cracks, tears, or pinholes, which may result from congenital membrane defects or from improper membrane electrode assembly (MEA) fabrication processes. The local areas corresponding to the interface between the lands and channels of the flow field or the sealing edges in a PEM fuel cell, which are subjected to excessive or nonuniform mechanical stresses, are also vulnerable to small perforations or tears. During fuel cell operation, the overall dimensional change due to nonhumidification [1], low humidification [2–4], and RH cycling [5–7] are also detrimental to mechanical durability. The constrained membrane in an assembled fuel cell experiences in-plane tension resulting from shrinkage under low RH, and in-plane compression during swelling under wet conditions. Recent experiments have demonstrated that the dynamic humidity in the fuel cell may generate humidity-induced stress as high as 2.3 MPa and a dimensional change of 11% toward free-standing PEMs [5]. The migration and accumulation of the catalysts and the decomposition of the seal into the membrane, as described in Sections 11.3.2.1 and 11.3.5, also negatively affect membrane conductivity and mechanical strength and significantly reduce ductility. A physical breach of the membrane due to local pinholes and perforations can result in a crossover of reactant gases into their respective reverse electrodes. If this happens, the highly exothermic direct combustion of the oxidant and reductant occurs on the catalyst surface and consequently generates local hot points. A destructive cycle of increasing gas crossover and pinhole production is then established, which undoubtedly accelerates the degradation of the membrane and the entire cell. The results of Huang et al. [8] suggested that mechanical failure of the membrane starts as a random, local imperfection that propagates to a catastrophic failure.

11.3.1.1.2. Thermal Degradation of the Membrane

To maintain well-hydrated PFSA membranes, the most favorable working temperature of a PEM fuel cell is usually from 60 to 80 °C. Conventional PFSA membranes are subject to critical breakdown at high temperatures due to the glass transition temperatures of PFSA polymers, which are around 80 °C. However, rapid startup, stable performance, and easy operation in subfreezing temperatures are necessary capabilities for fuel cell technologies to achieve before commercialization in vehicles and portable power supply applications. On the other hand, much effort has been made recently to develop PEM fuel cells that operate at temperatures >100 °C, to enhance electrochemical kinetics, simplify water management and cooling systems, and improve system CO tolerance. Membrane protonic conductivity drops significantly with decreasing water content when the fuel cell is operated at high temperatures and under low humidity.

Several studies have addressed the issue of thermal stability and thermal degradation of PFSA membranes. The PTFE-like molecular backbone gives Nafion® membranes their relative stability until 150 °C due to the strength of the C–F bond and the shielding effect of the electronegative fluorine atoms [9]. At higher temperatures, Nafion® begins to decompose via its side sulfonate acid groups. For example, the thermal stability of Nafion® was investigated by Surowiec and Bogoczek [10] using thermal gravimetric analysis (TGA), differential thermal analysis (DTA), as well as Fourier transform infrared (FTIR) spectroscopy, and only water was detected at a temperature <280 °C. At temperatures >280 °C, sulfonic acid groups were spilled off. In studies on the effects of heating Nafion® onto Pt in air, Chu et al. [11] found that sulfonic acid groups were lost after heating at 300 °C for 15 minu, whereas Deng et al. [12] measured small amounts of sulfur dioxide up to 400 °C. Detailed mechanisms for PFSA thermal degradation were proposed by Wilkie et al. [9] and Samms et al. [13], which included initial rupture of the C–S bond to produce sulfur dioxide, OH· radicals, and a left carbon-based radical for further cleavage at higher temperatures.

Regarding fuel cell operation at subfreezing temperatures, several studies on the state of water in PFSA membranes below the freezing temperature have been conducted. It has been suggested that several different states of water exist in the membrane, including "free water," which is not intimately bound to the polymer chain and will freeze at <0 °C. In addition, Cappadonia et al. [14] and Sivashinsky and Tanny [15] found that only a part of the water present in Nafion® underwent freezing. Cho et al. [16] reported that the contact resistance between the membrane and the electrode could increase after thermal cycles, whereas membrane ionic conductivity itself was not affected. However, McDonald et al. [17] found that after 385 temperature cycles between +80 °C and –40 °C, the ionic conductivity, gas impermeability, and mechanical strength of Nafion® membranes were severely impaired, although no catastrophic failures were detected. Phase transformation and changes in water volume due to freeze/thaw cycles have a detrimental effect on the membrane's lifetime. To avoid this, mitigation strategies have been proposed, including gas purging and solution purging to remove residual water during fuel cell startup and shutdown.

Another noteworthy hazard to PEM fuel cell durability at subzero temperatures is the influence of phase transformation and water volume changes on the physical properties of the membrane/electrode interface and electrode structure, in addition to the membrane. Cho et al. [16] observed a performance degradation rate of about 2.3% per freeze/thaw cycle from 80 to –10 °C. The cell performance degradation seen with thermal cycles was attributed to the physical damage of the electrode structure and MEA integrity, resulting from ice expansion during freezing. The analytical results of McDonald et al. [17] demonstrated the relationship of temperature cycling between 80 and –40 °C to membrane structure, water management, ionic conductivity, gas permeability, and mechanical strength. A detailed summary of research on PEM fuel cell freeze and rapid startup can be found in Ref. [18]. Experimental results from Xie et al. [19] have also revealed changes in the hydrophobic characteristics of the CL over time due to the dissolution of Nafion® or PTFE, which detrimentally affect the water management and mass transport abilities of the electrode.

11.3.1.1.3. Chemical/Electrochemical Degradation of the Membrane

During fuel cell operation with a load, the rates of hydrogen and air crossover to opposite sides of the membrane are relatively slow and result in only a 1–3%

loss in fuel cell efficiency. However, the aforementioned highly exothermic combustion between H_2 and O_2 may lead to pinholes in the membrane, destroying the MEA and causing catastrophic problems. More severely, the chemical reactions on the anode and cathode catalysts can produce peroxide ($HO\cdot$) and hydroperoxide ($HOO\cdot$) radicals, which are commonly believed to be responsible for chemical attacks on the membrane and catalysts. Further investigation has also revealed that the generation of these radicals and the chemical degradation of the membrane are accelerated when the fuel cell is operated under OCV and low humidity conditions. Under H_2 circumstances, the polymer backbone of the PFSA membrane preferentially reacts as follows [20]:

$$-CF_2- + 2H_2 \rightarrow -CH_2- + 2HF \qquad (11.I)$$

Following this reaction, the radicals attack the resulting $-CH_2-$ groups. Several mechanisms have been proposed, with conflicting views on whether the radicals are formed at the anode, at the cathode, or on both sides of the membrane. Pozio et al. [21] and other researchers [22,23] have provided evidence of predominant cathode degradation. Meanwhile, Mattsson et al. [24] observed no noticeable difference between the anode and cathode sides. Because H_2O_2 is so reactive, and decomposes relatively easily into water on the Pt surface, it can be assumed that there is little chance for H_2O_2 to accumulate on the Pt surface before it diffuses into the membrane. However, this assumption can be applied only to the cathode CL because a monolayer of hydrogen on the Pt surface at the anode inhibits H_2O_2 decomposition. For this reason, H_2O_2 formation due to oxygen gas crossover occurs mainly at the anode and progresses toward the cathode [25,27]. Therefore, more researchers believe that the loss of ionic groups begins at the anode side of the membrane, as shown in Fig. 11.1.

OCV and low humidification have been widely used as test methods to accelerate and diagnose membrane chemical degradation. Under OCV operation, H_2O_2 formed at the anode diffuses into the cathode and is removed by the decomposition on the Pt surface or evaporation into the air. The concentration of H_2O_2 at the anode and the cathode sides of the membrane depends on the amount of H_2O_2 generated at the anode and its removal rate at the cathode. Because H_2O_2 has a dipole moment of 2.2 debye, which is larger than that of water (1.85 debye), it can also be dragged by moving protons [28]. Therefore, similar to water, H_2O_2 transport across the membrane can be described by two physical mechanisms: electro-osmotic drag and diffusion. Even small currents can reduce the H_2O_2 concentration at the anode side of the membrane due to electro-osmotic drag, and the total amount of H_2O_2 inside the membrane will then decrease dramatically because the accumulation of H_2O_2 at the cathode can expedite the removal rate, which explains the accelerated degradation that occurs at OCV. As for the serious effect of low humidification on membrane

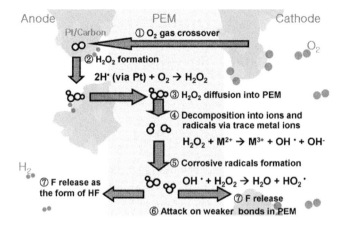

FIGURE 11.1 Mechanism of chemical degradation of a PEM membrane [26].

chemical degradation, this can be explained by H_2O_2 having a higher dipole moment than water. With low water content, H_2O_2 is also constrained more strongly than water by sulfonate heads in the membrane, due to its larger polarity. So most of the H_2O_2 contributes to the formation of reactive oxygen radicals, and has little chance of diffusing out to the cathode.

Recent research has shown that contamination by trace metal ions, such as Fe^{2+} and Cu^{2+}, originating from the corrosion of the metal bipolar plates or end plates can strongly accelerate membrane thinning and performance decay in a PEM fuel cell by catalyzing radical formation reactions. These cations show a stronger affinity than H^+ to the sulfonic acid group in PFSA membranes. During fuel cell operation, active sites are preferentially occupied by these multivalent ions and consequently, membrane bulk properties such as membrane ionic conductivity, water content, and H^+ transference numbers change in proportion to the cationic charge. This effect is not normally serious unless the contamination concentration goes beyond 50% of the sulfonic acid groups in the membrane. The second possible mode of membrane deterioration due to contaminant ions comes from the altered water flux inside the membrane, and in this case, just 5% of the contaminant is sufficient. The displacement of H^+ by foreign cations also results in an attenuated water flux and proton conductivity, leading to a much faster or more extensive membrane dehydration, especially near the anode. Contamination by trace metal ions, such as Fe^{2+} and Cu^{2+}, originating from the corrosion of the metal bipolar plates or end plates, can strongly accelerate membrane thinning and perfor- mance decay in a PEM fuel cell by catalyzing radical formation reactions, as shown in the following equations [29]:

$$H_2O_2 + Fe^{2+} \rightarrow HO\cdot + OH^- + Fe^{3+} \tag{11.II}$$

$$Fe^{2+} + HO \cdot \rightarrow Fe^{3+} + OH^- \qquad (11.\text{III})$$

$$H_2O_2 + HO \cdot \rightarrow HO_2 \cdot + H_2O \qquad (11.\text{IV})$$

$$Fe^{2+} + HO_2 \cdot \rightarrow Fe^{3+} + HO_2^- \qquad (11.\text{V})$$

$$Fe^{3+} + HO_2 \cdot \rightarrow Fe^{2+} + H^+ + O_2 \qquad (11.\text{VI})$$

The most recent research has demonstrated that the ranking of the four transition metals tested in terms of the greatest reduction in fuel cell performance was in the order $Al^{3+} \gg Fe^{2+} > Ni^{2+}$, Cr^{3+} [30].

Depending on the type of membrane, the HO· and HOO· radicals generated during the reaction can attack the α-carbon of an aromatic group, the ether links, or the branching points of the polymer. In PFSA membranes, the few carboxylate end groups with H-containing terminal bonds, which are inevitably formed during the polymer manufacturing process, are regarded as the inducing agent for membrane chemical decay due to their susceptibility to radical attack. One generally accepted mechanism, the unzipping reaction, initiates the abstraction of hydrogen from the end groups, releases HF and CO_2, and forms new carboxylate groups at the chain ends. An example of radical attack on the end group $-CF_2COOH$ is shown below [31].

$$R_f - CF_2COOH + \cdot OH \rightarrow R_f - CF_2 \cdot + CO_2 + H_2O \qquad (11.\text{VII})$$

$$R_f - CF_2 \cdot + \cdot OH \rightarrow R_f - CF_2OH \rightarrow R_f - COF + HF \qquad (11.\text{VIII})$$

$$R_f - COF + H_2O \rightarrow R_f - COOH + HF \qquad (11.\text{IX})$$

As the process repeats, the attack may propagate along the main chain, and eventually the polymer decomposes into low-molecular-weight compounds. Another possible mechanism is the scission of the polymer main chains, which suggests that the ether linkages are the side chain sites most susceptible to radical attack, producing vulnerable $-COOH$ groups. As a result, the average molecular weight of the polymer decreases, whereas the number of $-COOH$ groups increases with time. Even without susceptible end groups, under exposure to H_2, the polymer backbone of the PFSA membrane may preferentially react according to Eqn (11.I). Following this reaction, the radicals attack the resulting $-CH_2-$ groups. The rate of fluoride loss is typically considered to be an excellent measurement of PFSA membrane degradation.

11.3.1.2. Mitigation Strategies for Membrane Degradation

To prevent membrane mechanical failure, the MEA and flow field structures must be carefully designed to avoid local drying of the membrane, especially at the reactant inlet area. A membrane reinforced with e-PTFE, developed by Gore Fuel Cell Technologies, exhibited a lifetime order of magnitude longer than a non-reinforced membrane of comparable thickness [32], as shown in Fig. 11.2. Similar results for the enhanced membrane mechanical strength were reported by

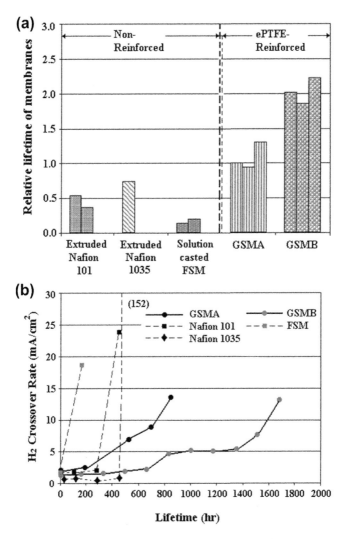

FIGURE 11.2 Comparison of Gore reinforced membranes and nonreinforced membranes. (a) Lifetime of various membranes in accelerated fuel cell conditions; (b) H_2 crossover rate as a function of time [32].

Wakizoe et al. [33] and Xu et al. [34] by using reinforced Aciplex® membranes and Nafion®–Teflon®–phosphotungstic acid composite membranes, respectively. The addition of carbon annotates is another method to improve the physical stability and mechanical strength of a PEM [35,36]. A well-designed interface between the nanotubes and the polymer also helps to improve the mechanical properties of a composite membrane [37]. More recently, Nafion®/silica nano-composite membrane [38] and TiO_2-nanowire-reinforced Nafion® composite membrane [39] with improved mechanical properties were developed by self-assembly of positively charged Nafion–silica nanoparticles or TiO_2 nanowires and negatively charged Nafion® molecules under low pH conditions.

In recent years, much effort has been devoted to PEM development and fabrication approaches aimed at achieving prolonged durability at temperatures >100 °C. The membranes developed so far can be classified into three groups: (1) modified PFSA membranes with enhanced thermal stability and water retention properties, which are impregnated with nonvolatile solvents or incorporate hydrophilic oxides and solid inorganic proton conductors, such as SiO_2, ZrO_2, TiO_2, and zirconium phosphate; (2) alternative sulfonated poly-mers and their composite membranes, such as SPSF (sulfonated polysulfones), SPEEK (sulfonated poly(ether ether ketone)), PBI (polybenzimidazole), and poly(vinylidene fluoride); and (3) acid–base polymer membranes, such as phosphoric-acid-doped Nafion®–PBI composite membranes.

With respect to PEM fuel cell freeze and rapid startup issues, two main strategies have been proposed to mitigate the fuel cell performance degrada-tion, based on whether the system uses extra energy during parking or startup. The first solution, the "keep-warm" method, is to consume power from a continuous or intermittent low-power energy source (an extra battery or hydrogen fuel converter) to keep the system above a certain threshold temperature during the parking period. The other option is to heat the fuel cell system to raise its temperature above the freezing point of water at startup. For this method, a higher power heat source is required, and it is strongly suggested that the method be combined with effective removal of residual water to save energy and alleviate physical damage to the MEA due to ice expansion. The possible methods for getting rid of the water include gas purging or washing it away with an antifreeze solution before fuel cell shutdown.

With respect to chemical and electrochemical degradation of the membrane, development of membranes that are chemically stable against peroxy radicals has drawn particular attention. One solution is to develop novel membranes with higher chemical stability, such as radiation-grafted FEP (fluoro-ethylene-propylene)-g-polystyrene membrane [40], in which polystyrene is used as a sacrificial material owing to its low resistance to radicals. Free-radical stabilizers and inhibitors such as hindered amines or antioxidants also have the potential to be mingled during membrane fabrication. Increased chemical stability can also be realized by modifying the structure of the available membrane. Curtin et al. [41] suggested that the radical attack on the residual

H-containing terminal bonds of the main chain of the PFSA membrane was the primary degradation mechanism. Elimination of the unstable end group significantly enhanced the chemical stability. In a third solution, damage caused by hydrogen peroxide can be suppressed by redesigning the MEA. For example, a composite membrane suggested by Yu et al. [23], in which a thin recast Nafion® membrane was bonded with a polystyrene sulfonic acid (PSSA) membrane, when positioned at the cathode of the cell could successfully prevent the oxidation degradation of the PSSA membrane. A fourth solution is to introduce peroxide-decomposition catalysts, such as heteropoly acids into the membrane, thereby moderating or eliminating membrane deterioration due to peroxide [42,43]. However, the advantage of this approach is partially counteracted by a decrease in membrane stability and conductivity, caused by the mixture of the catalysts. Last but not the least, the development and implementation of new metal coatings with improved corrosion resistance and of catalysts that produce less hydrogen peroxide are long-term goals for membrane durability enhancement.

11.3.1.3. Membrane Failure Testing and Diagnosis

Various diagnostic tools for the accurate analysis of PEM fuel cell and stack degradation have been developed. At present, characterization of the property changes in PEM fuel cell components is mainly concentrated on the following issues: (1) mass distribution, especially water distribution over the active electrode, including detection of flooding that leads to low catalyst utilization, (2) resistance diagnosis and membrane drying detection, which is closely related to membrane conductivity, (3) optimization of electrode structures and components, fuel cell design, and operating conditions, (4) current density distribution in dimensionally large-scale fuel cells, (5) temperature variation resulting from nonuniform electrochemical reaction and contact resistance in a single cell, and different interconnection resistances for a stack, and (6) flow visualization for direct observation of what is occurring within the fuel cell. Due to the complexity of the heat and mass transport processes occurring in fuel cells, there are typically a multitude of parameters to be determined. For all the previous reasons, it is important to examine the operation of PEM fuel cells or stacks using suitable techniques that allow for evaluation of these parameters separately and determination of the influence of each on the global fuel cell performance. The tools frequently used to characterize the membrane degradation are described in this chapter.

11.3.1.3.1. Hydrogen Crossover Rates

The hydrogen and oxygen that permeate through the membrane are consumed through the generation of heat and water without generating useful power, leading to fuel inefficiency. As discussed above, the gas crossover rate can be increased by both chemical and mechanical degradation.

When different concentrations of hydrogen or oxygen gas exist across a gas-permeable membrane, the gases can cross through the membrane due to the partial pressure gradient. The gas pressure applied at each side of the membrane equilibrates to the solubility coefficient (H_i) of the facing side of the membrane, creating a concentration gradient, which drives gas to permeate from one side of the membrane to the other. The gas permeation rate (N_i) of species i through a membrane can be expressed as follows:

$$N_i = D_i \frac{H_i^h p_i^h - H_i^l p_i^l}{l} \tag{11.1}$$

Here, if the solubility coefficient is assumed to be a function of temperature, and both membrane surfaces are at the same temperature, then the permeability coefficient of species i can be defined as follows:

$$k_i = D_i H_i = \frac{N_i l}{\Delta p_i} \tag{11.2}$$

Therefore, the permeability coefficient can be expressed as the product of the diffusion coefficient and the solubility coefficient. It can also be estimated by measuring the permeation rate through the membrane for a given gas pressure difference.

Many techniques have been developed to measure the gas permeation rate of membranes, including the volumetric method, the time-lag technique, the gas chromatography (GC) method, and electrochemical monitoring techniques. Among these, the most direct is in situ electrochemical measurement of gas permeation, which is measured as a mass transfer limited current. When hydrogen crossover measurement is performed, as discussed in Chapter 6, fully humidified hydrogen and nitrogen are introduced at the anode and the cathode at constant flow rates, and the potential of the cathode is set about 0.4 V higher than that of the anode. Hydrogen that diffuses through the membrane from the anode to the cathode is oxidized, and the protons produced are transported to the fuel cell anode, where they are reduced to hydrogen again. The current is directly proportional to the crossover rate of hydrogen through the membrane. Hydrogen crossover can also be measured using a steady-state electrochemical method, in which a fixed cathode potential >0.4 is applied to the fuel cell cathode. For a Nafion®-112 membrane of a 50-μm thickness, a crossover current of 1 mA cm^{-2} at atmospheric conditions was reported at the beginning of life, which corresponds to 2.6×10^{-13} mol$_{H2}$ cm^{-1} kPa^{-1} s^{-1}. End-of-life conditions were considered to correspond with values in the order of 13 mA cm^{-2} [44]. For Gore™ PRIMEA® series 57 catalyst coated membrane, the crossover current increased from 1.84 mA cm^{-2} at the beginning of life to 20.17 mA cm^{-2} at the end of life, around 1200 h [45]. Oxygen crossover is usually not measured in situ, and its permeability is typically half that of hydrogen.

11.3.1.3.2. F-ion Tests

The rate of release of F^- ions in the exhaust gases and product water is correlated to the rate of chemical degradation of fluorinated membranes, which is related to the decomposition of the PSFA chain induced by crossleakage of reactant gases. F^- ion analysis can be conducted offline at regular time intervals by ion chromatography with a conductivity detector or an ion-selective electrode. A fluoride-ion-selective electrode has also been used for continuously monitoring F^- ion changes.

Fluorinated membranes and ionomers undergo degradation during cell operation to produce degradation products such as F^- ions, sulfates, and small polymer end groups. It has already been established by UTC Power and DuPont that PEM fuel cell lifetime is directly related to the F^- ion release rate. For PFSA membranes, an average release rate of 0.01 $\mu g_F\, cm^{-2}\, h^{-1}$ was measured under mild saturated conditions. Under harsh conditions of low humidity, the fluoride release rate might increase to 3 $\mu g_F\, cm^{-2}$. For a 25-μm membrane with a density of 2 $g\, cm^{-3}$ and a fluoride content of 75%, the initial overall fluoride content of the membrane is approximately 3.8 $mg_F\, cm^{-2}$ [46]. These results therefore imply that under mild conditions, 2% of the total F content of the membrane is lost after 6000 h, whereas if the membrane is degraded under harsh conditions, loss of almost the entire membrane and complete failure will occur after about 1200 h. Data on the fluoride release rate have shown considerable scatter and are closely associated with ionomer changes in the membrane [7].

11.3.1.3.3. Gas Leakage Tests

Generally, stack manufacturers perform gas leakage tests to verify the correct assembly of the fuel cell stack. These tests are conducted primarily at the beginning of stack life, before any hydrogen use and any electrical performance test, to ensure that the stack's gas leakage and crossover rates are maintained within the specified tolerated limits. One cause of leakage that might occur during fuel cell degradation testing is the sudden rupture of the MEA in one or more cells of the stack. MEA rupture can result from the stress of an improper sealing gasket design or imbalanced pressure of the reactive gases on the anode and cathode sides. Another possible reason for MEA rupture is improper gas purging through the stack reactant outlet, which causes gas pressure fluctuations and periodic strikes on the MEA. A third cause is defective gasket assembly during fabrication or irregular thickness of the sealing gasket due to gasket degradation.

Different sorts of diagnosis methodologies have been used to investigate and highlight the specific problems of external or internal gas leakage. In the case of a single cell, some electrochemical voltammetry methods can be used during lifetime tests to generate slow-scan linear sweep voltammograms and to enable the estimation of the hydrogen crossover current density. The voltammetry diagnosis methods are quite different if a fuel cell stack composed of several single cells is considered. More importantly, when hydrogen is used inside an aged stack, safety issues for operators and materials must be taken into account. Pressure tests, which are performed with nitrogen or air, are generally used to highlight external or internal gas leakages. To estimate the exterior leakage rate, the anode, cathode, and cooling compartments are connected together, pressurized, and sealed off. The dropdown rate of the inside pressure is related to the leakage between the stack compartments and the external surround. Similarly, the leakage rate inside the fuel cell stack, in particular between the anode and the cathode, can be evaluated by pressurizing one compartment of the stack and monitor the pressure change of the other compartments [47]. However, it is difficult for these leakage check methods to locate which cell or cells have failed, let alone where the fault has occurred. Recently, a rapid solution based on OCV changes under different operating conditions has been proposed to determine the exact location of defective cells in an aged stack. The failed cells show abnormal performance and voltage patterns when compared with the other cells in the stack [48].

11.3.1.3.4. Surface Morphology Characteristics

Microscopic analysis can indicate cracks and especially membrane thinning that has occurred during operation. Moreover, EDX (energy-dispersive X-ray microanalysis) or chemical analysis can be used to determine the presence of contaminants in the membrane [49]. ^{19}F NMR has been used to establish changes in the chemical composition of PFSA membranes and degradation products. Characterization by X-ray diffraction (XRD) can reveal morphological changes that occurred in the membrane during aging under different operating conditions, for example, humidification and temperature. It is believed that a high crystallinity corresponds to open ion channels, so dehydrated and collapsed channels indicate a decrease in crystallinity. XPS (X-ray photoelectron spectroscopy) analysis can yield important information about the membrane structure and membrane degradation mechanisms. It was found that Nafion® decomposed in the hydrogen potential region of the fuel cell, through the interaction of the hydrophobic $(CF_2)_n$ groups of the membrane with H and/or C atoms [25].

11.3.1.3.5. OCV Test

The OCV test is an easy method to detect increases in the reactant crossover rate. However, it is less selective than hydrogen crossover current measurement and is far less quantitative. As shown in Chapter 7, differences and changes in OCV are caused mainly by two factors: the mixed potential of the Pt/PtO catalyst surface and hydrogen crossover. The OCV is a suitable measure of the state of the membrane, and it follows a negative trend that coincides with the fluoride release rate. Under Endoh's test conditions, a failing membrane showed an OCV decline from 1 V at the beginning of life to 0.7 V, and at the same time, a fluoride rate of 60 $\mu g \, day^{-1} \, cm^{-2}$ was measured [50]. A stabilized MEA shows a nearly stable OCV and a negligible fluoride release rate.

11.3.1.3.6. High-Frequency Resistance

As described in Chapter 3, the substantial part of the high-frequency resistance of a PEM fuel cell is mainly attributable to the membrane resistance. Additional contributions come from the GDLs, bipolar plates, and contact resistances. Monitoring of the frequency resistance is feasible during fuel cell degradation tests by in situ AC impedance spectroscopy. Although a high-frequency resistance certainly indicates changes in the membrane resistance that occur on degradation, it has been found to be a very inaccurate method for detecting degradation. Chemical degradation can lead to membrane thinning, which not only causes a lower resistance but also results in a reduced ionic conductivity that compensates for this effect. Further, corrosion products from metal plate oxidation or carbon corrosion may have larger effects on the total increase in resistance. So, although high-frequency resistance is a valuable parameter to monitor, its usefulness for diagnosing membrane degradation seems to be limited.

11.3.2. Electrocatalyst and Catalyst Support Failure

Pt and binary, ternary, or even quaternary Pt-transition metal alloys, such as PtCo, Pt–Cr–Ni, and Pt–Ru-Ir–Sn, placed on conductive supports have been proposed and implemented as electrocatalysts in PEM fuel cells. Commonly used supports include high-surface-area carbon materials, such as Vulcan-XC 72, Ketjen black, and Black Pearls BP2000. Carbon is an excellent material for supporting electrocatalysts, and it allows facile mass transport of reactants and reaction products and provides good electrical conductivity and stability under normal conditions. These catalysts are in principle able to meet the performance and cost requirements for high-volume fuel cell applications. However, from a catalyst durability viewpoint, the performance of currently known materials is still unsatisfactory under harsh operating conditions, including low humidity, low pH values, elevated temperature, and dynamic loads in combination with an oxidizing or reducing environment.

11.3.2.1. Electrocatalyst and Catalyst Support Degradation Mechanisms

Considerable effort has been put into the detailed examination of Pt catalyst degradation mechanisms during long-term operation. First, a pure Pt catalyst may be contaminated by impurities that originate from supply reactants or the fuel cell system. Second, the catalyst may lose its activity due to sintering or migration of Pt particles on the carbon support, detachment and dissolution of Pt into the electrolyte, and corrosion of the carbon support. Several mechanisms have been proposed to explain coarsening of catalyst particle size during PEM fuel cell operation: (1) small Pt particles may dissolve in the ionomer phase and redeposit on the surface of large particles, leading to particle growth, a phenomenon known as Ostwald ripening. In contrast, the dissolved Pt species may diffuse into the ionomer phase and subsequently precipitate into the membrane via reduction of Pt ions by hydrogen crossover from the anode side, which dramatically decreases membrane stability and conductivity; (2) the agglomeration of Pt particles on the carbon support may occur at the nanometer scale due to random cluster–cluster collisions, resulting in a typical log-normal distribution of particles sizes with a maximum at smaller particle sizes and a tail toward the larger particle sizes; (3) catalyst particle growth may also take place at the atomic scale by minimization of the clusters' Gibbs free energy. In this case, the particle size distribution can be characterized by a tail toward the smaller particle sizes and a maximum at the larger particle sizes. However, there is still no agreement on which mechanism is predominantly responsible for catalyst particle growth. Coarsening of the catalyst due to movement of its particles and coalescence on the carbon support can cause the catalytically active surface area to decrease. Lastly, the formation of metal oxides at the anode or cathode side probably leads to an increase in particle sizes and ultimately results in a decrease in catalyst activity.

As previously discussed, corrosion of the catalyst carbon support is another important issue pertaining to electrocatalyst and CL durability that has attracted considerable attention lately in both academic and industry research. In PEM fuel cells and stacks, two modes are believed to induce significant carbon corrosion: (1) transitioning between startup and shutdown cycles and (2) fuel starvation due to the blockage of H_2 from a portion of the anode under steady-state conditions. The first mode, referred to as an air-fuel front, can be caused by the nonuniform distribution of fuel on the anode and the crossover of oxygen through the membrane, which is likely to occur during the startup and shutdown of the PEM fuel cell. In the second mode, fuel starvation in individual cells may result from uneven flow sharing between cells during high overall stack use, or from gas flow blockage attributed to ice formation when fuel cells work at subfreezing temperatures. In both cases, the anode electrode is partially covered with hydrogen and, under circumstances of hydrogen exhaustion, the anode potential will be driven negatively until water and carbon oxidation takes place, according to the following equations [20]:

$$2H_2O \leftrightarrow O_2 + 4H^+ + 4e^- \quad E^\circ = 1.229V \text{ vs. } RHE \qquad (11.X)$$

$$C + 2H_2O \rightarrow CO_2 + 4H^+ + 4e^- \quad E^\circ = 0.207V \text{ vs. } RHE \qquad (11.XI)$$

Despite the thermodynamic instability, carbon corrosion in a normal PEM fuel cell is negligible at potentials <1.1 V vs. RHE due to its slow kinetics. However, recent experiments have confirmed that the presence of electrocatalysts such as Pt/C or PtRu/C can accelerate carbon corrosion and reduce the potentials for carbon oxidation to 0.55 V (vs. RHE) or lower. When provided with sufficient water in the fuel cell, carbon is actually protected from corrosion by virtue of the H_2O oxidation process, unless the water in the electrode is depleted or the cell is subjected to a high current density not sustainable by water oxidation alone. According to Eqn (11.XI), cell reversal as a result of fuel starvation has a potential impact on the durability of the CL, the GDL, or even the bipolar plate. As a consequence, the relative percentage of conductive material in the electrode may drop and the contact resistance with the current collector, as well as the internal resistance of the cell, will eventually increase. More seriously, the number of sites available to anchor the catalyst decreases with carbon corrosion, which causes catalyst metal sintering and, in extreme circumstances, structural collapse of the electrode.

11.3.2.2. Mitigation Strategies for Electrocatalyst and Catalyst Support Degradation

Recent research studies have proposed and successfully used several strategies to enhance catalyst durability. A key starting point is that fuel-cell-operating conditions play a major role in catalyst degradation. Dissolution of Pt from the carbon support is less favorable at low electrode potentials, which makes Pt catalysts more stable at the anode electrode than that at the cathode side. The experimental results of Mathias et al. [51] showed that the loss in Pt active surface area associated with an increase in testing time could be significantly decreased by operating the cell at both low RH and low temperature, as shown in Fig. 11.3. Borup et al. [52] recently found that the carbon corrosion of the CL increased with decreasing RH. They also revealed that growth in the cathode Pt particle size was much greater during potential cycling experiments than during steady-state testing, and that it increased with increasing potential, findings that were recently used in developing an AST method to evaluate electrocatalyst stability.

Second, corrosion of the carbon support due to fuel starvation can be alleviated by enhancing water retention on the anode, such as through modifications to the PTFE and/or ionomer, the addition of water-blocking components such as graphite, and the use of improved preferable catalysts for water electrolysis, as demonstrated by Knights et al. [2] in Fig. 11.4.

Third, Pt-alloy catalysts such as PtCo and Pt-Cr-Ni have been claimed to possess better activity and stability compared with pure Pt catalysts. The increased sintering resistance offered by the alloying elements or the larger alloy particle sizes may explain the observed improvement. However, XRD analysis has revealed a skin consisting of a monolayer of pure Pt formed on the surface of the alloys after long-term testing. This indicates that the non-noble metals in the Pt-transition metal alloy catalysts are more susceptible to dissolving in the ionomer phase, partially counteracting the advantage of Pt-alloy

(a)

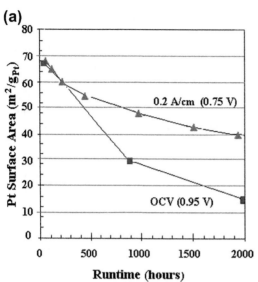

(b)

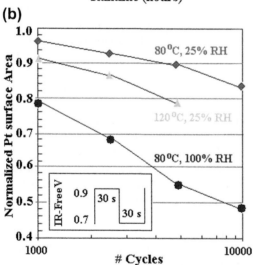

FIGURE 11.3 Impact of operating conditions on catalyst active surface area loss. (a) Pt surface area as a function of stack runtime; (b) impact of RH and high-temperature operation on Pt surface area loss of Pt/C as a function of potential cycles [51]. (For color version of this figure, the reader is referred to the online version of this book.)

catalysts. Metals such as Co, Cr, Fe, Ni, and V have already proven to be soluble in a fuel-cell-operating environment; Pt-Co/C has drawn more attention recently due to its superior stability compared with that of the other Pt-transition alloy catalysts. It is noteworthy that Adzic and coworkers [53] significantly improved Pt stability against dissolution under potential cycling regimes by modifying Pt nanoparticles with gold clusters. There were no obvious changes in the activity and surface area of Au-modified Pt under oxidizing conditions and potential cycling between 0.6 and 1.1 V after >30,000 cycles, in contrast to sizeable losses observed with Pt alone under the same conditions.

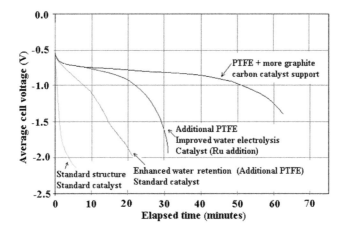

FIGURE 11.4 Comparison of different anode structures in severe failure testing. Each cell has an equivalent cathode (~0.7 mg cm^{-2}·Pt, supported on carbon). Testing conducted at 200 mg cm^{-2}, fully humidified nitrogen on anode. Anode loading at approximately 0.3 mg cm^{-2} Pt supported on carbon (varied materials and compositions) [2]. (For color version of this figure, the reader is referred to the online version of this book.)

The considerable improvement in the Au/Pt/C catalyst stability was mainly attributed to the existence of nondissolvable Au clusters.

As for the catalyst support, by strengthening the interaction between the metal particles and the carbon support, the sintering and dissolution of metal alloy catalysts can be alleviated. For example, Roy et al. [54] introduced a nitrogen-based carbon functionality to the carbon support surface by chemical modification and, consequently, the ability of the treated support to anchor metal particles as well as its catalytic activity showed obvious improvement. Another approach to improving Pt mass activity and durability is to study carbon nanomaterials as electrocatalyst supports. Multiwalled carbon nanotubes (CNTs) have demonstrated promise as catalyst supports in PEM fuel cell applications. Shao et al. [55] reported that the degradation rate of Pt/CNTs was nearly two times lower than that of Pt/C under the same accelerated durability testing conditions, which was attributed to the specific interaction between Pt and CNTs and to the higher resistance of the CNTs to electrochemical oxidation. Other carbon nanomaterials, such as carbon tubulen membrane, highly ordered nanoporous carbon, carbon nanohorns, carbon nanocoils, and carbon nanofibers have also been identified as potentially durable electrocatalyst supports for fuel cells [56–59]. In addition, a decrease in the support surface area or graphitization of the carbon support can also enhance the support's resistance to oxidation and carbon corrosion. However, the number of active surface sites on which to anchor metal particles correspondingly decreases, which is a potential detriment to the deposition of metal on the carbon support. Recent research studies have found that the primarily –OH and –COOH groups introduced to CNTs by oxidation treatment could enhance the reduction of Pt ionic species and further improve their dispersion and attachment properties. The Pt particles supported on oxidized CNTs displayed a durability superior to those on pristine CNTs or commercially available Pt/C [60].

11.3.2.3. Electrocatalyst and Catalyst Support Degradation Testing and Diagnosis

With the development of fuel cell technology, many different investigative tools, including electrochemical and physical/chemical methods, have become available that elucidate CL degradation. These methods provide valuable information on morphology (surface or cross-section of the CL, size distribution of the catalyst particles), elemental content and distribution, atomic structure of the local particles inside the CL, and electrochemical characteristics of the CL in fuel cell systems. At present, characterization of PEM fuel cell electrodes mainly concentrates on morphological characteristics (surface and microstructure), electrochemical diagnosis, and composition analysis.

11.3.2.3.1. Surface Morphology Characteristics

Extraordinary advancements have been achieved in characterizing the surface morphology of an electrode, among which microscopy has been a crucial tool in directly visualizing electrode and polymer morphology. The most common imaging techniques used in the analysis of PEM fuel cell materials and components are optical microscopy and electron microscopy. Optical microscopes use visible wavelengths of light to obtain an enlarged image of a small object, whereas electron microscopes use an electronic beam to examine objects on a very fine scale. Microscopy can provide information about not only surface morphology (the shape and size of the particles making up the sample) but also topography (surface features of the sample), composition (the elements and compounds that comprise the sample, and their relative amounts), and crystallography (atomic arrangement in a given zone). Two key types of electron microscopy are scanning electron microscopy (SEM) and transmission electron microscopy (TEM).

11.3.2.3.2. Optical Microscopy

The optical microscope, often referred to as the light microscope, uses visible light and a system of lenses to magnify images. A sample is usually mounted on a motorized stage and illuminated by a diffuse source of light. The image of the sample is projected via a condenser lens system onto an imaging system, such as the eye, a film, or a charge-coupled device. At very high magnification with transmitted light, point objects are seen as fuzzy disks surrounded by diffraction rings called Airy disks. The resolving power of a microscope is taken as the ability to distinguish between two closely spaced Airy disks. The limit of resolution depends on the wavelength of the illumination source, according to Abbe's theory:

$$d_{rs} = 0.612\lambda/NA_L \qquad (11.3)$$

where d_{rs} is the resolution, λ is the wavelength of light applied, and NA_L is the numerical aperture of the lens. Usually, a wavelength of 550 nm is considered to correspond to green light, and consequently, the limit of resolution of optical microscopy is about 200 nm. Optical microscopy has long been used to visually analyze the surface or cross-section of components used in PEM fuel cells. Recently, Liu et al. [61] used optical microscopy to measure the decrease in cross-sectional thickness of an MEA degraded by local fuel starvation. Through optical microscopy, Ma et al. [62] observed the adhesive effect that ionomer content in the CL had on different substrates. The results showed that if the ionomer content was <20 wt.%, the catalyst paste adhered poorly to the Nafion® membrane.

Although the standard optical microscope is easy to operate, one difficulty is the high-contrast image generated from completely or almost transparent samples, such as proton exchange membranes. To overcome this, fluorescence, dark field, and phase-contrast optical microscopy techniques have been developed, leading to microscopic images with sufficient contrast and high information content. Compared with other types of microscopy for surface imaging, such as SEM and TEM, optical microscopy is restricted by the diffraction limit of visible light to 1000 × magnification and 200-nm resolution. One way to improve the lateral resolution of the optical microscope is to use shorter wavelengths of light, such as ultraviolet. Another way is to use scanning near-field optical microscopy, which has been developed into a powerful surface analytical technique with a spatial resolution of ≥100 nm.

11.3.2.3.3. Scanning Electron Microscopy

The main difference between optical microscopy and electron microscopy is the substitution of an electron beam and electromagnetic coils for the light source and condenser lens. SEM can characterize a sample at several nanometers, which makes it suitable to detect the surface condition, thickness, and interfacial changes of the CL, as well as the elemental distribution changes when combined with energy-dispersive X-ray spectroscopy (EDS). SEM imagery can also be used to quantify morphological changes in PEM fuel cell components before and after degradation tests. Zhang et al. [63] studied the effect of open circuit operation on membrane and CL degradation in PEM fuel cells. SEM images of the cross-sectioned MEA before and after the degradation test are shown in Fig. 11.5. Compared with the "fresh" MEA (Fig. 11.5a), the degraded MEA (Fig. 11.5b) displays an obvious degree of thinning, with PEM thickness decreasing from 23 to 16 μm on both sides of the reinforced layer.

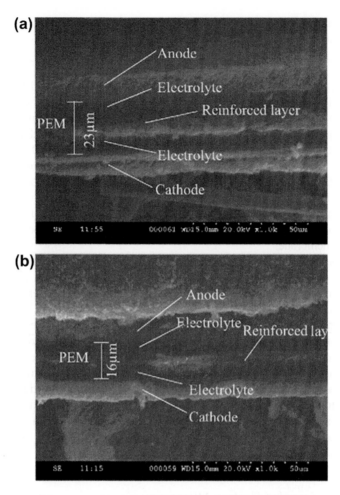

FIGURE 11.5　SEM images of the MEA (a) before and (b) after OC operation [63]. (For color version of this figure, the reader is referred to the online version of this book.)

Meanwhile, as shown in the circled area of Fig. 11.5b, membrane thinning does not occur evenly throughout the MEA. The SEM images show that the PEM thinning was a direct cause of the increased hydrogen crossover that resulted in an unrecoverable decrease in OCV and cell performance.

　The limitation of this technique is that samples for SEM need to be electrically conducting so that charge built up on the surface from the incident electron beam can be conducted away. This problem can be overcome by insulating the samples with a covering of thin, conducting coatings, typically of gold or platinum. To image an MEA cross-section, the sample is usually prepared by cooling the MEA in liquid nitrogen before fracturing, to avoid deformation.

11.3.2.3.4. Scanning Probe Microscopy

Another option for characterizing the morphology and topography of fuel cell electrodes is scanning probe microscopy (SPM). SPM forms images of the sample surface by scanning an atomically sharp, needle-like probe across the surface. As the probe scans, the probe–surface interaction as a function of position is recorded and variations in the topography of the sample surface can be observed. The resolution of SPM is not limited by diffraction, as in electron microscopy, but only by the size of the probe–sample interaction volume, which can be as small as a few picometers. The two main variants of SPM commonly used for morphological measurement are scanning tunneling microscopy (STM) and atomic force microscopy (AFM). For STM, a small electrical current is applied between the probe and the surface of a conducting sample to monitor variations in the surface topography. For AFM, the surface information is detected via the interatomic force between atoms on the probe and those on the surface of either an insulating or a conducting sample.

Ma et al. [62] used AFM and Kelvin probe microscopy (KPM) to examine the effects of different ionomer content (20–40 wt.%) on the morphology and surface potential mapping of CLs. In their AFM images, grain enlargement, pore reduction, and decreased surface roughness were observed with increased ionomer content in the CL. The KPM images showed an increase in the surface potential as ionomer content increased, which implies that the protonic conduction network reduced but did not prevent electrical conduction. Inoue et al. [64] investigated the effects of the overall mass of a CL and the mass ratio of electrolyte in a CL on PEM fuel cell performance, and evaluated the surface roughness of CLs with AFM. The experimental results suggested that the roughness was influenced by the mass ratio of electrolyte but not by the overall mass of a CL. The structure of the CL changed significantly at the optimum mass ratio.

Unlike electron microscopy, which provides a two-dimensional (2D) image of a sample, SPM provides a three-dimensional (3D) surface profile with an even higher resolution. Additionally, an electron microscope needs a vacuum environment for proper operation, whereas SPM can work perfectly in ambient air at standard temperature, or even in a liquid environment. In general, it is easier to use SPM than electron microscopy because SPM samples require minimal preparation. However, the disadvantages of SPM are that image acquisition is time consuming and the maximum image size is generally small.

11.3.2.4. Microstructure Analysis

11.3.2.4.1. Transmission Electron Microscopy

With the help of TEMs, the morphological changes of particles at the micro-scale and nanoscale can be observed after both in situ and ex situ aging processes. Owing to the low wavelength of incident electrons, TEM is capable of imaging a sample at a significantly high resolution, >0.2 nm, which makes it suitable for nanoscale characterization in PEM fuel cell research, such as the determination of the particle size and distribution. Particle imaging via improved TEM variants, such as high-angle annular dark field scanning TEM and high-resolution TEM, can characterize the MEA structure at extremely high resolution, <0.1 nm.

However, one major drawback of TEM is the limited penetration of electrons through the sample. Therefore, samples must be extremely thin (~0.1 μm). For the characterization of PEM fuel cell electrodes, this can be achieved by (1) lightly dispersing the powder (as-processed or scraped from an MEA electrode) across a thin (holey/lacy) carbon film or (2) preparing an intact cross-section of a three-layer MEA using diamond-knife ultramicrotomy. With the former method, information about particle size and distribution of the catalyst and its support can be obtained at the expense of complete destruction of the electrode structure. The latter method has been used successfully by preparing epoxy-embedded TEM cross-sections from thin three-layer MEAs. When the electrode structure is completely embedded with epoxy, imaging of the Pt catalyst particles and carbon support within the electrode is straightforward. Unfortunately, the presence of epoxy makes it virtually impossible to identify and characterize the continuous ionomer network that surrounds the catalyst network.

The recently developed ultramicrotomy sample preparation method based on partially embedded electrodes has enabled the direct imaging of the intact ionomer, carbon/Pt, and pore network surfaces within the MEA [65]. This technique has been used to characterize the differences between catalyst particles before and after degradation, including by determining where inside the MEA structure the growing particles are located. Figure 11.6a and b present, respectively, images of a freshly prepared MEA and an aged MEA that

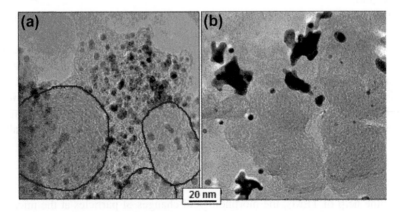

FIGURE 11.6 (a) TEM of freshly prepared MEA; (b) TEM of MEA after potential cycling at 80 to 1.2 V for 1500 cycles [66]. (For color version of this figure, the reader is referred to the online version of this book.)

underwent repeated high-potential cycling to 1.2 V [66]. In Fig. 11.6a, Pt particles of the fresh sample are evident in the ionomer region of the MEA, which suggests that the catalyst particles have separated from the carbon support material. In Fig. 11.6b, large particle agglomerates have formed in the ionomer region after aging.

11.3.2.4.2. X-ray Imaging

Although previously mentioned conventional imaging tools, such as optical microscopy, SEM, and AFM, are adequate to visualize the surface structures of fuel cell CLs, it has been difficult to accurately characterize their internal 3D arrays, porosity, and functionalities. To do that, destructive sample preparation through physical or chemical cross-section is always performed, which is tedious and introduces artifacts. Optical and confocal microscopy suffers from diffraction limits with a spatial resolution not >200 nm. Although electron microscopy can achieve spatial resolution in the nanometer scale, sample preparation can be very elaborate, including the need to be compatible with high vacuum conditions and be electrically conductive. Moreover, conventional imaging modalities will not easily characterize functional and structural changes of materials and sensors in 3D at the multiscale level. The basic principle of X-ray imaging is that the intensity of an X-ray beam is attenuated as it traverses through a material. The transmitted radiation, received by an array of detectors, produces a 2D or 3D map based on the variation in X-ray adsorption throughout the sample. X-ray computed tomography (CT) offers the capability to nondestructively resolve the 3D structure of porous materials with a high spatial resolution, using X-ray radiographs from many angles to computationally reconstruct a 3D image of the material.

In medicine and biology, nano-CT techniques have been used for imaging variations in bone density at an approximately 100-nm resolution, and for visualizing metal nanoparticles in cells at a 40-nm resolution. In the field of PEM fuel cell characterization, morphological imaging is relatively difficult because most of the solid volume is composed of low-phase-contrast materials (carbon and fluorocarbon ionomers), although the electrode also contains small (3–5 nm), dispersed Pt nanoparticles that occupy roughly 1% of the electrode volume. Recently, by using specialized X-ray lenses with a Fresnel zone plate objective and Zernike phase-contrast imaging, a high resolution of 50 nm has been realized in the electrode microstructure diagnosis. Post mortem nano-CT has been used to obtain 2D images of macroscopic Pt redistribution in PEM fuel cell electrodes, to study CL degradation after different ASTs [67]. Most recently, the 3D microstructures of PEM fuel cell electrodes were reconstructed, and they provide important information on the size and form of catalyst particle agglomerates and pore spaces, as shown in Fig. 11.7 [68]. The computational reconstructions, size distributions, and computed porosity obtained with nano-CT can be used for evaluating electrode preparation,

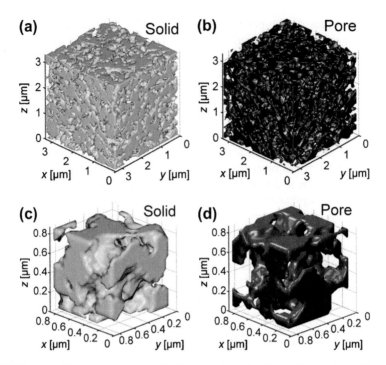

FIGURE 11.7　3D reconstructions of PEM fuel cell electrodes: 3D solid (a) and pore (b) phase of electrode, which includes the ionomer and primary pores; magnified views of the solid (c) and pore (d) phase reconstructions of the electrode. The cube dimensions in images (a) and (b) are 3.25 μm × 3.25 μm × 3.25 μm, and the porosity of the reconstructed electrode cube is 43% [68].

performing pore-scale simulation, and extracting effective morphological parameters for large-scale computational models.

11.3.2.5. Composition Analysis

11.3.2.5.1. Energy-Dispersive Spectrometry

When an electron beam interacts with a sample, secondary electrons, back-scattered electrons, X-rays, and other signals are generated as a result of collisions between the incident electrons and the electrons within the atoms that make up the sample. These signals carry information about the sample and provide clues to its composition, X-rays and backscattered electrons being most commonly used for investigating a sample's composition. X-ray emission occurs when an electron in a shell of an atom absorbs some energy from an incident electron and ejects to a higher energy level to create a vacancy (hole) in the original shell. An electron then drops back down to recombine with this vacancy, and an X-ray photon or Auger electron is emitted, equal in energy to the difference between the higher and lower energy levels. The wavelength of the X-ray radiation is generally regarded as the characteristic of the atom. Hence, after being detected and analyzed by an energy-dispersive spectrometer (EDS) or wavelength-dispersive spectrometer attached to the SEM system, a spectrum of the emitted X-ray wavelengths is commonly used for elemental and compositional analysis. Backscattered electrons are the result of the above-mentioned beam electrons being scattered back out of the sample. The percentage of beam electrons that become backscattered electrons is dependent on the atomic number, which makes the percentage a useful signal for analyzing the sample composition. An Everhart–Thornley detector or a solid-state detector can be used to collect these backscattered electrons and form an image that indicates the compositional information.

As a typical example, Pt dissolution and deposition processes were investigated by Bi et al. [69] under H_2/air and H_2/N_2 potential cycling conditions. In their study, the Pt distributions of (1) a fresh MEA and (2) MEAs after potential cycling were determined by SEM–EDS, as shown in Fig. 11.8. No significant amount of Pt was found in the membrane for either the fresh MEA or the H_2/N_2 cycled MEA, whereas a clear Pt band formed in the membrane for the H_2/air cycled MEA. The Pt deposition mechanism in the membrane was explored based on SEM-EDS and other diagnostic results. The dissolved Pt species may diffuse into the ionomer phase and subsequently precipitate in the membrane via reduction of Pt ions by hydrogen crossover from the anode side, thereby dramatically decreasing membrane stability and conductivity.

11.3.2.5.2. Thermal Gravimetric Analysis

Thermal analysis comprises a group of techniques in which a physical property of a substance is measured as a function of temperature while the substance is subjected to a controlled temperature program. It generally covers three different experimental techniques in PEM fuel cell research: TGA, DTA, and differential scanning calorimetry (DSC). DSC is a thermoanalytical technique in which the different amounts of heat required to increase the temperature of

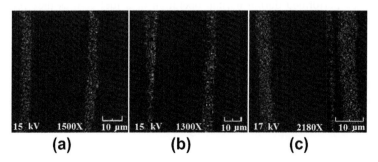

FIGURE 11.8   Pt distribution maps in three MEAs (left: anode; right: cathode): (a) fresh, (b) H_2/ N_2 potential cycled, and (c) H_2/air potential cycled [69].

a sample and of a reference material are measured as a function of temperature. In DTA, the temperature difference between a sample and an inert reference material is measured when both are subjected to identical heat treatments. The main application of DSC and DTA is the determination of phase transitions in various atmospheres, such as exothermic decompositions, that involve energy changes or heat capacity changes.

TGA is performed on samples to determine weight changes in relation to temperature changes and is commonly used in PEM fuel cell degradation research to characterize material composition. For example, TGA diagrams of the PEM can provide much information about degradation temperatures, absorbed moisture content, levels of inorganic and organic components, and solvent residues in the membrane [70]. By coupling TGA with gas analysis, FTIR spectroscopy, and/or mass spectrometry (MS), mass losses and volatile species of decomposition can be determined simultaneously, which will significantly improve the system's analytic capability.

11.3.2.6. Electrochemical Diagnosis

11.3.2.6.1. Polarization Curves

Analysis of a plot of cell potential against current density under a set of constant operating conditions, known as a polarization curve, is the most frequently used tool for characterizing the performance of fuel cells (both single cells and stacks). Although a steady-state polarization curve can be recorded in the potentiostatic or galvanostatic regions, nonsteady-state polarization is analyzed using a rapid current sweep. By measuring polarization curves, certain parameters such as the effects of the composition, flow rate, temperature, and RH of the reactant gases on cell performance can be characterized and compared systematically. This nondestructive tool has also commonly been used to evaluate fuel cell degradation over time.

Polarization curves provide information on the performance of the cell or stack as a whole. However, these measurements fail to produce much information about the performance of individual components within the cell and cannot be performed during normal operation of a fuel cell, as they require significant amounts of time to finish. In addition, they fail to differentiate between different mechanisms; for example, flooding and drying inside a fuel cell cannot both be distinguished in a single polarization curve. Resolving time-dependent processes occurring in the fuel cell and the stack is another important problem. For the latter purpose, current interruption, electrochemical impedance spectroscopy (EIS) measurements, and other electrochemical approaches are preferable.

11.3.2.6.2. Current Interrupt

In general, the current interrupt method is used to measure the ohmic losses in a PEM fuel cell. The principle is that the ohmic losses vanish much more

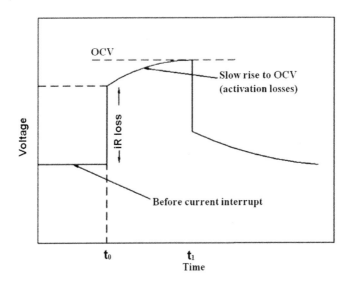

FIGURE 11.9   Ideal voltage transient in a PEM fuel cell after current interruption. The cell is operated at a fixed current. At $t = t_0$, the current is interrupted and the ohmic losses vanish almost immediately. After the current interruption, overpotentials start to decay and the voltage increases exponentially toward OCV. At $t = t1$, the current is again switched on [71].

quickly than do the electrochemical overpotentials when the current is interrupted. As shown schematically in Fig. 11.9, a typical current interrupt result is obtained by recording the transient voltages on interruption of the current after the fuel cell has been operated at a constant current. The ohmic losses disappear almost immediately, whereas the electrochemical (or activation) overpotentials decline to the open circuit value at a considerably slower rate. Therefore, a rapid acquisition of the transient data is of vital importance to adequately differentiate the ohmic and activation losses. Using the current interrupt method, Mennola et al. [71] determined the ohmic resistances in the individual cells of a PEM fuel cell stack. This was achieved by producing voltage transients and monitoring them with a digital oscilloscope connected in parallel with the individual cell. Their results showed a good agreement between the ohmic losses in the entire stack and the sum of the ohmic losses in each individual cell.

Compared with other methods, such as impedance spectroscopy, the current interrupt method has the advantage of relatively straightforward data analysis. However, one of the weaknesses of this method is that the information obtained for a single cell or stack is limited. Another issue is the difficulty in determining the exact point at which the voltage jumps instantaneously.

11.3.2.6 3. Electrochemical Impedance Spectroscopy

In contrast to linear sweep and potential step methods, where the system is far from equilibrium, EIS applies a small ac voltage or current perturbation/signal (of known amplitude and frequency) to the cell, and the amplitude and phase of the resulting signal are measured as a function of frequency. Basically, impedance is a measure of the ability of a system to impede the flow of electrical current; thus, EIS is another powerful technique that can resolve various sources of polarization loss in a short time, and it has been widely applied to PEM fuel cells. As introduced in Chapter 3, the high-frequency arcs of EIS reflect a combination of the double-layer capacitance in the CL, the effective charge transfer resistance, and the ohmic resistance, through which the latter can be directly compared with the data obtained from current interrupt measurements. The low-frequency arc reflects the impedance due to mass transport limitations.

Common uses of EIS for CL investigations are to study the ORR, to characterize transport (diffusion) losses, to evaluate ohmic resistance and electrode properties such as charge transfer resistance and double-layer capacitance, to optimize the MEA, and to evaluate fuel cell degradation.

This dynamic method can yield more information than do steady-state experiments and can provide diagnostic criteria for evaluating PEM fuel cell performance and degradation. The main advantage of EIS as a diagnostic tool for evaluating fuel cell behavior is its ability to resolve, in the frequency domain, the individual contributions of the various factors that determine overall PEM fuel cell power losses: ohmic, kinetic, and mass transport. Such a separation provides useful information for both optimization of fuel cell design and analysis of CL degradation.

11.3.2.6.4. Cyclic Voltammetry

Cyclic voltammetry (CV) is a commonly used in situ approach in fuel cell research, especially to assess catalyst activity, and has proven quite valuable for ascertaining the ECSA of gas diffusion electrodes. The ECSA of the electrode is estimated based on the relationship between the surface area and the H_2 adsorption charge on the electrode, as determined from CV measurement.

The disadvantage of this technique for assessing supported electrocatalysts is that the carbon features mask the H_2 adsorption and desorption character-istics—for example, the double-layer charging and redox behavior of surface active groups on carbon. To avoid carbon oxidation, the anodic limit is always set to <1.0 V (vs. DHE).

11.3.2.6.5. CO Stripping Voltammetry

CO stripping voltammetry is another common technique for determining the ECSA of electrodes through the oxidation of adsorbed CO at room temperature, operating under the same principle as CV. The CO stripping peak charge can also provide information about the active surface sites of the CL. Experimental results have demonstrated that the CO stripping peak potential can provide information on the composition of an unsupported metal alloy surface and is useful for exploring the reaction mechanism of a metal alloy with enhanced CO tolerance. This technique has also been used to investigate the effect of different electrode fabrication procedures on the structural properties of MEAs. Moreover, it has been found that exposing CO to platinum, and the subsequent removal of that CO by electrochemical stripping, is an excellent method of cleaning and activating Pt. The fuel cell achieves its maximum performance after several CO adsorption/CO_2 desorption cycles.

11.3.2.6.6. Other Analytical Techniques

Except for morphological observation, composition analysis, and electrochemical diagnosis, elemental content and atomic structural analysis techniques are useful diagnostic tools to quantitatively characterize microstructural/macrostructural changes in the CL during degradation. For elemental content analysis, inductively coupled plasma and atomic adsorption spectroscopy can be used to investigate Pt content changes in the CL. Guilminot et al. [72] also used ultraviolet spectroscopy successfully to detect the presence of Pt^{z+} ionic species. In terms of carbon support characterization, MS and GC are both effective tools for estimating the total amount of surface oxygen on carbon, when combined with the thermal desorption method. For atomic structural analysis, XRD and XPS are the most commonly used techniques to characterize the average Pt particle size and the surface electronic structural changes during the degradation process. According to a recent report by Yoshida et al. [73], X-ray absorption spectroscopy was carried out to obtain crucial information about the atomic/electronic structure of the surface Pt. The results revealed that the local Pt structure of a Pt/C catalyst was dependent on the particle size, a vital parameter that should lead to a difference in the electrochemical properties of the Pt catalyst. The authors also derived the local structural parameters, coordination number, and Pt–Pt bond distance from extended X-ray absorption fine structure oscillations. In addition, laser Raman spectroscopy [74] has also been conducted to detect the carbon structural disorder degree in research on CL degradation in PEM fuel cells. Except for the broad band at approximately 1600 cm^{-1}, assigned to ideal graphite, the presence of another band at approximately 1350 cm^{-1} proved the existence of disordered graphite in the CL after a high-potential holding test.

11.3.3. Catalyst/ionomer Interface Failure

11.3.3.1. Catalyst/ionomer Interface Degradation Mechanisms

For the membrane in an operating fuel cell, higher current density will intensify the chemical degradation of the membrane due to the much higher electrochemical reaction rates. Higher reaction rates include a higher rate of oxidative radical generation and a higher proton flux through the membrane. For the same reasons, the ionomer in the CL experiences similar degradation, because the electrochemical reactions take place at the interface between the catalyst clusters and the ionomer network. Meanwhile, high liquid water content may facilitate the movement of ions and particles, and hence the agglomeration or growth of catalyst nanoparticles. As the microstructural changes arising from catalyst, catalyst support, and ionomer degradation inside the CL accumulate with time, they will undoubtedly lead to decreased connection between different solid phases, and even mechanical damage to the MEA. Delamination can cause increased resistance, loss of apparent catalytic activity, and development of flooded areas and pinholes. Guilminot et al. [75] reported an obvious separation and cracks at aged cathode/membrane interfaces during testing over 529 h of constant power ($0.12 \ W \ cm^{-2}$). They proved that delamination and cracks between the CL and the PEM or GDL occurred more easily due to RH and temperature changes during load cycles. These interfacial degradations are considered unrecoverable and permanent compared with the recoverable and temporary changes due to water content fluctuations.

Water phase transformations and volume changes due to freeze/thaw cycles severely impair ionic conductivity, gas impermeability, and the mechanical strength of the membrane, eventually having a detrimental effect on the membrane's lifetime. In addition, another serious effect is interfacial degradation during subzero startups and freeze/thaw cycling. During cold startups in subzero environments, water produced in the CL may freeze instantaneously in the pore systems, covering the electrochemical active sites and thereby reducing reaction capability and damaging the interface structure. Yang et al. [76] characterized cross-sectional samples of aged MEAs after 110 cold startup cycles. By using TEM and XRD, they confirmed that interfacial delamination between the CL and PEM, as well as cathode CL pore collapse, was among the degradation mechanisms resulting from cold startups. CL delamination from both the PEM and the GDL was also observed by Yan et al. [77] when the cell cathode temperature fell to $<-5\ °C$ during cold startup studies. Similarly, under frequent freeze/thaw cycles, a shear force induced by phase transition between water and ice will cause uneven mechanical stress for different components, resulting in interfacial delamination and damage.

Because ice formation is the main reason for structural and performance degradation during exposure to subzero temperatures, proposed mitigation strategies include gas purging and solution purging to remove residual water during fuel cell startup and shutdown. Delamination between the CLs and the baer membrane can result in the development of flooded areas, increased resistance in the MEA, the development of pinhole areas, loss of apparent catalytic activity, and the development of areas susceptible to erosion.

11.3.3.2. Catalyst/Ionomer Interface Failure Testing and Diagnosis

As discussed above, interfacial delay and delamination between the CL and PEM, resulting from variations in operating conditions such as load cycling, freeze/thaw cycling, or reactant starvation (mostly fuel starvation), can severely impair ionic conductivity and eventually result in MEA failure. For the catalyst/ ionomer interface, SEM/EDX and X-ray CT imaging have often been used to characterize and diagnose failure modes and degradation mechanisms. Novel X-ray CT in particular has several advantages, including the fact that it is noninvasive, requires little or no sample preparation, does not require the sample to be conductive, and does not disturb the microstructure as compared with TEM or SEM. Moreover, novel CT instruments operate under ambient conditions rather than under high vacuum, thus producing fewer morphological changes due to severe dehydration of the ionomer. X-rays generate much less sample beam damage for polymers compared with the charged particles used in electron imaging techniques. The micro-CT sliced images of three MEAs after different ASTs, at 0.7-micron pixel resolution, are shown in Fig. 11.10. For the fresh, unused membrane (#1), the sample electrode density is uniform and the CL/ionomer interface is smooth. In the case of the degraded membrane (#2), after a drive cycle without stop until failure, the electrode interface became rough with membrane thinning, and significant carbon corrosion occurred at the cathode, with Pt redistribution. Membrane #3 underwent a drive cycle with start stops until failure. Its cathode electrode interface was severely damaged, and its membrane decay and thinning were the worst of the three [67].

11.3.4. Gas Diffusion Layer Failure

11.3.4.1. Gas Diffusion Layer Degradation Mechanisms

In a typical PEM fuel cell, The GDL is a dual-layer, carbon-based porous material, including a macroporous carbon fiber paper or carbon cloth substrate

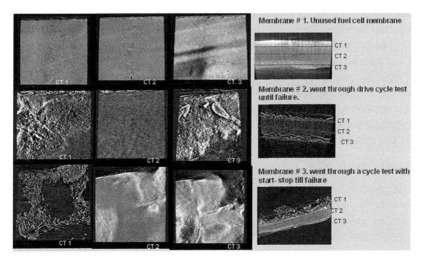

FIGURE 11.10   Micro-CT sliced images of three MEAs after different ASTs [67]. (For color version of this figure, the reader is referred to the online version of this book.)

covered by a thinner MPL, and it consists of carbon black powder and a hydrophobic agent. In previous studies of GDLs, the focal point has been the impact of GDL materials and design on PEM fuel cell performance, rather than durability. However, increased GDL surface hydrophilicity has been clearly observed after 11,000 h of operation and cold start conditions, which unquestionably indicates that further investigation of the GDL is warranted. To date, only a limited number of studies have focused on the degradation mechanisms of GDLs or on the relationship between GDL properties and fuel cell performance decay. Moreover, these studies have mainly used ex situ GDL aging procedures to avoid the possible confounding effects from adjoining components such as the CL and bipolar plate.

As the fuel cell operates, the PTFE and carbon composite of the GDLs are susceptible to chemical attack (i.e. OH· radicals, as electrochemical byproducts) and electrochemical (voltage) oxidation. Loss of PTFE and carbon results in changes in the physical properties of a GDL, such as decreases in GDL conductivity and hydrophobicity, which further lower MEA performance and negatively affect the durability of the whole fuel cell. Several GDL degradation mechanisms have been proposed, including carbon oxidation [78,79], PTFE decomposition [80,81], and mechanical degradation as a result of compression [82]. The first two mechanisms cause hydrophobicity loss and changes in the GDL pore structure, resulting in an increase in the water content of the GDL and MPL and thus altering the water balance in the MEA and limiting reactant mass transport. XPS investigations of the polymers in GDLs have demonstrated that the PTFE can be partially decomposed by electrochemical stressing. But the mechanism of PTFE decomposition in the GDL due to fuel cell operation has not been clearly understood [80,81]. Recently, a novel Nafion/MPL/polyimide barrier was developed and sandwiched between the anode and cathode GDLs to investigate the in situ degradation behavior of commercial GDLs [83]. The experimental results suggested that material loss plays an important role in GDL degradation mechanisms, whereas excessive mechanical stress before degradation weakens the GDL structure and changes its physical properties, consequently accelerating material loss in the GDL during aging. It should be noted that the carbon particles or carbon fibers in the GDL are more stable than the carbon particles in the CL, due to the absence of Pt that could otherwise catalyze the electrochemical oxidation of carbon.

11.3.4.2. Mitigation Strategies for GDL Degradation

Little information about mitigating GDL degradation is available from the literature. To improve GDL oxidative and electro-oxidative stability, Borup et al. [84] suggested using graphitized fibers during GDL preparation. Borup also proposed that a higher PTFE loading could benefit the water management ability of aged GDLs, as shown in Fig. 11.11. By incorporating the graphitized carbon material Pureblack® in the MPL, Owejan et al. [85] found a 25% improvement in the start/stop degradation rate at $1.2 \, A \, cm^{-2}$.

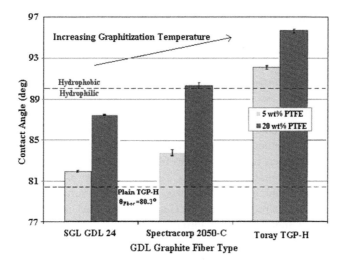

FIGURE 11.11   Effect of GDL graphite fiber type and PTFE loading on contact angle [85]. (For color version of this figure, the reader is referred to the online version of this book.)

11.3.4.3. Gas Diffusion Layer Failure Testing and Diagnosis

For GDL failure diagnosis, changes in the physical characteristics of the GDL before and after degradation tests have always drawn much attention, including porosimetry, permeability and gas diffusivity, contact angle, and conductivity.

11.3.4.3.1. Porosimetry

From a mass transport perspective, the proper porosities of the CL and GDL are critical parameters for optimizing PEM fuel cell performance. Due to the electrochemical reactions occurring inside the matrix CL, the reactants must pass through the GDL into the CL, and in the meantime, the residual water needs to be removed from the electrodes. Therefore, a porous structure is necessary in both the CL and the GDL. If the thickness and real weight of the CL or GDL are known, the porosity can easily be calculated by dividing the volume of the solid phase by the total volume of the electrode, as expressed in the following equation [86]:

$$\varepsilon_p = 1 - \frac{W_A}{\rho_{real}d_{th}} \tag{11.4}$$

where ε_p is the porosity, W_A is the real weight (g cm^{-2}), ρ_{real} is the solid-phase density (for carbon-based materials, ρ_{real} varies between 1.6 and 1.9 g cm^{-3}), and d_{th} is the thickness (either compressed or uncompressed).

Otherwise, the porosity of the electrode needs to be determined by diagnostic methods. So far, several methods have been proposed and implemented to comprehensively characterize porosity, including mercury or gas porosimetry, capillary flow porosimetry, and standard contact porosimetry.

With regard to mercury porosimetry, the applied pressure for injecting mercury into the sample is inversely proportional to the pore radius, in accordance with the Washburn equation, and consequently, the pore size distribution can be evaluated. Mercury porosimetry can be used to characterize larger pore sizes (3 nm–300 μm), whereas gas porosimetry is suited to smaller sizes (0.5–100 nm). As an example, mercury porosimetry measurements were conducted by Wu et al. [83] to determine changes in the total porosity and pore size distribution of GDL samples before and after degradation. However, one limitation of this method is that the sample has to be cut into small pieces and stacked in the holder, requiring relatively large volumes of sample and thus compromising the measuring accuracy. Another concern is the high pressure required for determining small pore sizes, which may result in the collapse of the electrode's pore structure. Jena and Gupta [87] reported using capillary flow porosimetry by monitoring the through-plane and in-plane flow of a gas through dry and wet two-layer porous electrodes. Measurement using this method can yield microstructural information such as largest pore diameter, mean flow pore diameter, cumulative flow percentage, and pore size distribution.

Standard contact porosimetry was recently developed by Volfkovich et al. [88] to characterize the porous electrodes. In this method, disks of the sample and two porous standards are first filled with a low contact angle liquid (e.g. octane or decane) and weighed. Then, the sample is sandwiched between the standards and held in compression to attain capillary equilibrium. By heating and/or vacuum treatment or by a flow of dry inert gas through the sandwich, a certain amount of wetting liquid is removed, and the liquid in the sample is allowed to reach capillary equilibrium again with the standard. The disks are then disassembled and weighed. This process is repeated until all the liquid is completely evaporated from the sample. Standard contact porosimetry with appropriate standard samples can be used to measure pore sizes in the range of $1-3 \times 10^6$ nm. The main drawback is that these measurements are time consuming.

11.3.4.3.2. Permeability and Gas Diffusivity

When the porosity (ε_p) of the electrode is known, the effective diffusivity of the gas phase in this porous media, D_{eff}, can be calculated according to the Bruggeman correlation:

$$D_{eff} = D^0[\varepsilon_p(1 - S_{ls})]^\tau \tag{11.5}$$

where D^0 is the diffusion coefficient, S_{ls} is the liquid saturation, and τ is the tortuosity. As for the liquid phase, the driving force, capillary pressure (P_c), is related to the porosity and the surface tension (σ_{st}), as shown in [89]:

$$P_c = \sigma_{st}\cos(\theta_c)(\varepsilon_p/K_{perm})^{1/2}J(S) \tag{11.6}$$

where θ_c is the contact angle, K_{perm} represents the absolute permeability, and $J(S)$ is the Leverett function. Based on the fluid flow direction through the porous electrode, the permeability coefficients are normally defined as in-plane permeability (x, y directions) and through-plane permeability (z direction). Experimental determination of permeability has been reported by several research groups using their homemade apparatuses with manometers. The principle of measuring the permeability (k_{perm}) is based on the Darcy formula:

$$v_f = \frac{k_{\text{perm}}}{\mu_{\text{fv}}} \frac{\Delta P}{l_{\text{th}}} \tag{11.7}$$

where v_f is the fluid velocity, ΔP is the pressure drop, l_{th} is the thickness of the electrode, and μ_{fv} is the fluid viscosity. Then, the permeability of the porous electrode can be calculated by transforming :

$$k_{\text{perm}} = \mu_{\text{fv}} v_v \frac{l_{\text{th}}}{\Delta P} \tag{11.8}$$

Williams et al. [90] reported that the through-plane permeability was $0.8–3.1 \times 10^{-11}$ m^2 for baer carbon paper and that the addition of microporous layers decreased this value by approximately two orders of magnitude. Prasanna et al. [91] reported a permeability of $1–8 \times 10^{-11}$ m^2 for carbon paper, which decreased significantly as the Tefion® loading was increased.

11.3.4.3.3. Contact Angle

The hydrophobicity and hydrophilicity of the GDL and CL play complex and critical roles in water management within PEM fuel cells. Adequate hydrophilicity is necessary for better PEM fuel cell performance and extended lifetime. If the amount of water in the membrane is too low, the membrane conductivity will decrease, as will the fuel cell performance. However, if an electrode is too hydrophilic, the excess water cannot be removed efficiently; liquid water floods the electrodes, interfering with mass transport of the reactant. Significant effort has been put into investigating water transport and water balance within PEM fuel cells. Research has found that the hydrophobic properties of electrodes can be controlled by the choice of carbon, the ionomer/carbon ratios, the content of the hydrophobic agent, and the pretreatment and fabrication procedures.

The wettability of a PEM fuel cell electrode is normally characterized according to its contact angle to water, which can be divided into two categories: external surface contact angle and internal contact angle. The surface is said to be hydrophilic when its contact angle to water is $<90°$, and hydrophobic if the contact angle is $>90°$. Most methods in use today, such as goniometry (the sessile drop method), the capillary meniscus height method, or the Wilhelmy plate gravimetric method [92] aim to determine the external surface contact angle. Goniometry is the most common method of measuring the contact angle of a liquid on a solid surface, and involves placing a small liquid droplet on a GDL substrate. The contact angle can be determined by measuring the angle between the tangent of the droplet surface at the contact line and the surface. In this method, the droplet size should be small enough to eliminate the influence of the droplet's weight. For the capillary meniscus height method, the sample is first dipped into water, then an optical technique is used to directly record and measure the capillary meniscus height, as shown in Fig. 11.12 [93].

Considering the force balance between gravity and surface tension through a meniscus line, the contact angle (θ_c) between water and the sample has the following relationship with the meniscus height [93]:

$$\sin\theta_c = 1 - \frac{\Delta\rho \cdot gh^2}{2\sigma_{st}} \tag{11.9}$$

where $\Delta\rho$ is the difference between the densities of water and vapor, g is the gravitational acceleration, h is the meniscus height, and σ_{st} is the liquid–gas surface tension of water. For the Wilhelmy plate method, the sample is dipped into liquid, while the force on the sample due to wetting (F_m) is measured via a tensiometer or microbalance. The contact angle between water and the sample can be expressed as follows:

$$\cos\theta_c = \frac{F_m}{2l_w \cdot \sigma_{st}} \tag{11.10}$$

where l_w is the wetted length of the sample. The advantages of the Wilhelmy plate method include that the angle values obtained represent averages over the sample's entire wetted length, and that the temperature of the liquid can be precisely monitored. These methods for determining the external surface contact angle are suitable for materials with smooth surfaces, such as proton

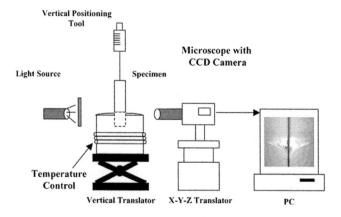

FIGURE 11.12 Experimental setup of the capillary meniscus height method for measuring the external contact angle [93]. (For color version of this figure, the reader is referred to the online version of this book.)

exchange membranes, bipolar plates, or CLs prepared on the membrane surface. Brack et al. [94] studied the membrane properties by means of goniometry and found that the contact angle increased with dehydration of the membrane.

However, when these methods were applied to characterize the porous materials with rough surfaces, such as GDLs, the values of the external contact angle as reported in the literature were between 120° and 140°, or even higher [93]. Because the contact angle to water of pure PTFE is only 108°, these large values cannot be explained by the presence of a hydrophobic agent inside the GDL pores but rather by the contribution of GDL surface roughness. For a rough structured surface, these methods do not provide a true measure of the interfacial properties of the material. Instead, microstructural aspects dominate the observed behavior through such phenomena as droplet pinning. To overcome the limitation of these methods for measuring the external surface contact angle of porous electrodes, Parry et al. used the Washburn method to characterize the internal wetting properties of different GDLs with low PTFE loadings [95]. Capillary rise experiments were performed in a tensiometer by submerging the GDL samples in water. Each sample was held by a metal clamp attached to a microbalance. The mass of liquid water absorbed by the sample was recorded as a function of time. When inertia and gravity forces are negligible, the internal contact angle can be calculated according to the Lucas–Washburn equation:

$$\cos\theta_c = \frac{m_L^2}{t_a} \frac{\eta_L}{C_W \rho_L^2 \gamma_{LV}} \tag{11.11}$$

where m_L is the mass of the liquid absorbed by the sample in time t_a; C_W is the Washburn constant of the GDL sample; and η_L, ρ_L, and γ_{LV} are the liquid viscosity, density, and surface tension, respectively. However, the Washburn method is only applicable to hydrophilic materials. Recently, Gurau et al. [92] combined the Washburn method with the Owens–Wendt theory to estimate the internal contact angle to water of hydrophobic GDLs. In their experiments, the Washburn method was first conducted with a set of wetting fluids to find their internal contact angles to the GDL material. The Owens–Wendt theory was used next to extrapolate the data obtained with the Washburn method and estimate the contact angle to water of the GDL material.

11.3.4.3.4. Conductivity

From the viewpoint of PEM fuel cell performance, high electrical and protonic conductivities are very desirable for optimizing the electrode structure and components. To measure lateral electrical conductivity, a 4-probe technique provides more accurate measurements compared with the 2-probe method. Both local and large-scale conductivity can be determined, depending on the location and separation of the probes. Although this is not a mainstream characterization method, it is useful as a quality control or development tool. The 4-probe technique can yield valuable information on crack formation and membrane deterioration. Through-plane electrical conductivity can be measured by sandwiching the electrode sample between two gold-coated copper plates and subsequently compressing the assembly under a certain pressure [83]. A fixed current (I) is applied with a DC power supply, and the resulting voltage drop between two gold-coated plates (ΔV) is measured with a sensitive multimeter. The through-plane electrical resistance (R) can be described by

$$R = \frac{\Delta V}{I} = \frac{\rho_{el} \cdot L_{th}}{A_{el}} \tag{11.12}$$

where ρ_{el} is the through-plane electrical resistivity, A_{el} is the area of the sample, and L_{th} is the thickness of the electrode sample under a compressive load.

11.3.5. Sealing Gasket Failure

11.3.5.1. Sealing Gasket Degradation Mechanisms

As described in Chapter 2, the sealing material is placed between the bipolar plates, not only to prevent gas and coolant leakage and crossover but also to function as electrical insulation and stack height control. Typically, elastomeric materials are used as seals, as they are relatively inexpensive and easy to fabricate. Although there is a substantial literature discussing chemical or thermal degradation of elastomeric seal materials, only a few are concerned with such a degradation and its mechanisms in a PEM fuel cell environment. Typical sealing materials used in PEM fuel cells include fluorinated elastomers, ethylene propylene diene monomer rubber (EPDM), silicone, and glass-reinforced PTFE. All these materials offer good chemical resistance; however, there are problems associated with each system. Fluoroelastomers offer excellent sealing properties but are poor when it comes to gasketing and disassembly of the cell. Silicone offers good thermal and chemical resistance but has a high value of compression set. PTFE-based systems are only suited for one-time use due to their poor compression set. Material selection concerns include outgassing, degradation, and extraction of the seal components, which may contribute to contaminant migration into the membrane. Another key concern is the potential for internal and/or external leakage of fuel cell reactant gases when seal materials lose their compression set capabilities, and thus lose their "sealing contact" with the adjoining substrates.

Tan et al. [96,97] studied the degradation characteristics of commercial gasket materials, including silicone, fluoroelastomers, and EPDM, in a simulated fuel cell environment consisting of solutions containing hydrogen fluoride and sulfuric acid, and found significant leaching of fillers from seal material, which contaminated the GDL and the MEA. Cleghorn et al. [44] and St-Pierre and

Jia [98] observed the complete degradation of a glass-reinforced silicone seal in a fuel cell stack, with silicon from the seal detectable throughout the MEA. Schulze et al. [99] detailed the degradation of silicon-based seals, and using XPS detected residues of silicone in the anode CL and cathode GDL. They concluded that the direction of movement of the silicone traces was from the anode to the cathode, due to the electrical field, and that it was blocked by the PEM. However, traces of decomposition products from the sealing material in both the membrane and electrodes were detected by Du et al. [100], as shown in Fig. 11.13.

During fuel cell operation, the seal material is subjected to mechanical stress in the presence of variable temperature and chemical environments, which pose significant durability concerns for polymeric materials in terms of accelerated degradation. Thermal cycling alone may result in an enhanced stress relaxation and compression set, leading to reduced stack pressure and adversely affecting the fuel cell performance. Compression of the seal material results in a direction-dependent stress state where compressive and tensile stresses are present. Residual stresses induced during seal material curing or due to shrinkage against the bipolar plate may be tensile in nature and lead to failure over time. The presence of tensile stresses and a corrosive environment may lead to the formation of a crack or the growth of a pre-existing crack, which may lead to seal failure. The seal failure modes can be grouped into two categories: (1) seal cracking/fracture due to material degradation in the fuel cell operating environment and (2) reduction in sealing contact stress due to stress relaxation.

The process causing stress relaxation may be physical or chemical in nature, and in a fuel cell both the processes can occur simultaneously. Physical relaxation involves the motion of molecular chains toward new configurations

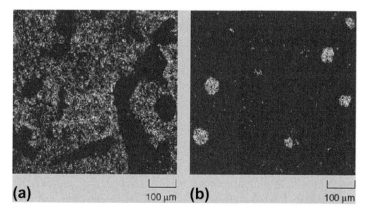

(a) 100 µm **(b)** 100 µm

FIGURE 11.13 SEM-EDX images of an embrittled membrane sample. The holes and tears resulted from reduced mechanical integrity caused by the crystallization of (a) silicon-containing and (b) calcium-containing particles from the degradation of incompatible sealing materials inside and on the surface of a membrane [100].

in equilibrium at the new strained state, and may involve the movement of entanglements and dangling ends, which is generally believed to be reversible in cross-linked systems on removing the strain from the system. Chemical relaxation involves primarily scission and cross-linking events, resulting from the breakage and formation of covalent bonds, respectively. Chemical processes may occur, either in the absence of oxygen (thermal degradation) or in its presence (oxidative degradation), both leading primarily to chain scission reactions. The chemical component of relaxation is typically irreversible. At normal to low temperatures and/or for short periods of time, stress relaxation is dominated by the physical processes, whereas at high temperatures and/or for prolonged periods, the chemical processes are often dominant. Cracks may be present on the surface of the seal as a result of manufacturing or handling and may also develop during service due to seal material degradation. A pre-existing crack on the seal surface may grow and reach a critical size, leading to seal failure. Also, the aggressively acidic environment within a PEM fuel cell, together with thermal stressing or hydrogen embrittlement, may significantly reduce the energy required for the crack to propagate.

The degradation of seals results in the loss of their force retention and can lead to compression loss, external leaks of coolant, gas crossover, or plate electrical shorting, eventually accelerating fuel cell performance degradation. The migration and accumulation of sealing materials within the electrodes will also negatively change the electrodes' hydrophobic character and probably poison the Pt catalysts. Further, traces from the seal may diffuse into the membrane phase and consequently lead to a decrease in the membrane conductivity and a reduction in the mechanical integrity of the membrane, both of which would severely impair the fuel cell lifetime. Seal selection through ex situ and in situ screening processes should be based on the overall chemical and mechanical properties of the materials. Recently, the Fuel Cell and Hydrogen Energy Association summarized and recommended the standard test protocols for screening fuel cell gaskets material, including tension testing, brittleness temperature, chemical resistance, and outgassing [101]. To the best of our knowledge, with regard to seal material degradation, no publications relevant to mitigation strategies are yet available.

11.3.5.2. Sealing Gasket Failure Testing and Diagnosis

The degradation of seal material due to aging in the presence of a corrosive environment, elevated temperatures, and mechanical stress may result in the loss of sealing force, leading to external leaks of coolant, gas crossover, or plate electrical shorting. It is therefore very important to develop and fabricate durable seals that can last the desired lifetime of a fuel cell unit. To be able to choose the best-suited sealing material or to reliably judge whether a given material is suitable at all for sealing applications, seal material characterization must be conducted and an accelerated means to degrade seal material must be adopted so that the performance of such gaskets or seals under service conditions can be accurately predicted.

11.3.5.2.1. Chemical Characterization

Optical microscopy is always used to reveal topographical changes on a sealing gasket surface during degradation tests, showing the progression from an initially smooth to a rough and cracked surface, and finally to crack propagation. Fillers are required to enhance the mechanical properties of elastomeric materials for sealing gasket applications, for example, tensile strength, hardness, and resistance to compression set. As discussed above, some of the filler materials, such as silicon dioxide and calcium carbonate, can be attacked in a PEM fuel cell operating environment. Atomic absorption spectrometry has been used to analyze the elements leached from sealing gaskets, such as silicon, calcium, and magnesium. Attenuated total reflection FTIR spectroscopy can be used to study the changes in the gasket surface chemistry, and it has been found that the chemical degradation of sealing gaskets is likely due to de-crosslinking via hydrolysis of cross-link sites and chain scission in the backbone over time. XPS is a surface-sensitive analysis method to elucidate surface chemicals and has always been used to determine qualitative and quantitative information about elements on the gasket surface before and after degradation tests [102].

11.3.5.2.2. Physical Characterization

Sealing gaskets are normally amorphous polymers above the glass transition temperature. They exhibit the ability to elongate to a large extent under an applied force and subsequently return to their original shape. To obtain the material properties of bulk specimens, standard tests such as uniaxial tension, trouser tear, and compression stress relaxation can be conducted. These tests are typically performed according to the procedures outlined in the American Society for Testing and Materials (ASTM) standards such as ASTM D412 for tension and ASTM D624 for tear properties [103], and provide the baseline properties of the neat material, as summarized in Table 11.3.

11.3.5.2.2.1. Compression Stress Relaxation The definition of compression stress relaxation is that when a constant strain is applied to the gasket sample, the force necessary to maintain that strain is not constant but decreases with time; this behavior is called stress relaxation. The test apparatus used for compression stress relaxation measurements is the Wykeham Farrance device. It provides information for the prediction of the service life of materials by measuring the sealing force decay of a sealing gasket sample as a function of time, temperature, and environment.

The device precisely measures the counterforce exerted by a sample maintained at constant strain between two stainless steel plates inside the compression jig over a period of time. The decay force is then plotted against

TABLE 11.3 The Properties and Test Methods of Sealing Gaskets [101,103]

Test	Standard	Properties of Interest	Suggested Equipment
Compression stress relaxation	ASTM D6147-97	Calculation of sealing force in the seal in various environments and temperatures	Custom-designed jig
Tensile	ASTM D638	Tensile strength, elongation to break	Tensile testing equipment
Tear	ASTM D624 Type T	Seal material resistance to crack propagation	Tear testing equipment
Compression set	ASTM D395-03	Residual deformation after removal of compressive stress	Custom-designed jig
Durometer	ASTM D2240-04e1	Indentation hardness, elastic modulus	Microindenter
Mass uptake/ weight change	ASTM D570	Diffusion coefficient, solubility	Analytical balance
Outgassing	-	Volatile organic components liberated with heat and time	GC–MS or TGA

time to generate the stress–relaxation curve, which provides a valuable tool in failure diagnosis and new gasket materials screening.

11.3.5.2.2.2. Tensile Test For characterization of the tensile properties of an elastomer, tensile strength, strain to break, tension set, and modulus at 100% strain are typically used. In the measurement of tensile stress–strain properties, a test piece is stretched to the breaking point, and the force and elongation are measured at regular intervals. The modulus is relevant where stiffness of the product is important. Tensile strength can be used to evaluate aging performance, though tensile strength decreases more slowly than does elongation to break. Owing to the differences in aging characteristics of tensile strength and elongation to break, the most important property criterion for lifetime prediction must be selected based on the particular application.

11.3.5.2.2.3. Tear Test In measuring tensile strength, the material has to completely break through the cross-section in the absence of a defect, whereas measurement of the tear strength indicates the material resistance to the propagation of a defect. Tear properties are commonly determined at a constant rate, but such properties can also be determined at various rates due to the rate-dependent behavior of elastomeric materials. The tear strength of non-crystallizing elastomers depends on the rate of tearing and temperature. These variations parallel closely the variation of viscoelastic properties with rate and temperature, that is, the tear strength increases with increasing viscoelastic energy dissipation. Tearing in noncrystallizing elastomers often proceeds in a steady, time-dependent manner, whereby the force in a trouser tear test carried out at a constant rate remains relatively constant.

11.3.5.2.2.4. Microindentation Test The microindentation test has been widely applied to measure the mechanical properties of solids such as metals and ceramics because of the ease and speed with which it can be done. It has been performed on rubber coatings and elastomer films, and in recent years, indentation tests have also been conducted on gasket materials to assess the mechanical property changes due to gasket degradation, such as hardness and elastic modulus. In a microindentation test, a diamond or stainless steel indenter of specific geometry is impressed into the surface of the test samples using a known applied load. The microindenter monitors and records the load and displacement of the indenter and obtains an indentation load–depth curve. The indentation load at the peak indentation depth can be used as a manifestation of the surface hardening of the samples. Hertz contact theory is often used to obtain the elastic modulus from the indentation load–indentation depth curves. Based on the Hertz theory of elastic contact, considering the contact between a rigid sphere (the indenter tip) and a flat surface (the gasket sample), the relationship between the total displacement of both the indenter and the sample, δ, and the load, P, can be written as follows:

$$\delta = \left(\frac{9P^2}{16RE^2} \right)^{\frac{1}{3}}$$ (11.13)

where R is the radius of the indenter, whereas E is a combination of the modulus of the indenter and the sample; E can be given by

$$\frac{1}{E} = \frac{1 - \vartheta^2_{\text{indenter}}}{E_{\text{indenter}}} + \frac{1 - \vartheta^2_{\text{sample}}}{E_{\text{sample}}}$$ (11.14)

where E_{indenter} and E_{sample} are the elastic modulus and $\vartheta_{\text{indenter}}$ and $\vartheta_{\text{sample}}$ are Poisson's ratios of the indenter and the sample, respectively. Equation (11.14) can be rewritten as follows:

$$E = \frac{3}{4\sqrt{R}} P\delta^{-1.5}$$ (11.15)

When a rigid indenter compresses a soft flat sample such as a gasket sample, d is the depth of the indentation because the diamond or steel indenter's deformation is negligible relative to that of the sample. Based on Eqn (11.14) and experimental indentation load and indentation depth, the elastic modulus of the sample can be obtained.

11.3.5.2.2.5. Mass Uptake/weight Change When monitoring gasket degradation, the weights of the sealing gasket samples before and after degradation tests are always recorded by a microelectronic balance. The percent weight loss, W_L, is calculated by the following equation [104]:

$$W_L(\%) = \frac{W_2 - W_1}{W_1} \times 100 \qquad (11.16)$$

where W_1 is the initial weight of the sample in air, and W_2 is the weight of the aged sample in air.

11.3.5.2.2.6. Outgassing Gasket materials that are "cured-in-place" or "formed-in-place" are known to outgas harmful volatile compounds during curing (e.g. solvents, cross-linking agents). In addition, fully cured gaskets can also emit species (e.g. low molecular reaction products) that may negatively affect membrane health. Therefore, it is imperative to identify the type and concentration of potential contaminants using ex situ testing methods. This evaluation can involve measurement of volatile organic content using gas chromatography–MS (GC–MS) or by performing thermogravimetric (TGA) analysis under simulated nominal fuel cell operating conditions.

11.4. ACCELERATED STRESS TEST METHODS AND PROTOCOLS

Traditional durability data analysis in engineering involves analyzing times-to-failure data obtained under normal operating conditions to quantify the life characteristics of the product, system, or component. In the case of fuel cells, such times-to-failure data are always very difficult to obtain due to the issues mentioned above: prolonged test periods and high costs. More importantly, a fuel cell stack is a complicated system comprising various components for which the degradation mechanisms, component interactions, and effects of operating conditions need to be fully understood before establishing fuel cell commercial viability. As mentioned earlier, several fuel cell developers have implemented various ASTs to analyze the failure modes of fuel cell components, to increase sample throughput and reduce experimental time. Currently, the US Department of Energy (DOE) and the US Fuel Cell Council (USFCC) are each trying to establish PEM fuel cell durability testing protocols with the intent of providing a standard set of test conditions and operating procedures for evaluating new cell component materials and structures. An accelerated stress testing method should not only activate the targeted failure mode of the specific component, but it should also minimize the confounding effects from other components. For instance, the AST protocol for catalyst supports is different from that for electrocatalysts because the components experience different degradation mechanisms under different conditions. Similarly, the AST for mechanical degradation of the membrane should distinguish and isolate the effects of chemical degradation of the membrane. These protocols help prevent the prolonged test periods and high costs associated with real-time tests, assess the performance and durability of PEM fuel cell components, and ensure that the generated data can be compared.

11.5. CHAPTER SUMMARY

It is universally acknowledged that cost, durability, and reliability are delaying the commercialization of PEM fuel cell technology. Above all, durability is the most critical issue and influences the other two issues. Various efforts have been made to investigate the degradation mechanisms of fuel cell systems and components in an attempt to enhance the durability of fuel cells. However, the current understanding of the degradation mechanisms of PEM fuel cell components is still insufficient. Continuing efforts are critical to propose necessary mitigation strategies and eventually to facilitate the move toward commercialization of PEM fuel cell technology.

With respect to membrane durability, great achievements have been made by modifying the membrane structure to improve its chemical/electrochemical stability and by using a PTFE-reinforced membrane to enhance its mechanical stability. However, further improvements in preventing crossover and preserving stability are necessary for a successful operation in the rugged environment of automotive applications, rather than only under mild steady-state conditions. Similarly, catalysts must also survive the harsh transient operating conditions of a vehicle, such as load and RH cycles. Unfortunately, the catalyst decay that occurs with present-day materials is still too high to meet DOE performance targets. Further optimization of materials and an improved understanding of degradation mechanisms are needed to alleviate Pt dissolution and carbon corrosion. Pt-alloy catalysts such as PtCo or Pt–Cr–Ni loaded on materials highly resistant to electrochemical oxidation, such as CNTs, are suggested for further work. The limited research on GDLs and sealing materials is based mainly on ex situ analysis. Accelerated stresses include mechanical press and/or chemical oxidation. More work in these areas is needed to improve the fuel cell stability in the long term.

The durable physical features of an electrode usually include surface morphology, microstructural properties, physical characteristics, chemical characteristics, and composition. Very often, it becomes important to know as much as possible about the microstructure of an electrode to determine how to improve its efficiency and lifetime in carrying out the relevant electrochemical reaction. In this chapter, the commonly used techniques for characterizing PEM fuel cell electrode components have been addressed. Our discussion has focused on the merits and limitations of each technique when it is applied to PEM fuel cell failure diagnosis research.

The establishment of AST protocols will provide a standard set of test conditions and operating procedures for evaluating new cell component materials and structures. The present AST protocols developed individually by DOE and USFCC are still limited to the component level (electrocatalyst, catalyst support, membrane, and MEA). It is worth noting that, even though these two protocols generally agree with each other, they still differ in a few areas. Achieving the completion and unanimity of AST protocols for the PEM fuel cell as a whole, in addition to those for the components, is imperative for the near future.

REFERENCES

[1] Büchi FN, Srinivasan S. J Electrochem Soc 1997;144:2767–72.

[2] Knights SD, Colbow KM, St-Pierre J, Wilkinson DP. J Power Sources 2004;127:127–34.

[3] Yu J, Matsuura T, Yoshikawa Y, Islam MN, Hori M. Electrochem Solid State Lett 2005; 8:A156–8.

[4] Endoh E, Terazono S, Widjaja H, Takimoto Y. Electrochem Solid State Lett 2004; 7:A209–11.

[5] Tang HL, Shen PK, Jiang SP, Fang W, Mu P. J Power Sources 2007;170:85–92.

[6] Kusoglu A, Karlsson AM, Santare MH, Cleghorn S, Johnson WB. J Power Sources 2007;170:345–58.

[7] Vengatesan S, Fowler MW, Yuan XZ, Wang HJ. J Power Sources 2011;196:5045–52.

[8] Huang X, Solasi R, Zou Y, Feshler M, Reifsnider K, Condit D, Burlatsky S, Madden T. J Polym Sci 2006;16:2346–57.

[9] Wilkie CA, Thomsen JR, Mittleman ML. J Appl Polym Sci 1991;42:901–9.

[10] Surowiec J, Bogoczek R. Studies on the thermal stability of the perfluorinated cation-exchange membrane nafion-417. J Therm Anal Calorim. 1998;33:1097–102.

[11] Chu D, Gervasio D, Razaq M, Yeager EB. J Appl Electrochem 1990;20:157–62.

[12] Deng Q, Wilkie CA, Moore RB, Mauritz KA. Polymers 1998;39:5961–72.

[13] Samms SR, Wasmus S, Savinell RF. J Electrochem Soc 1996;143:1498–504.

[14] Cappadonia M, Erning JW, Stimming U. J Electroanal Chem 1994;376:189–93.

[15] Sivashinsky N, Tanny GB. J Appl Polym Sci 1981;26:2625–37.

[16] Cho EA, Ko JJ, Ha HY, Hong SA, Lee KY, Lim TW, Oh IH. J Electrochem Soc 2004;151:A661–5.

[17] McDonald RC, Mittelsteadt CK, Thompson EL. Fuel Cells 2004;4:208–13.

[18] http://www.nrel.gov/hydrogen/pdfs/pem_fc_freeze_milestone.pdf.

[19] Xie J, Wood III DL, Wayne DM, Zawodzinski TA, Atanassov P, Borup RL. J Electrochem Soc 2005;152:A104–13.

[20] Wilkinson DP, St-Pierre J. Durability. In: Vielstich W, Gasteiger HA, Lamm A, editors. Handbook of fuel cells: fundamentals, technology and applications, vol. 3. New York: John Wiley & Sons; 2003. p. 611–26.

[21] Pozio A, Silva RF, De Francesco M, Giorgi L. Electrochim Acta 2003;48:1543–9.

[22] Stucki S, Scherer GG, Schlagowski S, Fischer E. J Appl Electrochem 1998;28:1041–9.

[23] Yu JR, Yi BL, Xing DM, Liu FQ, Shao ZG, Fu YZ, Zhang HM. Phys Chem Chem Phys 2003;5:611–5.

[24] Mattsson B, Ericson H, Torell LM, Sundholm F. Electrochim Acta 2000;45:1405–8.

[25] Huang C, Tan KS, Lin J, Tan KL. Chem Phys Lett 2003;371:80–5.

[26] Jung M, Williams KA. J Power Sources 2011;196:2717–24.

[27] Scherer GG. Phys Chem 1990;94:1008–14.

[28] Ohma A, Suga S, Yamamoto S, Shinohara K. J Electrochem Soc 2007;154:B757–60.

[29] Inaba M, Kinumoto T, Kiriake M, Umebayashi R, Tasaka A, Ogumi Z. Electrochim Acta 2006;51:5746–53.

[30] Sulek M, Adams J, Kaberline S, Ricketts M, Waldecker JR. J Power Sources 2011;196:8967–72.

[31] Zhou C, Savant D, Ghassemi H, Schiraldi DA, Zawodzinski Jr TA. Fuel cells-proton-exchange membrane fuel cells membrane: life-limiting considerations. In: Garche J, editor. Encyclopedia of electrochemical power sources. New York: Elsevier Academic Press; 2009. p. 755–63.

[32] Liu W, Ruth K, Rusch G. J New Mater Electrochem Syst 2001;4:227–31.

[33] Wakizoe M, Murata H, Takei H. Asahi chemical aciplex-S membrane for PEMFC. In: Proceedings of fuel cell seminar. Portland, USA; 1998. p. 487–90.

[34] Xu H, Wu M, Liu Y, Mittal V, Vieth R, Kunz HR, Bonville LJ, Fenton JM. ECS Trans 2006;3:561–8.

[35] Ijeri V, Cappelletto L, Bianco S, Tortello M, Spinelli P, Tresso E. J Memb Sci 2010;363:265–70.

[36] Liu S, Liu Y, Cebeci H, de Villoria RG, Lin JH, Wardle BL, Zhang QM. Adv Funct Mater 2010;20:3266–71.

[37] Tonpheng B, Yu JC, Andersson BM, Andersson O. Macromolecules 2010;43:7680–8.

[38] Tang H, Wan Z, Pan M, Jiang SP. Electrochem Commun 2007;9:2003–8.

[39] Wang Z, Tang H, Pan M. J Memb Sci 2011;369:250–7.

[40] Gubler L, Gürsel SA, Scherer GG. Fuel Cells 2005;5:317–35.

[41] Curtin DE, Lousenberg RD, Henry TJ, Tangeman PC, Tisack ME. J Power Sources 2004;131:41–8.

[42] Ramani V, Kunz HR, Fenton JM. J Power Sources 2005;152:182–8.

[43] Haugen GM, Meng F, Aieta N, Horan JL, Kuo MC, Frey MH, Hamrock SJ, Herring AM. ECS Trans 2006;3:551–9.

[44] Cleghorn SJC, Mayfield DK, Moore DA, Moore JC, Rusch G, Sherman TW, Sisofo NT, Beuscher U. J Power Sources 2006;158:446–54.

[45] Wu JF, Yuan XZ, Martin JJ, Wang HJ, Yang DJ, Qiao JL, Ma JX. J Power Sources 2010;195:1171–6.

[46] Healy J, Hayden C, Xie T, Olson K, Waldo R, Brundage R, Gasteiger H, Abbott J. Fuel Cells 2005;5:302–8.

[47] Tian G, Wasterlain S, Endichi I, Candusso D, Harel F, François X, Péra M, Hissel D, Kauffmann J. J Power Sources 2008;182:449–61.

[48] Ashraf Khorasani M, Asghari S, Mokmeli A, Shahsamandi MH, Faghih Imani B. Int J Hydrogen Energy 2010;35:9269–75.

[49] Kelly MJ, Fafilek G, Besenhard JO, Kronberger H, Nauer GE. J Power Sources 2005;145:249–52.

[50] Endoh E. Fuel cell seminar abstracts, courtesy associates. Palm Springs; 2005. p. 180–183.

[51] Mathias MF, Makharia R, Gasteiger HA, Conley JJ, Fuller TJ, Gittleman GJ, Kocha SS, Miller DP, Mittelsteadt CK, Xie T, Yan SG, Yu PT. Electrochem Soc Interface 2005; 14:24–35.

[52] Borup RL, Davey JR, Garzon FH, Wood DL, Inbody MA. J Power Sources 2006;163:76–81.

[53] Zhang J, Sasaki K, Sutter E, Adzic RR. Science 2007;315:220–2.

[54] Roy SC, Harding AW, Russell AE, Thomas KM. J Electrochem Soc 1997;144:2323–8.

[55] Shao YY, Yin GP, Gao YZ, Shi PF. J Electrochem Soc 2006;153:A1093–7.

[56] Joo SH, Choi SJ, Oh I, Kwak J, Liu Z, Terasaki O. Nature 2001;414:470–2.

[57] Yoshitake T. Physica B Condens Matter 2002;323:124–6.

[58] Guha A, Zawodzinski TA, Schiraldi DA. J Power Sources 2010;195:5167–75.

[59] Li W, Waje M, Chen Z, Larsen P, Yan Y. Carbon 2010;48:995–1003.

[60] Chiang YC, Ciou JR. Int J Hydrogen Energy 2011;36:6826–31.

[61] Liu ZY, Brady BK, Carter RN, Litteer B, Budinski M, Hyun JK, Muller DA. J Electrochem Soc 2008;125:B979–84.

[62] Ma S, Solterbeck CH, Odgaard M, Skou E. Appl Phys A 2009;96:581–9.

[63] Zhang SS, Yuan XZ, Hin JNC, Wang HJ, Wu JF, Friedrich KA, Schulze M. J Power Sources 2010;195:1142–8.

[64] Inoue H, Daiguji H, Hihara E. The structure of catalyst layers and cell performance in proton exchange membrane fuel cell. JSME Int J Ser B 2004;47-2:228–34.

[65] More KL, Reeves KS. Microsc Microanal 2005;11:2104–5.

[66] Borup R, Meyers J, Pivovar B, Kim YS, Mukundan R, Garland N, Myers D, Wilson M, Garzon F, Wood D, Zelenay P, More K, Stroh K, Zawodzinski T, Boncella J, McGrath JE, Inaba M, Miyatake K, Hori M, Ota K, Ogumi Z, Miyata S, Nishikata A, Siroma Z, Uchimoto Y, Yasuda K, Kimijima K, Iwashita N. Chem Rev 2007;107:3904–51.

[67] Lau SH, Chiu WKS, Garzon F, Chang H, Tkachuk A, Feser M, Yun W. JPCS 2009;152 (Article No. 012059).

[68] Epting WK, Gelb J, Litster S. Adv Funct Mater 2012;22:555–60.

[69] Bi W, Gray GE, Fuller TF. PEM fuel cell Pt/C dissolution and deposition in nafion electrolyte. Electrochem Solid State Lett 2007;10:B101–4.

[70] Adjemian KT, Dominey R, Krishnan L, Ota H, Majsztrik P, Zhang T, Mann J, Kirby B, Gatto L, Velo-Simpson M, Leahy J, Srinivasan S, Benziger JB, Bocarsly AB. Chem Mater 2006;18:2238–48.

[71] Mennola T, Mikkola M, Noponen M. J Power Sources 2002;112:261–72.

[72] Guilminot E, Corella A, Charlot F, Maillard F, Chatenet M. J Electrochem Soc 2007;154:B96–B105.

[73] Yoshida H, Kinumoto T, Iriyama Y, Uchimoto Y, Ogumi Z. ECS Trans 2007;11:1321–9.

[74] Yoda T, Uchida H, Watanabe M. Electrochim Acta 2007;52:5997–6005.

[75] Guilminot E, Corcella A, Chatenet M, Maillard F, Charlot F, Berthome G, Iojoiu C, Sanchez JY, Rossinot E, Clauded E. J Electrochem Soc 2007;154:B1106–14.

[76] Yang XG, Tabuchi Y, Kagami F, Wang CY. J Electrochem Soc 2008;155:B752–61.

[77] Yan Q, Toghiani H, Lee YW, Liang K, Causey H. J Power Sources 2006;160:1242–50.

[78] Stevens D, Dahn J. Carbon 2005;43:179–88.

[79] Cai M, Ruthkosky MS, Merzougui B, Swathirajan S, Balogh MP, Oh SH. J Power Sources 2006;160:977–86.

[80] Schulze M, Christenn C. Appl Surf Sci 2005;252:148–53.

[81] Schulze M, Wagner N, Kaz T, Friedrich K. Electrochim Acta 2007;52:2328–36.

[82] Lee C, Mérida W. J Power Sources 2007;164:141–53.

[83] Wu JF, Martin JJ, Orfino FP, Wang HJ, Legzdins C, Yuan XZ, Sun C. J Power Sources 2010;195:1888–94.

[84] Borup R, Davey JR, Wood DL, Garzon FH, Inbody M. In: Wohlers C, editor. Proceedings of international symposium of fuel cell durability. Massachusetts: Knowledge Press; 2005.

[85] Owejan JE, Yu PT, Makharia R. ECS Trans 2007;11:1049–57.

[86] Barbir F. PEM fuel cells: theory and practice. Burlington, MA: Elsevier Academic press; 2005.

[87] Jena A, Gupta K. J Power Sources 2001;96:214–9.

[88] Volfkovich YM, Bagotzky VS, Sosenkin VE, Blinov IA. Colloids Surf A 2001; 187–188:349–65.

[89] Wang CY, Cheng P. Adv Heat Transf 1997;30:93–196.

[90] Williams MV, Begg E, Bonville L, Kunz HR. J Electrochem Soc 2004;151:A1173–80.

[91] Prasanna M, Ha HY, Cho EA, Hong SA, Oh IH. J Power Sources 2004;131:147–54.

[92] Gurau V, Bluemle MJ, De Castro ES, Tsou YM, Mann Jr JA, Zawodzinski Jr TA. J Power Sources 2006;160:1156–62.

[93] Lim C, Wang Y. Electrochim Acta 2004;49:4149–56.

[94] Brack HP, Slaski M, Gubler L, Scherer GG, Alkan S, Wokaun A. Fuel Cells 2004;4:141–6.

[95] Parry V, Appert E, Joud JC. Appl Surf Sci. 2010;256:2474–8.

[96] Tan J, Chao YJ, Van ZJW, Lee WK. Mater Sci Eng A 2007;445–446:669–75.

[97] Lin CW, Chien CH, Tan J, Chao YJ, Van Zee JW. J Power Sources 2011;196:1955–66.

[98] St-Pierre J, Jia N. J New Mater Electrochem Syst 2002;5:263–71.

[99] Schulze M, Knori T, Schneider A, Gulzow E. J Power Sources 2004;127:222–9.

[100] Du B, Guo Q, Pollard R, Rodriguez D, Smith C, Elter J. J Oral Microbiol 2006;58:45–9.

[101] http://www.fchea.org/core/import/PDFs/FCHEA_Gasket_Report_2011.pdf.

[102] Li G, Tan J, Gong J. J Power Sources 2012;205:244–51.

[103] http://www.astm.org/.

[104] Tan J, Chao YJ, Yang M, Williams CT, Van Zee JW. J Mater Eng Perform 2008;17:785–92.

Chapter 12

Electrochemical Half-Cells for Evaluating PEM Fuel Cell Catalysts and Catalyst Layers

Chapter Outline

PEM Fuel Cell Testing and Diagnosis. http://dx.doi.org/10.1016/B978-0-444-53688-4.00012-7

12.1. INTRODUCTION

As described in Chapter 1, the electrochemical reactions in a PEM fuel cell include two half-cell reactions: the fuel oxidation reaction, such as the hydrogen oxidation reaction (HOR), occurs at the anode while the oxygen reduction reaction (ORR) proceeds at the cathode. When investigating one reaction and its associated catalysis mechanism, as well as the effects that operating conditions have on this reaction, possible interference from the other reaction is normally eliminated through a half-cell method. In addition, for quick downselection of electrode materials and components, such as the catalyst and its associated catalyst layer, an ex situ approach using a half-cell setup is the quickest and most cost-effective method. Half-cell testing is usually conducted in a three-electrode system containing working, counter electrode (CE) and reference electrode (RE). Cyclic voltammetry (CV), rotating disk electrode (RDE), rotating ring-disk electrode (RRDE), and electrochemical impedance spectroscopy are the typical half-cell testing techniques to investigate a catalyst's characteristics in terms of the HOR and ORR. Besides their utility for investigating these two reactions, some special half-cell designs also allow testing of other operating conditions, such as catalyst layer/membrane electrode assembly (MEA) designs, temperature, pressure, humidity, as well as fuel and air-flow rates.

12.2. CONVENTIONAL THREE-ELECTRODE HALF-CELL

12.2.1. Half-Cell Design and Fabrication

Figure 12.1 shows the design for a conventional three-electrode half-cell. The three electrodes in this half-cell (or electrochemical cell) are (1) the catalyst-coated working electrode (WE), for example, an RDE made of either carbon, such as glassy carbon (GC), or other stable metal materials, such as Au or Pt, or an RRDE whose disk is made of carbon material (GC), Au or Pt, and whose ring is made of Pt located at the outside edge of the disk electrode; (2) the CE (e.g. Pt wire or Pt foil); and (3) the RE, such as a standard or normal hydrogen electrode (SHE or NHE), a reversible hydrogen electrode (RHE), an Ag/AgCl electrode, a saturated calomel electrode (SCE), an $Hg/HgSO_4$ electrode, or another kind. Note that the difference between an SHE and an NHE is that the former uses a theoretical Pt electrode/1.0 M H^+ aqueous solution interface and assumes that the H^+ ions do not interact with other ions (a condition that is not physically attainable), whereas the latter is actually constructed using a practical Pt electrode/1.0 M H^+ aqueous solution (such as 0.5 M H_2SO_4) interface. At standard conditions, the SHE's potential is defined as 0.000 V and is used as the standard RE potential. However, in practice, only the NHE is physically attainable, but its potential is almost the same as that of the SHEs. The difference between the NHE and the RHE is that the latter uses the same electrolyte solution as does the measurement cell, rather than

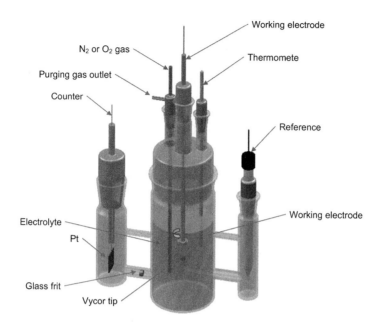

N₂ or O₂ gas

Purging gas outlet

Counter

Working electrode

Thermomete

Reference

Working electrode

Electrolyte

Pt

Glass frit

Vycor tip

FIGURE 12.1   Schematic of conventional three-electrode half-cell. (For color version of this figure, the reader is referred to the online version of this book.)

a 1.0 M H$^+$ aqueous solution, as in the NHE. For example, if the electrolyte solution in the measurement cell is 0.05 M H$_2$SO$_4$, the electrolyte solution in the RHE electrolyte chamber is also 0.05 M H$_2$SO$_4$. Therefore, there is a large potential difference between the NHE and the RHE, and only when both electrolyte solutions are 1.0 M H$^+$ aqueous solution can they have the same electrode potential.

As shown in Fig. 12.1, the three-electrode cell has an inlet and an outlet for gas purging. For surface CV measurement, N$_2$ gas is used to purge the electrolyte solution for 30 min to remove dissolved O$_2$. For ORR measurement, pure O$_2$ gas or air is used to purge the electrolyte solution and introduce dissolved O$_2$ into the solution. For CO-stripping experiments, a CO/N$_2$ mixed gas is used to purge the solution and introduce dissolved CO into the solution. In addition, there is a port for a thermometer to monitor the temperature of the electrolyte solution. To control the temperature, the whole cell is emerged in a thermal bath in which the temperature of the liquid can be adjusted to the desired level.

RDE and RRDE techniques have been widely used for studying the kinetics of the ORR [1–10] and HOR [11–17] in PEM fuel cells. Particularly in ex situ evaluations of catalysts and catalyst layers, the conventional half-cell is used to measure the electrochemical Pt surface area (EPSA), ORR mass activity, catalyst stability, as well as non-noble metal catalyst activity and stability, as described in later sections of this chapter.

12.2.2. Pt-Based Catalyst Measurements for the ORR

A necessary step in conducting these measurements is preparation of the WE. A well-developed procedure for electrode preparation is described by Mayrhofer et al. [3]. A small amount of catalyst power (Pt- or Pt alloy-based catalyst) is first mixed ultrasonically with deionized water, followed by the addition of alcohol or isopropanol (~1.0 ml alcohol to 5 mg catalyst) and 5 wt.% Nafion$^{®}$ ionomer solution (~1/40 volume ratio with the alcohol) to form a well-mixed catalyst ink. Then a small amount of catalyst ink is pipetted and coated onto a disk electrode surface such as GC or gold electrode, with a geometric area of 0.2–0.5 cm^2. The coated electrode is then left to air dry. The total catalyst loading on a GC electrode can be adjusted to 0.02–0.2 mg cm^{-2}.

Another way to prepare the electrode is to make a catalyst ink that does not contain Nafion$^{®}$ ionomer solution. The ink is pipetted and coated onto the GC or gold surface, and then, the required amount of Nafion$^{®}$ ionomer solution is dropped onto the top of the catalyst coating to form a catalyst layer coated electrode.

The following is an example of measuring EPSA, ORR mass activity, and stability [18]. In this instance, the WE was coated with a Pt-based catalyst and then put into an electrochemical half-cell containing N$_2$-saturated 0.1 M HClO$_4$ or H$_2$SO$_4$ solution for CV testing. The temperature of this electrolyte solution was controlled at 20–30 °C. Before the data were recorded, the WE was pre-conditioned by using CV in the potential range of 0.0–1.4 V vs. RHE at a sweep rate of 100 mV s^{-1} for 20 cycles. Then, the potential was scanned in the potential range of 0.05–1.2 V at a scan rate of 20 mV s^{-1}. Twenty scans were taken for each experiment, and the stable twentieth cyclic voltammogram was used to calculate the EPSA. Figure 12.2 shows a typical example of a cyclic voltammogram recorded using a catalyst-coated GC electrode.

The EPSA was calculated by averaging the charges of the hydrogen adsorption/desorption peaks (Fig. 12.2) in the potential range of 0.05–0.40 V, corrected for double layer charging. The recognized value for the monolayer hydrogen adsorption charge on a smooth Pt electrode surface, 0.21 mC cm^{-2}(Pt), was used for the calculation. By averaging the seven electrode measurements in Fig. 12.2, an EPSA of 88 m^2 g^{-1} can be obtained, which is close to Gasteiger's value of 80 m^2 g^{-1} [19].

To confirm the EPSA data obtained by the CV method, CO-stripping experiments were also carried out in CO-saturated 0.1 M HClO$_4$ solution. The electrolyte solution was first purged with nitrogen gas for at least 30 min before preconditioning the WE. The catalyst-coated WE was preconditioned by CV in the potential range of 0.0–1.4 V vs. RHE at a sweep rate of 100 mV s^{-1} for 20 cycles. After preconditioning, CO was absorbed by purging the solution with CO gas at a flow rate of 200 ml min^{-1}, while holding the WE potential at 0.05 V vs. RHE. By keeping the potential at the same value, the purging gas was switched to nitrogen for 30 min to remove CO traces from the solution.

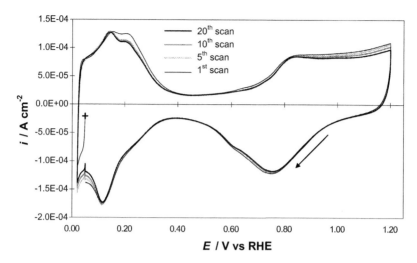

FIGURE 12.2   Cyclic voltammogram of baseline 47 wt.% Pt/C catalyst (purchased from Tanaka Kikinzoku Kogyo Co. Ltd) coated GC electrode, recorded in N_2-saturated 0.1 M $HClO_4$. Potential scan rate: 20 mV s^{-1}; Pt/C catalyst loading: 50 μg cm^{-2} [18].

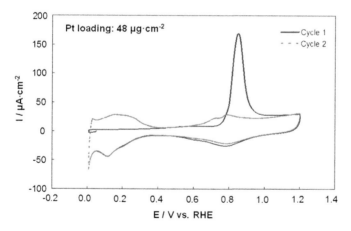

FIGURE 12.3   A CO-stripping voltammogram on 47 wt.% Pt/C catalyst-coated GC electrode, recorded in N_2-saturated 0.1 M $HClO_4$. Potential scan rate: 20 mV s^{-1}; Pt/C catalyst loading: 48 μg cm^{-2} [18]. (For color version of this figure, the reader is referred to the online version of this book.)

Then, the potential was scanned in the range of 0.05–1.2 V at a rate of 20 mV s^{-1}. Figure 12.3 shows a typical example of a CO-stripping cyclic voltammogram.

The EPSA was then calculated by integrating the charge of the CO peak and subtracting the background charge arising from double layer charging and

oxide formation. This calculation was based on the procedure used by
Vidakovic et al. [20]. The obtained EPSA was 77 $m^2\,g^{-1}$, which is consistent
with the value obtained from the H_2 adsorption/desorption peaks in the
following scans (75 $m^2\,g^{-1}$), and close to the EPSA of 88 $m^2\,g^{-1}$obtained by
CV in Fig. 12.2.

Mass activity of the catalyst toward the ORR can also be evaluated with
a conventional half-cell by using the RDE technique. Normally, mass activity is
measured using the RDE technique in O_2-saturated 0.1 M $HClO_4$ solution
at 30 °C. Figure 12.4 shows the voltammetric curves in the potential range of
1.2–0.4 V at a scan rate of 5 mV s^{-1}, for different rotation rates. A diffusion-
limiting current can be observed at a potential <0.8 V, depending on the
rotation rate. The kinetic current of the catalytic ORR (i_k) was calculated by
using the current at 0.9 V vs. RHE on the voltammetric curve, at an electrode
rotation rate of 1600 rpm, followed by a calculation by using the following
equation:

$$i_k = \frac{i_d \times i}{i_d - i}$$ (12.1)

where i is the current measured at 0.9 V and i_d is the diffusion-limiting current.
The mass activity, with a unit of A mg^{-1} (Pt), was calculated by dividing the i_k
by the Pt loading. For example, for a commercially available Pt/C catalyst,
a typical mass activity of 0.11 A mg^{-1} (Pt) at 0.9 V vs. RHE and 30 °C can be
obtained.

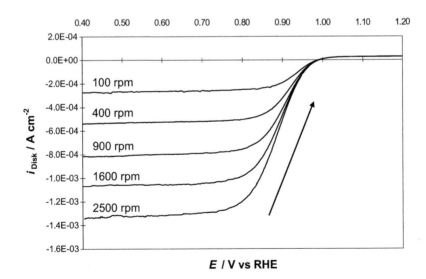

FIGURE 12.4 Current–voltage curves recorded on a rotating disk GC electrode coated with
47 wt.% Pt/C catalyst, measured in O_2-saturated 0.1 M $HClO_4$. Potential scan rate: 5 mV s^{-1}, Pt/C
catalyst loading: 50 μg cm^{-2} [18].

Catalyst stability can also be measured with a half-cell by using either a potential square wave scan or CV cycling. In a typical experiment, before durability testing, the EPSA and ORR mass activity are measured using the same procedure described above. Then, potential square wave scans are performed, 30 s each at 0.6 or 1.2 V vs. RHE for 1000 cycles, after which both the EPSA and the ORR mass activity are measured and then compared with the values obtained before the durability test. For a commercially available Pt/C catalyst, the EPSA and ORR mass activity losses after 1000 cycles are 38% and 33%, respectively.

CV cycling can also be used to test the stability. In typical CV cycling, the potential is scanned between 0.05 and 1.2 V vs. RHE for 1000 cycles. The EPSA and ORR mass activity are measured before and after durability testing. For a commercially available Pt/C catalyst, the EPSA and ORR mass activity losses after 1000 cycles are approximately 20% and 40%, respectively.

12.2.3. Non-noble Metal Catalysts for the ORR [2]

The half-cell shown in Fig. 12.1 can also be used for evaluating non-Pt-based catalysts. For example, Lee et al. [2] investigated the ORR kinetics and mechanisms on carbon-supported cobalt polypyrrole (Co–PPy/C) catalysts by using both RDE and RRDE techniques. Figure 12.5 presents the ORR current–voltage (I–V) curves in 0.5 M H$_2$SO$_4$ with several Co–PPy/C catalysts, which shows that the catalyst heat treated at 800 °C can yield the highest catalytic

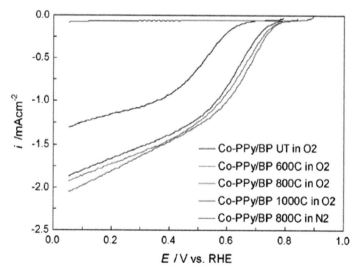

FIGURE 12.5   Polarization curves of unpyrolyzed and pyrolyzed Co–PPy/C catalysts at 25 °C in 0.5 M H$_2$SO$_4$ under saturated N$_2$ and O$_2$, respectively. Potential scan rate: 5 mV s^{-1}. Electrode rotation rate: 400 rpm. Co–PPy/C loading: 0.122 mg cm^{-2} [2]. (For color version of this figure, the reader is referred to the online version of this book.)

activity in terms of both ORR onset potential and limiting current. Lee et al. [2] also recorded the ORR I–V curves at various electrode rotation rates to evaluate the catalyzed ORR mechanism. According to the Koutecky–Levich theory, the current density, i, of the ORR on an RDE can be expressed as in Eqn (12.2):

$$\frac{1}{i} = \frac{1}{i_k} + \frac{1}{i_d}$$ (12.2)

Here, i_d is the diffusion-limiting current density and can be expressed as in Eqn (12.3):

$$i_d = 0.62nFC_{O_2}D_{O_2}^{2/3}v^{-1/6}\omega^{1/2}$$ (12.3)

where n is the number of electrons transferred during the overall reduction process, C_{O_2} is the concentration of O_2, D_{O_2} is the diffusion coefficient of O_2, v is the kinetic viscosity of the electrolyte, and ω is the electrode rotation rate. It is worth pointing out that the RDE theory expressed by Eqn (12.2) should be modified if a porous layer electrode rather than a smooth electrode is used, as shown in Eqn (12.8) (section 12.2.4). The effect arises from the thickness of the Nafion® ionomer porous layer, which can limit the diffusion of O_2 within the electrode [21]; in that study, certain calculations indicated that the effect of the ionomer layer thickness on the i_k values was insignificant, so this effect was ignored.

According to Eqns (12.2) and (12.3), the reciprocal of the current density, i_d, is plotted against the reciprocal of the square root of the rotation rate, ω, to obtain the Koutecky–Levich plots, as shown in Fig. 12.6. It can be seen that the slope of the Koutecky–Levich plot with an unpyrolyzed catalyst is close to that of the theoretical 2-electron transfer reaction, and the slopes with pyrolyzed catalysts are close to that of the theoretical 4-electron transfer reaction, indicating that the ORR proceeds with a 2-electron transfer mechanism on an unpyrolyzed catalyst and with a 4-electron transfer mechanism on pyrolyzed catalysts.

The authors also conducted their mechanism study using an RRDE technique and calculated the electron transfer number and the percentage of produced H_2O_2 according to Eqns (12.4) and (12.5) [2]:

$$n = \frac{4 - 2(I_r/I_d)}{N}$$ (12.4)

$$\%H_2O_2 = \frac{(I_r/I_d)}{N} \times 100$$ (12.5)

where I_r and I_d are the currents at the ring and disk electrodes, respectively, and N is the RRDE collection efficiency, which was determined by the authors in this study to be 0.21 [2]. The electron transfer numbers calculated from their RRDE analysis were close to those calculated based on the Koutecky–Levich

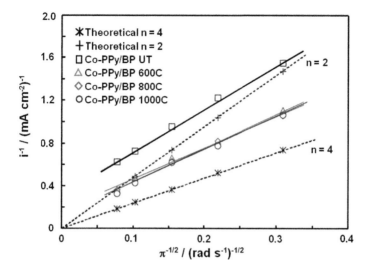

FIGURE 12.6 Koutecky–Levich plots for the ORR on unpyrolyzed and pyrolyzed Co–PPy/C catalysts at 0.3 V vs. RHE in 0.5 M H_2SO_4 under saturated O_2. The current densities were normalized to the geometric area. Co-PPy/C loading: 0.122 mg cm^{-2} [2]. (For color version of this figure, the reader is referred to the online version of this book.)

plot. From the electron numbers obtained, they concluded that the ORR proceeds via different mechanisms on unpyrolyzed and pyrolzed catalysts.

In another study, on a Magneli phase Ti_4O_7 electrode for the ORR using a half-cell RDE/RRDE technique, Li et al. [22] identified that the ORR mechanism on a Ti_4O_7 electrode is a combination of 2- and 4-electron transfer pathways in KOH aqueous solutions. The ORR kinetic parameters—such as the chemical reaction rate constant between O_2 and the active reaction site of the catalyst before electron transfer (k_c), the electron transfer rate constant in the ORR rate-determining step (RDS) (k_e), the electron transfer coefficient in the RDS (α), and the ORR exchange current density—were obtained in alkaline solution, as listed in Table 12.1.

Regarding the ORR mechanism, the authors believed that the first step should be oxygen adsorption on the Ti_4O_7 electrode surface, which is then electrochemically reduced through several elementary steps. The H_2O_2 produced (in the form of HO_2^- in alkaline solution) can then be further reduced to water or can dissolve into the solution, depending on the electrode material used. If the HO_2^- is relatively stable, it can enter into the solution before being further reduced by another two electrons to H_2O. This dissolved HO_2^- will be detected by the ring electrode when the RRDE is rotated. To facilitate understanding, the authors offered the following hypothetical steps for the ORR mechanism:

$$Ti_4O_7 + O_2 \rightarrow O_2 - Ti_4O_7 \tag{12.I}$$

TABLE 12.1 Electron Transfer Rate Constant (k_e) and Electron Transfer Coefficient (α) in the ORR Rate-Determining Step, Calculated Based on the Intercept and the Slope Values Obtained Using Koutecky–Levich plots. The Electron Transfer Number in the ORR Rate-Determining Step was Taken to be 1 [22]

KOH Concentration in Aqueous Solution	1 M	4 M	6 M
k_e (cm s^{-1})	1.6×10^{-12}	8.1×10^{-10}	9.8×10^{-8}
i_o (mA cm^{-2})	3.8×10^{-10}	7.7×10^{-8}	4.8×10^{-6}
α	0.19	0.15	0.12

$$O_2 - Ti_4O_7 + e^- \rightarrow O_2^- - Ti_4O_7 \quad \text{(Rate-Determining Step)} \qquad (12.\text{II})$$

$$O_2^- - Ti_4O_7 + e^- + H_2O \rightarrow HO_2^- - Ti_4O_7 + OH^- \qquad (12.\text{III})$$

$$x(HO_2^- - Ti_4O_7) + xH_2O + 2xe^- \rightarrow xTi_4O_7 + 3x(OH^-) \qquad (12.\text{IV})$$

$$(1-x)(HO_2^- - Ti_4O_7) \rightarrow (1-x)Ti_4O_7 + (1-x)HO_2^- \qquad (12.\text{V})$$

In this mechanism, Reaction (12.I) is the chemical reaction to form the adduct, which has a reaction rate constant of 1.0×10^{-2} cm s^{-1}, as determined by the RDE measurements in this work. Reaction (12.II) is the ORR RDS on the Ti_4O_7 electrode surface, whose rate constants are given in Table 12.1. Reaction (12.III) represents the reactions for peroxide formation. After HO_2^- formation, HO_2^- can react in one of two ways: further 2-electron reduction to OH^- through Reaction (12.IV), or chemical desorption through Reaction (12.V) to form a free peroxide ion, which then enters into the bulk solution and can be detected by the ring electrode of the RRDE. The ORR on the Ti4O7 electrode has a mixed 2- and 4-electron transfer pathway and gives an overall electron transfer number of <4. The relative portion of Reaction (12.IV) can be expressed as x, and the portion of Reaction (12.V) can be expressed as $(1-x)$. When $x = 1$, the mechanism will follow a totally 4-electron transfer pathway, and when $x = 0$, the mechanism will be a totally 2-electron pathway. If the x value is >0 and <1, the ORR will have a mixed 2- and 4-electron transfer pathway. Note that this ORR mechanism is only hypothetical, to facilitate further discussion. More evidence is needed to validate the mechanism.

It is worth mentioning that kinetic parameters such as the electron transfer coefficient and the exchange current density can also be obtained by the Tafel method, based on RDE currents. For example, in a study by Li et al. [21] on the ORR catalyzed by a carbon composite non-noble metal catalyst, $La_{0.6}Ca_{0.4}CoO_3$/C (abbreviated as LCCO/C), the RDE current–voltage curves were recorded at different electrode rotation rates in solutions containing three different KOH concentrations. The RDE current–potential curves were analyzed using the following Tafel and Nernst equations:

$$E = E^o_{O_2/OH^-} + \frac{2.303RT}{\alpha n_\alpha F}\log(i^o_{O_2/OH^-}) - \frac{2.303RT}{\alpha n_\alpha F}\log\left(\frac{i i_d}{i_d - i}\right) \qquad (12.6)$$

$$E = E^o_{O_2/OH^-} - \frac{2.303RT}{F}\log(OH^-) \qquad (12.7)$$

where E is the electrode potential, $E^o_{O_2/OH^-}$ is the thermodynamic electrode potential of the ORR, R is the gas constant $(8.314 \, J\,K^{-1}\,mol^{-1})$, T is the absolute temperature (K), n_α and α are the electron transfer number and coefficient in the RDS of the catalyzed ORR, $i^o_{O_2/OH^-}$ is the ORR exchange current density, and both i and i_d have the same meaning as in Eqn (12.2).

The diffusion-current-corrected current densities in these regions (at 1600 rpm) were used to construct the Tafel plots shown in Fig. 12.7. It can be seen that well-defined linear plots of $\log(i i_d/i_d - i)$ vs. E were obtained in these potential ranges. The resulting values of the Tafel slope $(2.303RT/\alpha n_\alpha F)$ for the LCCO/C and LCCO electrodes in all three KOH concentrations were obtained, from which the values of the electron transfer coefficient, α, were calculated by assuming that the RDS is a 1-electron transfer process $(n_\alpha = 1)$. The exchange current density $i^o_{O_2/OH^-}$ was also determined from the intercept of the Tafel plots.

12.2.4. Pt-Based Catalyst Layer Measurements Using the HOR [23]

Schmidt et al. [23] used a conventional electrochemical half-cell to characterize the electrocatalytic properties of highly dispersed electrocatalysts in an RDE configuration. The WE was rotating GC coated by Pt/Vulcan catalyst powder that had been impregnated with a thin Nafion® ionomer film. RDE measurements on the catalyzed HOR showed that if the equivalent Nafion® film in the catalyst layer was too thick, the mass transfer process could be significantly affected. To take account of this effect, an alternative expression to Eqn (12.2) was written as follows:

$$\frac{1}{i} = \frac{1}{i_k} + \frac{1}{i_f} + \frac{1}{i_d} \qquad (12.8)$$

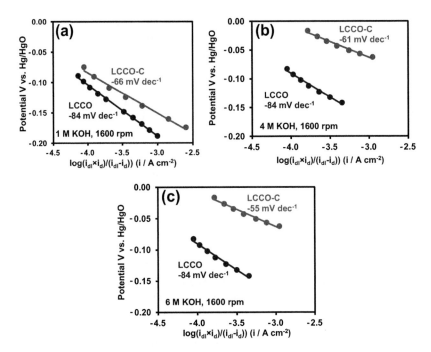

FIGURE 12.7 (Color online) ORR Tafel plots constructed using diffusion-current-corrected current densities recorded on LCCO/C (red) and LCCO (blue) electrodes in O_2-saturated (a) 1 M, (b) 4 M, and (c) 6 M KOH solutions. LCCO/C loading: 0.41 mg cm^{-2} (LCCO:C = 0.4:0.6 w:w) [21]. (For interpretation of the references to color in this figure legend, the reader is referred to the web version of this book.)

where i_f is diffusion-limiting current within the Nafion® ionomer layer, which can be expressed as in Eqn (12.9):

$$i_f = nFAC_{H_2}^f D_{H_2}^f L^{-1} \qquad (12.9)$$

where n, F, and A have the same meanings as in Eqn (12.3), $C_{H_2}^f$ and $D_{H_2}^f$ are, respectively, the concentration and diffusion coefficients of H_2 inside the Nafion® layer, and L is its thickness. Based on these equations and the measured data at different Nafion® film thicknesses, the effect of i_f was investigated, as shown in Fig. 12.8. Figure 12.8d shows that when the equivalent Nafion® film thickness is >0.5 mm, the effect of i_f on the RDE current should be taken into account when investigating the electrode reaction kinetics.

This method to correct the i_f effect was also used by Li et al. [21] for the ORR. In their study, the thickness of the Nafion® ionomer layer was calculated to be approximately 0.63 μm by assuming that the density of Nafion® film is 2 g cm^{-3}. Based on the assumption that the concentration and diffusion coefficients of oxygen in the Nafion® layer displayed no remarkable difference between 1 M H_2SO_4 and alkaline solutions, i_f was calculated to be 57 mA cm^{-2}

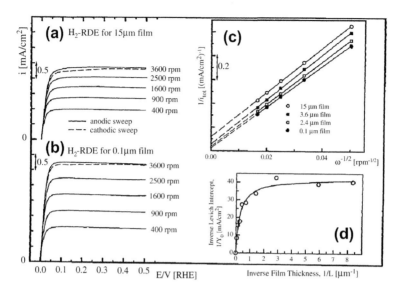

FIGURE 12.8 (a) Hydrogen oxidation on a 15-μm-thick Nafion® film; (b) hydrogen oxidation on a 0.1-μm-thick Nafion® film; (c) Koutecky–Levich plots based on the hydrogen oxidation data with different Nafion® film thicknesses; (d) inverse Levich intercepts obtained from (c) vs. inverse film thickness. $T = 25\,°C$; Pt loading $= 7\,\mu g\,cm^{-2}$; 0.5 M H_2SO_4; scan rate $= 10\,mV\,s^{-1}$ [23].

according to the reported values of the O_2 concentration diffusion coefficient. This i_f value was used to correct the kinetic parameters in their work. Their Koutecky–Levich data were plotted at different electrode potentials, from -0.20 to $-0.45\,V$, in three KOH concentrations. Each data group was extrapolated to $\omega^{-1/2} = 0$ to obtain the intercept $(1/i_k + 1/i_f)$, from which the ORR kinetic current densities (i_k) were obtained by subtracting the value of i_f. They found that the deviation could be as high as 20% after i_f correction, depending on both the electrode potential and the current range, especially at lower alkaline concentrations.

12.3. HALF-CELL DESIGN TO MIMIC FUEL CELL ELECTRODE SITUATION FOR LIQUID FUEL OXIDATION REACTION [24]

12.3.1. Half-Cell Design and Fabrication

To mimic fuel cell operating conditions, some novel half-cells have been designed and fabricated for evaluating catalysts and catalyst layers. For example, Fig. 12.9a shows the schematic setup of a half-cell designed for fuel cell electrode characterization and catalyst evaluation [24]. In this half-cell, the CE reservoir is on the left-hand side, and the WE and its holder are placed in the central chamber. Figure 12.9b shows the WE glass frit, and an upward oblique

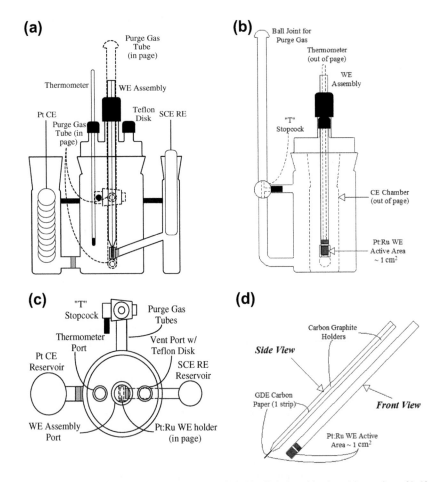

FIGURE 12.9 (a) Half-cell design front view, (b) half-cell design side view, (c) top view of half-cell assembly, and (d) side and front views of WE assembly with 1 cm^2 active area. Wound Teflon® tape not shown [24].

salt bridge that is connected to the RE chamber. This upward oblique orientation (rather than a horizontally positioned reference salt bridge with a glass frit) is designed to avoid gas accumulation inside the bridge tube. Experiments have shown that this special design feature is necessary because the accumulation of gas inside the RE bridge can isolate the electrolyte connection between the RE and WE, resulting in measurement failure. The WE holder has its back connected to the RE salt bridge glass frit. The front and top of the WE holder is open to the electrolyte, and the front is aligned to face straight right to the fritted glass port mouth of the CE reservoir, straight on the left-hand side. The WE electrode active surface, onto which the catalyst layer is coated, is made to face the outside from the holder. The back of the WE active surface, which is not catalyzed, is made to face the RE salt bridge frit. In this way, the current may only pass between the WE surface and the CE; consequently, the current distribution field can be effectively blocked by both the WE glass holder wall and the RE salt bridge tube wall. As a result, there should not be significant current distribution between the WE and RE, leading to minimal IR drop (current-electrolyte resistance drop) between the WE and RE. The WE assembly, as shown in Fig. 12.9d, consists of a strip of gas diffusion electrode carbon paper (GDE, 1 cm × 15 cm) sandwiched between two carbon graphite holders. All three pieces are held together by tightly wound Teflon® tape, to leave a 1.0 cm^2 (geometric) active area exposed on the WE tip. In Fig. 12.9a and b, there are three cap ports on the top cap of the cell. Two outer ports are sealed, and the third, with one holder, is for the thermometer. The central port holds the WE shaft via O-rings at the top and bottom. The O-rings keep the WE tip stationary and help prevent the tip from breaking. Fig. 12.9c gives a clear view of the three cap ports as well as the three-way valve ("T" stop-cock) and purge gas tube, both on the top of the drawing. The three-way valve ("T" stop-cock) and purge gas tube are also both shown on the left-hand side of Fig. 12.9b. The valve and tubes are used for bubbling gas from the bottom of the half-cell, or passing gas over the surface of the electrolyte, or isolating the cell from the gas flow altogether. The whole half-cell resides in a temperature controlled water bath covered with insulating plastic balls that are used to keep the temperature of the bath more stable and to minimize evaporation.

This half-cell was used for evaluating methanol oxidation catalysts and catalyst layers [24], as described in the following section. It is expected that this cell could also be suitable to mimic fuel cell operating conditions for other liquid fuel oxidation because the WE structure is similar to the structure of the anode in a direct liquid fuel cell.

12.3.2. Methanol Oxidation Under Mimicked Fuel Cell Conditions [24]

Figure 12.10 shows the CVs for methanol oxidation on a Pt/Ru catalyzed electrode at various temperatures. It can be seen that there is a significant increase in the current density with increasing temperature. From the data in this figure, the kinetic parameters of methanol oxidation can be estimated. For an electrochemical reaction controlled purely by electron transfer kinetics, if the reaction overpotential is large enough (>60 mV), the Butler–Volmer equation can be simplified to the form of a Tafel equation, which is similar to Eqn (12.6):

$$E = E^o_{MeOH/CO_2} - \frac{2.303RT}{\alpha n_\alpha F}\log(i^o_{MeOH/CO_2}) + \frac{2.303RT}{\alpha n_\alpha F}\log(i) \qquad (12.10)$$

where E is the anode potential, E^o_{MeOH/CO_2} is the standard electrode potential for methanol oxidation, α and n_α have the same meanings as in Eqn (12.6)

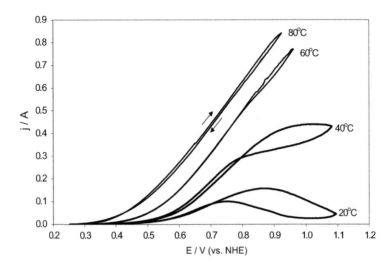

FIGURE 12.10 Cyclic voltammograms for methanol oxidation using 40 wt.% Pt:Ru (1:1 atomic ratio) on Vulcan XC-72 with a catalyst loading of 1.0 mg cm^{-2} in 1.0 mol dm^{-3} methanol and sulfuric acid at different temperatures, as marked on each curve. The anode geometric area was 1.0 cm^2. Potential scan rate: 5 mV s^{-1} [24].

except that they are for methanol oxidation, and i^o_{MeOH/CO_2} is the exchange current density of methanol oxidation. In Fig. 12.10, due to the slow potential scan rate, the corresponding forward and backward currents at each potential point during the CV scan can be averaged to obtain a steady-state current. The Tafel plots of log (i) vs. E can be used to determine the kinetic parameters for methanol oxidation, according to Eqn (12.10).

The Tafel plots based on the data in Fig. 12.10 at all four temperatures in the potential range of 0.31–0.75 V showed that when the electrode potential was increased, all curves began to deviate from the expected linear behavior; this indicates that the oxidation reaction was no longer a purely kinetically controlled process in this higher potential range. At low temperatures, such as <40 °C, deviation from linearity was mainly due to poisoning by the CO produced during methanol oxidation, which can be seen from the CVs at 20 and 40 °C in Fig. 12.10 (the "looped," scattered, forward, and backward currents at higher potentials). At higher temperatures, such as >60 °C, the deviation from linearity was believed to be caused by limited methanol diffusion through the catalyst layer to the reaction sites. The Tafel plots of log(i)~ vs. E in the potential range of 0.35–0.65 V (vs. NHE) were used to estimate the kinetic parameters of the Tafel slope b ($= 2.303RT/\alpha n_\alpha F$); i^o_{MeOH/CO_2}, and αn_α, as listed in Table 12.2. The Tafel slope values (~160 mV) are consistent with those reported in the literature [25].

Because the measurements were conducted on an electrode that mimics a direct methanol fuel cell (DMFC) anode electrode, the data shown in

TABLE 12.2 Kinetic Parameters at Different Temperatures, Estimated From the Tafel Plots of Methanol Oxidation on a Fuel Cell Anode Catalyzed by Carbon-Supported Pt/Ru (40%, 1:1 atomic ratio) Catalyst, 1.0 mol dm^{-3} Methanol Solution with 1.0 mol dm^{-3} H$_2$SO$_4$ as Supporting Electrolyte, and at Ambient Pressure

T (K)	293	313	333	353
b (mV)	159	161	163	169
i^o_{MeOH/CO_2} (A cm^{-2})	1.93×10^{-8}	3.69×10^{-8}	8.33×10^{-8}	2.02×10^{-7}
αn_α	0.36	0.39	0.40	0.41
E^o_{MeOH/CO_2} (V)	−0.0120	−0.0262	−0.0404	−0.0546

Table 12.2 may be treated as in situ data for methanol oxidation under DMFC operating conditions.

The methanol oxidation activation energy was also obtained according to Eqn (12.11):

$$Ln(i^o_{MeOH/CO_2}) = Ln(i^o_{o,MeOH/CO_2}) - \frac{E_a}{RT} \quad (12.11)$$

where E_a is the reaction activation energy and $i^o_{o,MeOH/CO_2}$ is the absolute reaction exchange current density of the reaction. The plot of the logarithm of the exchange current density vs. T^{-1} based on the data in Table 12.2 gave a slope of E_a/R, from which the reaction activation energy of methanol oxidation was obtained, which was 33.6 kJ mol^{-1}.

12.3.3. Electrochemical Impedance Spectroscopic Measurements for Methanol Oxidation [24]

The half-cell shown in Fig. 12.9 can also be used for measuring the AC impedance of methanol oxidation. The data obtained at different electrode potentials and different temperatures are shown in Figs 12.11 and 12.12, respectively. In Fig. 12.11, a small arc in the high-frequency (HF) region and a well-developed semicircle in the mid-frequency (MF) region are discernible, followed by an inductive arc in the low-frequency (LF) region.

The MF arc shrinks rapidly when the bias potential increases from 0.35 to 0.65 V. In Fig. 12.12, at approximately 0.3-V bias potential, an increase in the temperature from 20 to 80 °C also makes the MF arc shrink significantly. The curve intercepts at the HF end of the Z' axis normally represent the solution electrolyte resistance. The HF arc may be ascribed to the ionic resistance within

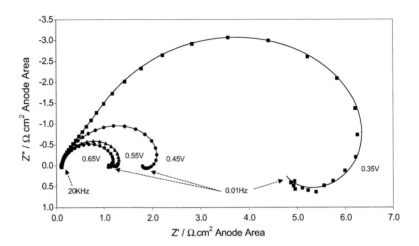

FIGURE 12.11 Nyquist plots of methanol oxidation at 80 °C and 0.35 V, 0.45 V, 0.55 V, and 0.65 V (vs. NHE) for an electrode with 40 wt.% Pt:Ru (1:1 atomic ratio) on Vulcan XC-72 in 1.0 mol dm^{-3} methanol and 1.0 mol dm^{-3} sulfuric acid. Experimental data (symbols) and fitted results for equivalent circuit model (ECM) curves (solid lines) [24]. (For color version of this figure, the reader is referred to the online version of this book.)

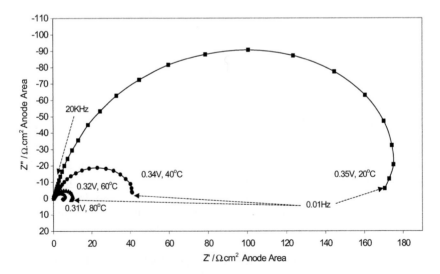

FIGURE 12.12 Nyquist plots of methanol oxidation at 20 °C, 40 °C, 60 °C, and 80 °C, at approximately 0.3 V (vs. NHE) for methanol oxidation with 40 wt.% Pt:Ru (1:1 atomic ratio) on Vulcan XC-72 in 1.0 mol dm^{-3} methanol and 1.0 mol dm^{-3} sulfuric acid. Experimental data (symbols) and fitted curves (solid lines) [24].

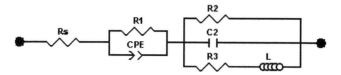

FIGURE 12.13   ECM for methanol oxidation on a Pt/Ru catalyzed fuel cell anode.

the catalyst layer, coupled with a pseudocapacitance within the GDE. The MF arc is usually associated with the electrochemical reaction charge transfer process across the catalyst/electrolyte interface during the electrochemical reaction, and the LF inductive loop may be caused by an adsorbed intermediate species, such as CO. This adsorbed CO can be removed by oxidation and results in an inductive loop on the LF end of the spectrum [26–28]. Normally, the bigger the arc radii, the slower the corresponding process will be. In Fig. 12.11, the increase in potential results in shrinkage of the MF arc radius; this indicating that the rate of charge transfer for methanol oxidation increases with increasing electrode potential, which is to be expected from the electrochemical kinetics [29].

For a more quantitative description of the behavior shown in Figs 12.11 and 12.12, an ECM was constructed by these authors, as shown in Fig. 12.13.

In the ECM (Fig. 12.13), R_s is the ohmic resistance of the solution electrolyte, while R_1 and C_1 (constant phase element, CPE), respectively, represent the ionic ohmic resistance and ionic capacitance in the catalyst layer; the capacitance (C_1) was replaced by a CPE to more accurately reflect the porous electrode behavior [30]. R_2 and C_2 represent methanol oxidation charge transfer resistance and interfacial double layer capacitance, respectively, and R_3 represents intermediate adsorbate resistance due to the increase in intermediate adsorbate coverage at the reaction site of the catalyst surface. An adsorbed intermediate, such as CO, can be oxidized to CO_2 above a critical potential to result in an inductance, L_1 [28]. The inductance of the instrument in the HF region was not pursued in this ECM.

Experimental data fitting using the ECM was performed with ZPlot's complex nonlinear least square fitting program. It was confirmed that the experimental data were fitted very well at all four temperatures and potentials, and this demonstrates that the proposed model in Fig. 12.13 is feasible for description of the methanol oxidation process. The charge transfer resistance (R_2) of the methanol anode oxidation was obtained from the ECM fitting at the four different temperatures and potentials, as plotted in Fig. 12.14.

From Fig. 12.14, it can be seen that the charge transfer resistance, R_2, decreases monotonically with increasing temperature for each potential. The decrease in this resistance with increasing electrode potential reflects

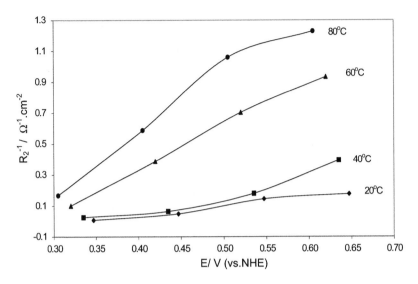

FIGURE 12.14　Reciprocals of R_2 as a function of temperature and electrode potential, obtained by fitting the impedance data by using the ECM shown in Fig. 12.13 [24].

that an increase in the electrode potential can speed up the methanol electro-oxidation kinetics, which is consistent with the CV results shown in Fig. 12.10.

12.4. HALF-CELL DESIGN TO MIMIC THE FUEL CELL ELECTRODE SITUATION FOR THE ORR AND HOR [31]

12.4.1. Half-Cell Design and Fabrication

Figure 12.15 shows a half-cell design consisting of three electrodes: the WE, CE (a platinum sheet), and RE (an SCE). A Teflon® container with both the CE and RE inlets and two gas vents serves as the electrolyte chamber, filled with 0.5 M H_2SO_4 solution. Before each electrochemical measurement, this electrolyte solution is bubbled with N_2 for 30 min to remove dissolved O_2. The WE (anode or cathode) and flow field plate containing gas channels are located inside the metal titanium current collector. The RE is located right beside the WE so that its tip touches the membrane of the catalyst-coated membrane electrode (CCME). In this way, the IR drop between the RE and the catalyst layer can be minimized. The gas diffusion layer (GDL, carbon paper) of the CCME is exposed to the feed gases, and the catalyst layer side—which has an active area of 1.0 mm² (i.e. 1.0 mm × 1.0 mm) covered by Nafion® 112 membrane—is exposed to the electrolyte solution through a small round window at the end of the electrolyte chamber. Note that because the membrane is in direct contact with the aqueous electrolyte solution, the situation

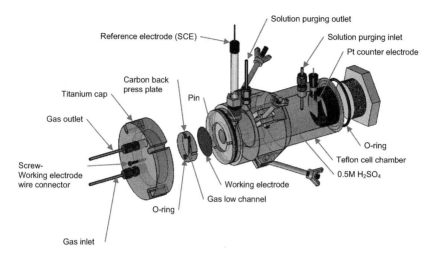

FIGURE 12.15 Schematic of a designed half-cell for PEM fuel cell anode or cathode catalyst/catalyst layer evaluation [31]. (For color version of this figure, the reader is referred to the online version of this book.)

resembles that of a fuel cell whose cathode MEA is totally flooded. Four metal pins are used to hold together the electrolyte chamber and the metal current collector.

12.4.2. Half-Cell Design Validation Using Both the ORR and the HOR [31]

In validating the half-cell shown in Fig. 12.15, the authors used 20 wt.% Pt/C and 20 wt.% $Pt_{0.5}Ru_{0.5}$/C catalysts to construct the cathode and anode catalyst layers for the ORR and HOR, respectively [31]. The CCME, constructed for cathode or anode, consisted of three layers: (1) the GDL on which (2) the catalyst layer was coated, and (3) the Nafion$^{®}$ membrane, which was bonded to the top of the catalyst layer using hot-pressing. When this CCME was assembled into the half-cell, the GDL side was made to face the channel in the flow field plate, and the membrane side was made to face and be in contact with the electrolyte solution.

With optimization of the design, in particular, the size of the electrolyte window, an overpotential of 47 mV at 1.0 A cm^{-2} for the HOR was achieved, which was less than the 50 mV observed in a fuel cell HOR. This demonstrates that this cell design is feasible.

To further confirm the feasibility of the half-cell, the Pt-based CCME was also used to test the ORR at the cathode. Figure 12.16 shows the ORR curves as a function of air-flow rate. It can be seen that an air-flow rate of

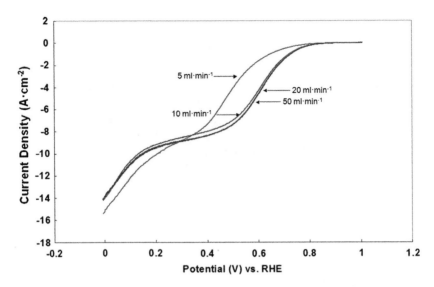

FIGURE 12.16 ORR curves at different air-flow rates, as marked. Pt/C (20 wt.%) was used to prepare the cathode catalyst layer, which had a Pt loading of 2.0 mg cm^{-2}. Potential scan rate: 20 mV s^{-1} [31]. (For color version of this figure, the reader is referred to the online version of this book.)

5 ml min^{-1} is not enough to support the ORR current; the air-flow rate must be >10 ml min^{-1}. The curves also show plateau current densities of approximately 7 A cm^{-2}; this value is limited by O$_2$ diffusion. When the electrode potential is more negative than 0.2 V, another electrode process starts, which is probably H$_2$O reduction to produce H$_2$. The result shown in Fig. 12.16 demonstrates that this half-cell can also be used for studying cathode ORR catalysts.

12.4.3. Half-Cell Testing for CO-Tolerant Anode Catalysts [31]

The half-cell shown in Fig. 12.15 was also used to study CO-tolerant catalysts [31]. In this study, both Pt-based and Pt$_{0.5}$Ru$_{0.5}$-based CCMEs were used for HOR testing in the presence of various CO concentrations.

Figure 12.17 shows the HOR performance catalyzed by these two catalysts in the presence of various CO concentrations (10, 50, 100, 300, and 500 ppm) in the H$_2$ stream. It can be seen that the Pt$_{0.5}$Ru$_{0.5}$/C catalyst has a lower ORR catalytic activity compared to the Pt/C catalyst. However, when the CO concentration is >100 ppm, Pt$_{0.5}$Ru$_{0.5}$/C shows a higher performance than Pt/C does, which suggests that Pt$_{0.5}$Ru$_{0.5}$/C has a higher CO tolerance. Note that in the presence of >100 ppm CO, when the electrode potential is more positive than approximately 0.55 V (vs. RHE), the current density increases rapidly on

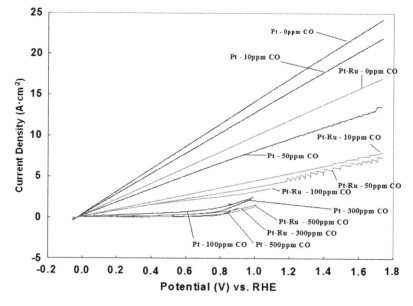

FIGURE 12.17 Comparison of H_2 oxidation performance with Pt-based and Pt–Ru-based electrodes in the presence of various CO concentrations. Pt/C (20 wt.%) was used to prepare the anode catalyst layer, with a Pt loading of 2.0 mg cm^{-2}. Potential scan rate: 20 mV s^{-1}; H_2 flow rate: 5 ml min^{-1} [31]. (For color version of this figure, the reader is referred to the online version of this book.)

the curves. This should indicate a surface CO oxidation reaction that releases more Pt sites for the HOR.

12.5. CHAPTER SUMMARY

For quickly downselecting fuel cell electrode materials and components such as the catalyst and its associated catalyst layer, an ex situ approach using an electrochemical half-cell setup is the quickest and most cost-effective method because it allows the investigation of one electrochemical reaction and its associated catalysis mechanism as well as the effects that operating conditions have on this reaction, without interference from another electrochemical reaction. By using novel half-cell designs that enable mimicking of fuel cell operating conditions, some evaluation is possible. In this chapter, three different kinds of half-cell design are presented; these include the conversional three-electrode half-cell and two novel half-cells that allow such mimicking. Several examples are also given to validate these half-cell designs, including the evaluation of (1) ORR catalysts in terms of their EPSA, mass activity, and stability; (2) the catalyst layer for the effect of its Nafion®

content; (3) catalysts for the methanol oxidation reaction; and (4) HOR catalysts in the presence of CO.

REFERENCES

[1] Gottesfeld S, Raistrick ID, Srinivasan S. J Electrochem Soc 1987;134:1455–62.

[2] Lee K, Zhang L, Lui H, Hui R, Shi Z, Zhang J. Electrochim Acta 2009;54:4704–11.

[3] Mayrhofer KJJ, Strmcnik D, Blizanac BB, Stamenkovic V, Arenz M, Markovic NM. Electrochim Acta 2008;53:3181–8.

[4] Vázquez-Huerta G, Ramos-Sánchez G, Rodríguez-Castellanos A, Meza-Calderón D, Antaño-López R, Solorza-Feria O. J Electroanal Chem 2010;645:35–40.

[5] Zhang L, Lee K, Bezerra CWB, Zhang J, Zhang J. Electrochim Acta 2009;54:6631–6.

[6] Zhang L, Lee K, Zhang J. Electrochim Acta 2007;52:3088–94.

[7] Ioroi T, Yasuda K, Paffett MT, Beery JG, Gottesfeld S, Liu GC-K, Sanderson RJ, Vernstrom G, Stevens DA, Atanasoski RT, Debe MK, Dahn JR, Gottesfeld S, Raistrick ID, Srinivasan S. J Electrochem Soc 2005;152:A1917–24.

[8] Paffett MT, Beery JG, Gottesfeld S, Liu GC-K, Sanderson RJ, Vernstrom G, Stevens DA, Atanasoski RT, Debe MK, Dahn JR, Gottesfeld S, Raistrick ID, Srinivasan S. J Electrochem Soc 1988;135:1431–6.

[9] Liu GC-K, Sanderson RJ, Vernstrom G, Stevens DA, Atanasoski RT, Debe MK, Dahn JR, Gottesfeld S, Raistrick ID, Srinivasan S. J Electrochem Soc 2010;157:B207–14.

[10] Song C, Zhang L, Zhang J, Wilkinson DP, Baker R. Fuel Cells 2007;7:9–15.

[11] Elezovic NR, Gajic-Krstajic L, Radmilovic V, Vracar L, Krstajic NV. Electrochim Acta 2009;54:1375–82.

[12] Esparbé I, Brillas E, Centellas F, Garrido JA, Rodríguez RM, Arias C, Cabot P-L. J Power Sources 2009;190:201–9.

[13] Innocente AF, Ângelo ACD. J Power Sources 2006;162:151–9.

[14] Pronkin SN, Bonnefont A, Ruvinskiy PS, Savinova ER. Electrochim Acta 2010;55:3312–23.

[15] Velázquez A, Centellas F, Garrido JA, Arias C, Rodríguez RM, Brillas E, Cabot P-L. J Power Sources 2010;195:710–9.

[16] Sheng W, Gasteiger HA, Shao-Horn Y. J Electrochem Soc 2010;157:B1529–36.

[17] Pietron JJ. J Electrochem Soc 2009;156:B1322–8.

[18] Baker R, Song C, Neburchilov V, Zhang J. Unpublished experiment results from institute for fuel cell innovation. National Research Council Canada; 2010.

[19] Gasteiger HA, Kocha SS, Sompalli B, Wagner FT. Appl Catal B Environ 2005;56:9–35.

[20] Vidakovic T, Christov M, Sundmacher K. Electrochim Acta 2007;52:5606–13.

[21] Li X, Qu W, Zhang J, Wang H. J Electrochem Soc 2011;158(5):A597–604.

[22] Li X, Zhu AL, Qu W, Wang H, Hui R, Zhang L, Zhang J. Electrochim Acta 2010;55:5891–8.

[23] Schmidt J, Gasteiger HA, Stab GD, Urban PM, Kolb DM, Behm RJ. J Electrochem Soc 1998;145:2354–8.

[24] Baker R, Xie Z, Zhang J, Wilkinson DP. 3–17. In: Ghosh D, editor. Fuel cell and hydrogen technologies, 44th annual conference of metallurgists of CIM. Calgary, Alberta, Canada: Published by MET-SOC (2005); 2005.

[25] Khazova OA, Mikhailova AA, Skundin AM, Tuseeva EK, Havránek A, Wippermann K. Fuel Cells 2002;2:99–108.

[26] Ciureanu M, Roberge R. J Phys Chem B 2001;105:3531–9.

[27] Muller JT, Urban PM. J Power Sources 1999;75:139–43.

[28] Muller JT, Urban PM, Holderich WF. J Power Sources 1999;84:157–60.
[29] Yuan X, Song C, Wang H, Zhang J. Electrochemical impedance spectroscopy in PEM fuel cells. London: Springer-Verlag London Ltd; 2010.
[30] Brug GJ, G Van Den Eeden AL, Sluyters-Rehbach M, Sluyters JH. J Electroanal Chem 1984;176:275–95.
[31] Zhang L, Kim J, Singh S, Tsay K, Zhang J. Inter J Green Energy 2012; online available, DOI:10.1080/15435075.2012.729170.

AAS	atomic adsorption spectroscopy
AB	acetylene black
AC	alternating current
A-CL	anode catalyst layer
ADT	accelerated degradation test *or* accelerated durability test
AECD	apparent exchange current density
AFM	atomic force microscopy
A-GDM	anode gas diffusion medium
A-MPL	anode microporous layer
AP	1-aminopyrene
APR	average pore radius
AST	accelerated stress test
ATR-FTIR	attenuated total reflection Fourier transform infrared spectroscopy
BM	benzyl mercaptan
BMI	1-butyl, 3-methylimidazolium
BOL	beginning of lifetime
BP	Black Pearls®
C-CL	cathode catalyst layer
CL	catalyst layer
CCM	catalyst-coated membrane
CCME	catalyst-coated membrane electrode
CE	counter electrode
C-GDM	cathode gas diffusion medium
CL	catalyst layer
C-MPL	cathode microporous layer
CNT	carbon nanotube
CPE	constant phase element
CT	computed tomography
CV	cyclic voltammetry
DC	direct current
DMAc	dimethylacetamide
DMFC	direct methanol fuel cell
DOE	United States Department of Energy
DSC	differential scanning calorimetry
DTA	differential thermal analysis
EPSA	electrochemical Pt surface area
EC-AFM	electrochemical atomic force microscopy
ECM	equivalent circuit model
EDS	energy dispersive X-ray spectroscopy
EDX	energy dispersive X-ray analysis
EIS	electrochemical impedance spectroscopy
EOL	end of lifetime
EPDM	ethylene propylene diene monomer
ESA	electrochemical surface area

EXAFS	extended X-ray absorption fine structure
FC	fuel cell
FEP	fluorinated ethylene propylene
FF	flow field
FRA	frequency response analyzer
FTIR	Fourier transform infrared spectroscopy
FWHM	full width at half maximum
GC	gas chromatography *or* glassy carbon
GC–MS	gas chromatography–mass spectroscopy
GDB	gas diffusion backing
GDE	gas diffusion electrode
GDHL	gas diffusion half-layer
GDL	gas diffusion layer
GDM	gas diffusion medium/media
HAADF-STEM	high-angle annular dark field scanning transmission electron microscopy
HF	high frequency
HFR	high-frequency resistance
HOR	hydrogen oxidation reaction
HR-TEM	high-resolution transmission electron microscopy
HT-PEM	high-temperature PEM
ICP	inductively coupled plasma
IECD	intrinsic exchange current density
IR	infrared
KPM	Kelvin probe microscopy
LCCO	$La_{0.6}Ca_{0.4}CoO_3$
LF	low frequency
LRS	laser Raman spectroscopy
LSV	linear sweep voltammetry
LT-PEM	low-temperature PEM
MEA	membrane electrode assembly
MF	midfrequency
MPL	microporous layer
MS	mass spectrometry
NDC	normal direction conductivity
NHE	normal hydrogen electrode
NMR	nuclear magnetic resonance
OCV	open circuit voltage
ORR	oxygen reduction reaction
PA	phosphoric acid
PAAM	polyacrylamide
PBI	polybenzimidazole
PCA	1-pyrenecarboxylic acid
PCB	printed circuit board
PEEK	poly(ether ether ketone)
PEEKK	poly(ether ether ketone ketone)
PEI	polyethylenimine
PEM	proton exchange membrane or polymer electrolyte membrane
PEMFC	polymer electrolyte membrane fuel cell *or* proton exchange membrane fuel cell
PEO	poly(ethylene oxide)
PFSA	perfluorosulfonic acid
PFSI	perfluorosulfonated ionomer *or* perfluorosulfonimide

PP	polyphenylenes
PSF	polysulfones
PSSA	polystyrene sulfonic acid
PTA	phosphotungstic acid
PTFE	polytetrafluoroethylene
PVA	polyvinyl alcohol
PVDF	polyvinylidene difluoride
RDE	rotating disk electrode
RRDE	rotating ring-disk electrode
RE	reference electrode
RH	relative humidity
RHE	reversible hydrogen electrode
RRDE	rotating ring-disk electrode
SAB	Shawinigan acetylene black
SCE	saturated calomel reference electrode
SEM	scanning electron microscopy
SEM-EDS	scanning electron microscopy–energy-dispersive X-ray microscopy
SHE	standard hydrogen electrode
SPEEK	sulfonated poly(ether ether ketone)
SPM	scanning probe microscopy
SPSF	sulfonated polysulfone
STM	scanning tunneling microscopy
TDC	tangential direction conductivity
TEM	transmission electron microscope/microscopy
TFCl	thin-film catalyst layer
TGA	thermal gravimetric analysis
TPSA	total Pt surface area
USFCC	US Fuel Cell Council
UVS	ultraviolet spectroscopy
VOC	volatile organic compound
WE	working electrode
WDS	wavelength-dispersive X-ray spectroscopy
XAS	X-ray absorption spectroscopy
XPS	X-ray photoelectron spectroscopy
XRD	X-ray diffraction

Index

Note: Page numbers with "f" denote figures; "t" tables.

Printed and bound by CPI Group (UK) Ltd, Croydon, CR0 4YY

08/05/2025

01864815-0002